Texas Instruments

TI-83 Plus
GRAPHING CALCULATOR
GUIDEBOOK

18 19 20

Important

Texas Instruments makes no warranty, either expressed or implied, including but not limited to any implied warranties of merchantability and fitness for a particular purpose, regarding any programs or book materials and makes such materials available solely on an "as-is" basis.

In no event shall Texas Instruments be liable to anyone for special, collateral, incidental, or consequential damages in connection with or arising out of the purchase or use of these materials, and the sole and exclusive liability of Texas Instruments, regardless of the form of action, shall not exceed the purchase price of this equipment. Moreover, Texas Instruments shall not be liable for any claim of any kind whatsoever against the use of these materials by any other party.

US FCC Information Concerning Radio Frequency Interference

This equipment has been tested and found to comply with the limits for a Class B digital device, pursuant to Part 15 of the FCC rules. These limits are designed to provide reasonable protection against harmful interference in a residential installation. This equipment generates, uses, and can radiate radio frequency energy and, if not installed and used in accordance with the instructions, may cause harmful interference with radio communications. However, there is no guarantee that interference will not occur in a particular installation.

If this equipment does cause harmful interference to radio or television reception, which can be determined by turning the equipment off and on, you can try to correct the interference by one or more of the following measures:

- Reorient or relocate the receiving antenna.
- Increase the separation between the equipment and receiver.
- Connect the equipment into an outlet on a circuit different from that to which the receiver is connected.
- Consult the dealer or an experienced radio/television technician for help.

Caution: Any changes or modifications to this equipment not expressly approved by Texas Instruments may void your authority to operate the equipment

This manual describes how to use the TI-83 Plus Graphing Calculator. Getting Started is an overview of TI-83 Plus features. Chapter 1 describes how the TI-83 Plus operates. Other chapters describe various interactive features. Chapter 17 shows how to combine these features to solve problems.

Chapter 6: Sequence Graphing

Chapter 7: Tables

Chapter 8: DRAW Operations

Chapter 9: Split Screen

Chapter 10: Matrices

Chapter 11: Lists

Chapter 12: Statistics

Chapter 13: Inferential Statistics and Distributions

Chapter 14: Applications

Chapter 15: CATALOG, Strings, Hyperbolic Functions

Chapter 16: Programming

Chapter 17: Activities

Flash – Electronic Upgradability

The TI-83 Plus uses Flash technology, which lets you upgrade to future software versions without buying a new calculator.

For details, refer to: Chapter 19

As new functionality becomes available, you can electronically upgrade your TI-83 Plus from the Internet. Future software versions include maintenance upgrades that will be released free of charge, as well as new applications and major software upgrades that will be available for purchase from the TI web site: **http:// www.ti.com/calc**

184K bytes of Memory

184K bytes of memory are built into the TI-83 Plus. About 24K of RAM (random access memory) are available for you to compute and store functions, programs, and data.

For details, refer to: Chapter 18

About 160K of user data archive allow you to store data, programs, applications, or any other variables to a safe location where they cannot be edited or deleted inadvertently. You can also free up RAM by archiving variables to user data

Applications

Applications can be installed to customize the TI-83 Plus to your classroom needs. The big 160K archive space lets you store up to ten (10) applications at one time. Applications can also be stored on a computer for later use or linked unit-to-unit.

For details, refer to: Chapter 18

Archiving

You can store variables in the TI-83 Plus user data archive, a protected area of memory separate from RAM. The user data archive lets you:

For details, refer to: Chapter 18

- Store data, programs, applications or any other variables to a safe location where they cannot be edited or deleted inadvertently.
- Create additional free RAM by archiving variables.

By archiving variables that do not need to be edited frequently, you can free up RAM for applications that may require additional memory.

Calculator-Based Laboratory™ (CBL 2™, CBL™) and Calculator-Based Ranger™ (CBR™)

The TI-83 Plus comes with the CBL/CBR application already installed. When coupled with the (optional) CBL 2/CBL or CBR accessories, you can use the TI-83 Plus to analyze real world data.

For details, refer to: Chapter 14

CBL 2/CBL and CBR let you explore mathematical and scientific relationships among distance, velocity, acceleration, and time using data collected from activities you perform.

CBL 2/CBL and CBR differ in that CBL 2/CBL allows you to collect data using several different probes analyzing temperature, light, volt type, or sonic (motion) data. CBR collects data using a built-in Sonic probe. CBL 2/CBL and CBR accessories can be linked together to collect more than one type of data at the same time. You can find more information on CBL 2/CBL and CBR in their user manuals.

Getting Started: Do This First!

Contents

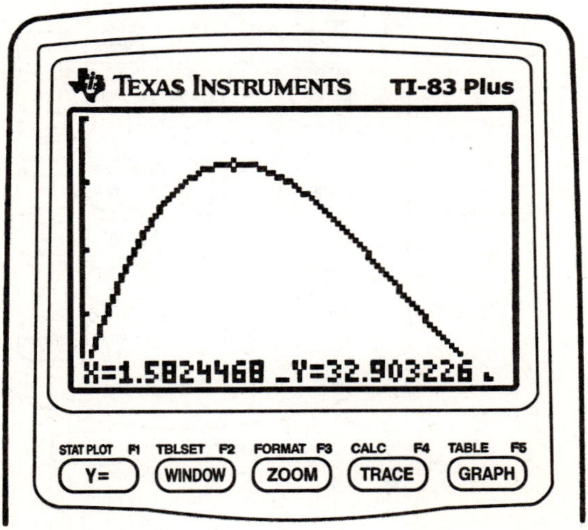

Generally, the keyboard is divided into these zones: graphing keys, editing keys, advanced function keys, and scientific calculator keys.

Keyboard Zones

Graphing keys access the interactive graphing features.

Editing keys allow you to edit expressions and values.

Advanced function keys display menus that access the advanced functions.

Scientific calculator keys access the capabilities of a standard scientific calculator.

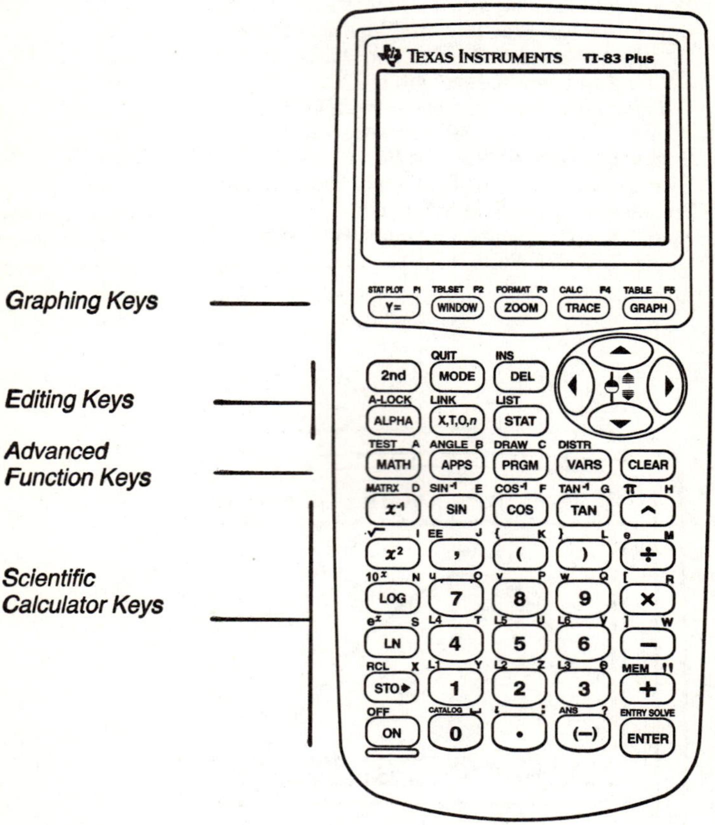

Using the Color-Coded Keyboard

The keys on the TI-83 Plus are color-coded to help you easily locate the key you need.

The light gray keys are the number keys. The blue keys along the right side of the keyboard are the common math functions. The blue keys across the top set up and display graphs. The blue [APPS] key provides access to applications such as the Finance application.

The primary function of each key is printed on the keys. For example, when you press [MATH], the MATH menu is displayed.

Using the [2nd] and [ALPHA] Keys

The secondary function of each key is printed in yellow above the key. When you press the yellow [2nd] key, the character, abbreviation, or word printed in yellow above the other keys becomes active for the next keystroke. For example, when you press [2nd] and then [MATH], the TEST menu is displayed. This guidebook describes this keystroke combination as [2nd] [TEST].

The alpha function of each key is printed in green above the key. When you press the green [ALPHA] key, the alpha character printed in green above the other keys becomes active for the next keystroke. For example, when you press [ALPHA] and then [MATH], the letter **A** is entered. This guidebook describes this keystroke combination as [ALPHA] [A].

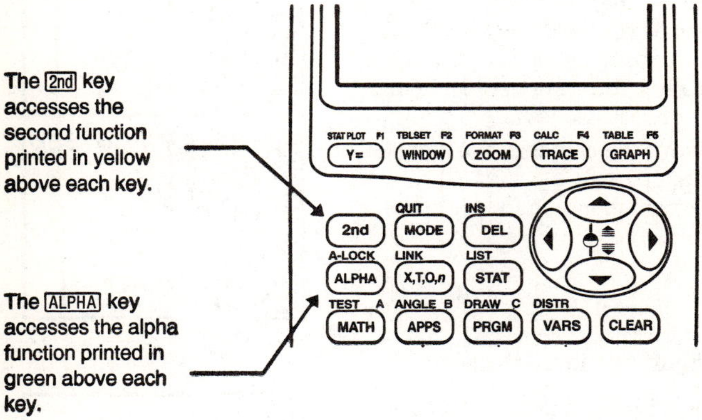

Displaying a Menu

While using your TI-83 Plus, you often will need to access items from its menus.

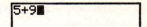

When you press a key that displays a menu, that menu temporarily replaces the screen where you are working. For example, when you press [MATH], the MATH menu is displayed as a full screen.

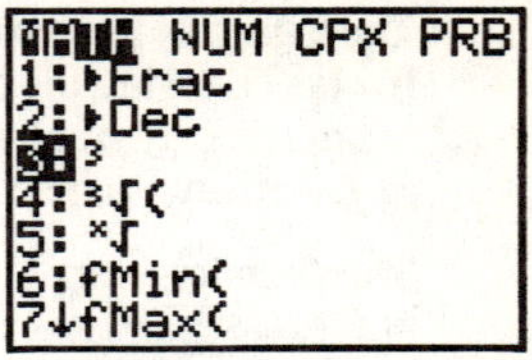

After you select an item from a menu, the screen where you are working usually is displayed again.

Moving from One Menu to Another

Some keys access more than one menu. When you press such a key, the names of all accessible menus are displayed on the top line. When you highlight a menu name, the items in that menu are displayed. Press [▶] and [◀] to highlight each menu name.

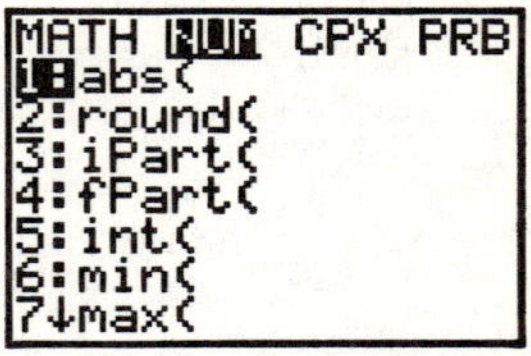

Selecting an Item from a Menu

The number or letter next to the current menu item is highlighted. If the menu continues beyond the screen, a down arrow (↓) replaces the colon (:) in the last displayed item. If you scroll beyond the last displayed item, an up arrow (↑) replaces the colon in the first item displayed.You can select an item in either of two ways.

- Press [▼] or [▲] to move the cursor to the number or letter of the item; press [ENTER].
- Press the key or key combination for the number or letter next to the item.

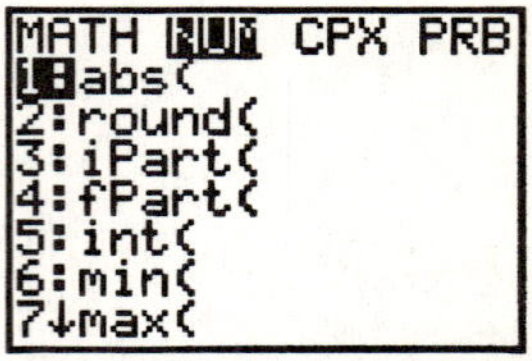

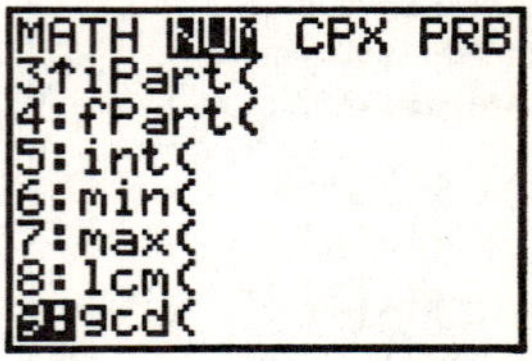

Leaving a Menu without Making a Selection

You can leave a menu without making a selection in any of three ways.

- Press [CLEAR] to return to the screen where you were.
- Press [2nd] [QUIT] to return to the home screen.
- Press a key for another menu or screen.

Before starting the sample problems in this chapter, follow the steps on this page to reset the TI-83 Plus to clear RAM memory. This ensures that the keystrokes in this chapter will produce the illustrated results.

To reset the TI-83 Plus, follow these steps.

1. Press [ON] to turn on the calculator.
2. Press and release [2nd], and then press [MEM] (above [+]).

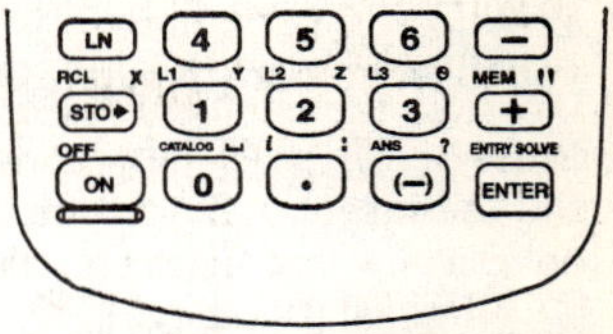

When you press [2nd], you access the operation printed in yellow above the next key that you press. [MEM] is the [2nd] operation of the [+] key.

The MEMORY menu is displayed.

3. Press **7** to select **7:Reset**.

   The RAM ARCHIVE ALL menu is displayed.

4. Press **1** to display the RESET RAM menu.
5. Press **2** to select **2:Reset**.

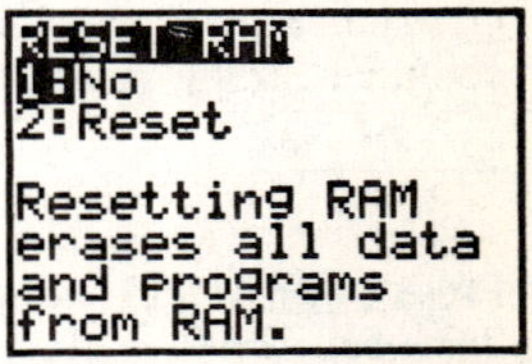

   RAM memory is cleared.

6. Press Enter to display the home screen.

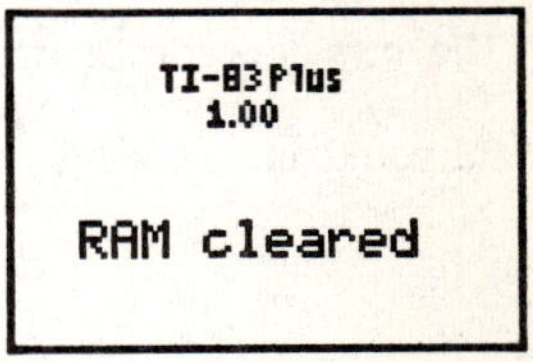

   When you reset the TI-83 Plus, the display contrast is also reset.

   - If the screen is very light or blank, press and release [2nd], and then press and hold [▲] to darken the screen.
   - If the screen is very dark, press and release [2nd], and then press and hold [▼] to lighten the screen.

Use the quadratic formula to solve the quadratic equations $3X^2 + 5X + 2 = 0$ and $2X^2 - X + 3 = 0$. Begin with the equation $3X^2 + 5X + 2 = 0$.

1. Press **3** STO▶ ALPHA [A] (above MATH) to store the coefficient of the X^2 term.

```
3→A:5→B:2→C■
```

2. Press ALPHA [:] (above .). The colon allows you to enter more than one instruction on a line.
3. Press **5** STO▶ ALPHA [B] (above APPS) to store the coefficient of the X term. Press ALPHA [:] to enter a new instruction on the same line. Press **2** STO▶ ALPHA [C] (above PRGM) to store the constant.
4. Press ENTER to store the values to the variables A, B, and C.

   The last value you stored is shown on the right side of the display. The cursor moves to the next line, ready for your next entry.

```
3→A:5→B:2→C
                2
■
```

5. Press ((-) ALPHA [B] + 2nd [√] ALPHA [B] x^2 − **4** ALPHA [A] ALPHA [C])) ÷ (**2** ALPHA [A]) to enter the expression for one of the solutions for the quadratic formula,

   $$\frac{-b \pm \sqrt{b^2 - 4ac}}{2a}$$

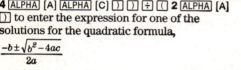

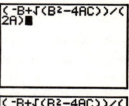

```
(-B+√(B²-4AC))/(
2A)■
```

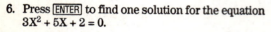

6. Press ENTER to find one solution for the equation $3X^2 + 5X + 2 = 0$.

   The answer is shown on the right side of the display. The cursor moves to the next line, ready for you to enter the next expression.

```
(-B+√(B²-4AC))/(
2A)
     -.6666666667
■
```

You can show the solution as a fraction.

1. Press [MATH] to display the MATH menu.

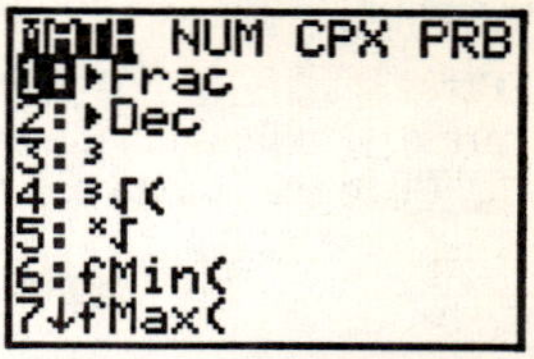

2. Press **1** to select **1:▸Frac** from the MATH menu.

   When you press **1**, **Ans▸Frac** is displayed on the home screen. **Ans** is a variable that contains the last calculated answer.

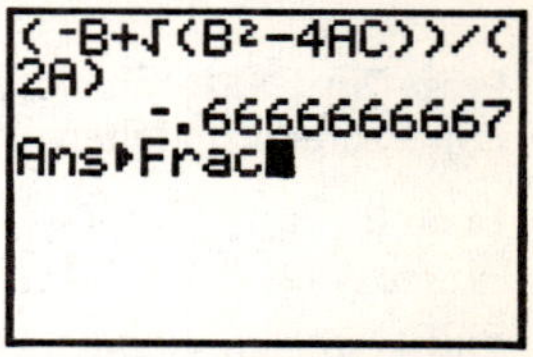

3. Press [ENTER] to convert the result to a fraction.

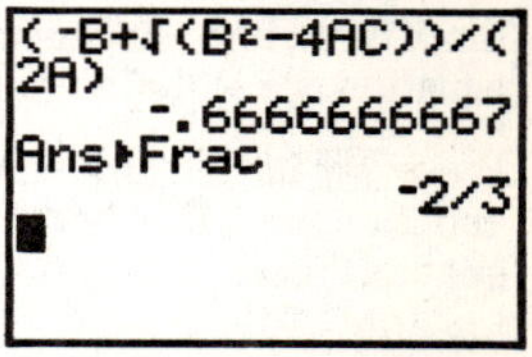

To save keystrokes, you can recall the last expression you entered, and then edit it for a new calculation.

4. Press [2nd] [ENTRY] (above [ENTER]) to recall the fraction conversion entry, and then press [2nd] [ENTRY] again to recall the quadratic-formula expression,

   $$\frac{-b+\sqrt{b^2-4ac}}{2a}$$

5. Press [▲] to move the cursor onto the **+** sign in the formula. Press [−] to edit the quadratic-formula expression to become:

   $$\frac{-b-\sqrt{b^2-4ac}}{2a}$$

6. Press [ENTER] to find the other solution for the quadratic equation $3X^2 + 5X + 2 = 0$.

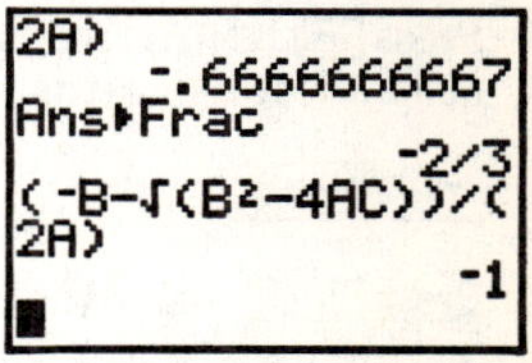

Now solve the equation $2X^2 - X + 3 = 0$. When you set **a+b*i*** complex number mode, the TI-83 Plus displays complex results.

1. Press [MODE] [▼] [▼] [▼] [▼] [▼] [▼] (6 times), and then press [▶] to position the cursor over **a+b*i***. Press [ENTER] to select **a+b*i*** complex-number mode.

2. Press [2nd] [QUIT] (above [MODE]) to return to the home screen, and then press [CLEAR] to clear it.

3. Press **2** [STO▸] [ALPHA] [A] [ALPHA] [:] [(-)] **1** [STO▸] [ALPHA] [B] [ALPHA] [:] **3** [STO▸] [ALPHA] [C] [ENTER].

   The coefficient of the X^2 term, the coefficient of the X term, and the constant for the new equation are stored to A, B, and C, respectively.

4. Press [2nd] [ENTRY] to recall the store instruction, and then press [2nd] [ENTRY] again to recall the quadratic-formula expression,

   $$\frac{-b-\sqrt{b^2-4ac}}{2a}$$

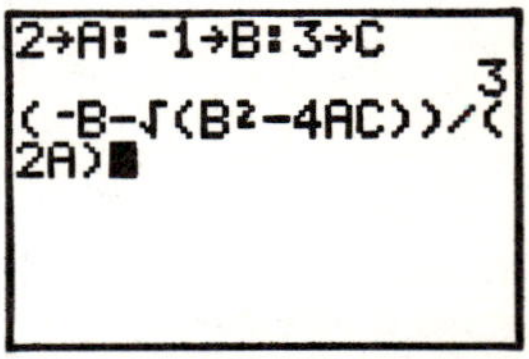

5. Press [ENTER] to find one solution for the equation $2X^2 - X + 3 = 0$.

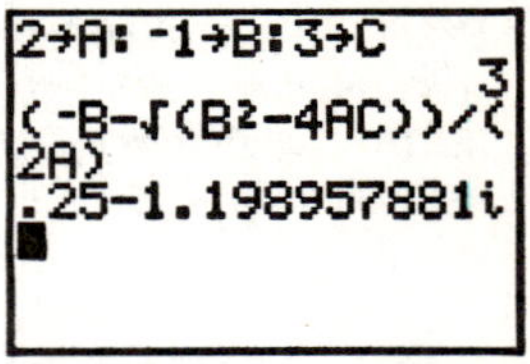

6. Press [2nd] [ENTRY] repeatedly until this quadratic-formula expression is displayed:

   $$\frac{-b+\sqrt{b^2-4ac}}{2a}$$

7. Press [ENTER] to find the other solution for the quadratic equation: $2X^2 - X + 3 = 0$.

Note: An alternative for solving equations for real numbers is to use the built-in Equation Solver (Chapter 2).

Take a 20 cm × 25 cm. sheet of paper and cut X × X squares from two corners. Cut X × 12.5 cm rectangles from the other two corners as shown in the diagram below. Fold the paper into a box with a lid. What value of X would give your box the maximum volume V? Use the table and graphs to determine the solution.

Begin by defining a function that describes the volume of the box.

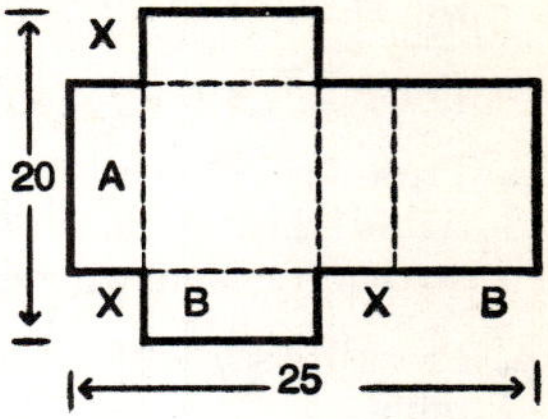

From the diagram: $2X + A = 20$

$2X + 2B = 25$

$V = A*B*X$

Substituting: $V = (20 - 2X)(25/2 - X)X$

1. Press [Y=] to display the Y= editor, which is where you define functions for tables and graphing.

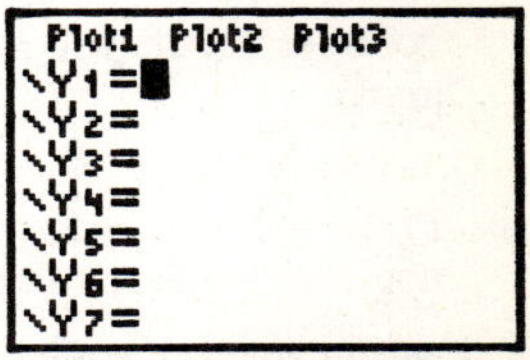

2. Press [(] 20 [-] 2 [X,T,Θ,n] [)] [(] 25 [÷] 2 [-] [X,T,Θ,n] [)] [X,T,Θ,n] [ENTER] to define the volume function as **Y1** in terms of **X**.

   [X,T,Θ,n] lets you enter **X** quickly, without having to press [ALPHA]. The highlighted **=** sign indicates that **Y1** is selected.

Defining a Table of Values: Box with Lid

The table feature of the TI-83 Plus displays numeric information about a function. You can use a table of values from the function defined on page 9 to estimate an answer to the problem.

1. Press [2nd] [TBLSET] (above [WINDOW]) to display the TABLE SETUP menu.

2. Press [ENTER] to accept **TblStart=0**.
3. Press **1** [ENTER] to define the table increment **ΔTbl=1**. Leave **Indpnt: Auto** and **Depend: Auto** so that the table will be generated automatically.
4. Press [2nd] [TABLE] (above [GRAPH]) to display the table.

   Notice that the maximum value for **Y1** (box's volume) occurs when **X** is about **4**, between **3** and **5**.

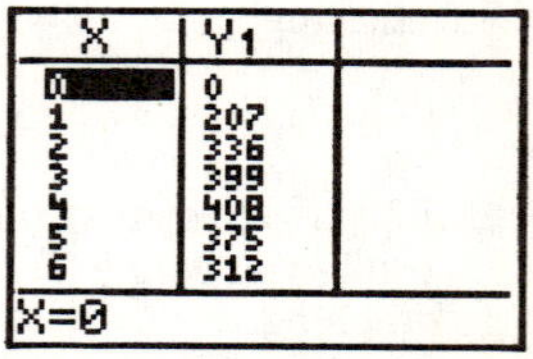

5. Press and hold [▼] to scroll the table until a negative result for **Y1** is displayed.

   Notice that the maximum length of **X** for this problem occurs where the sign of **Y1** (box's volume) changes from positive to negative, between **10** and **11**.

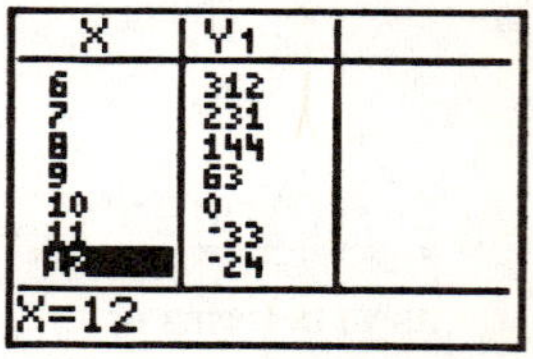

6. Press [2nd] [TBLSET].

   Notice that **TblStart** has changed to **6** to reflect the first line of the table as it was last displayed. (In step 5, the first value of **X** displayed in the table is **6**.)

You can adjust the way a table is displayed to get more information about a defined function. With smaller values for **ΔTbl**, you can zoom in on the table.

1. Press **3** [ENTER] to set **TblStart**. Press [.] **1** [ENTER] to set **ΔTbl**.

   This adjusts the table setup to get a more accurate estimate of **X** for maximum volume **Y1**.

2. Press [2nd] [TABLE].
3. Press [▼] and [▲] to scroll the table.

   Notice that the maximum value for **Y1** is **410.26**, which occurs at **X=3.7**. Therefore, the maximum occurs where **3.6<X<3.8**.

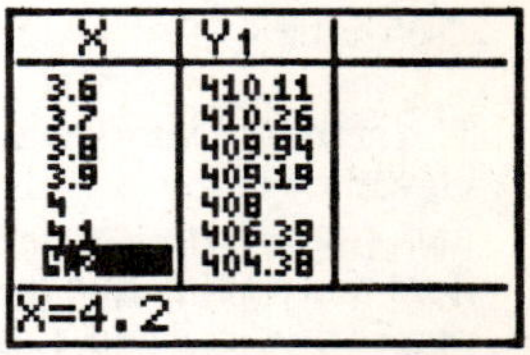

4. Press [2nd] [TBLSET]. Press **3** [.] **6** [ENTER] to set **TblStart**. Press [.] **01** [ENTER] to set **ΔTbl**.

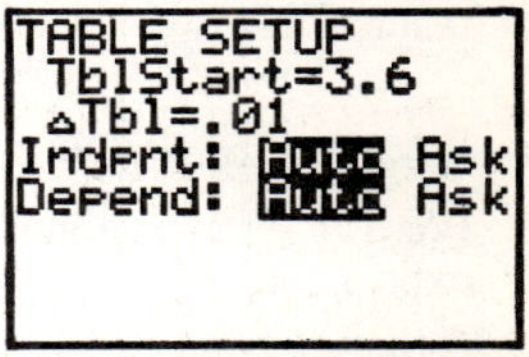

5. Press [2nd] [TABLE], and then press [▼] and [▲] to scroll the table.

   Four equivalent maximum values are shown, **410.26** at **X=3.67**, **3.68**, **3.69**, and **3.70**.

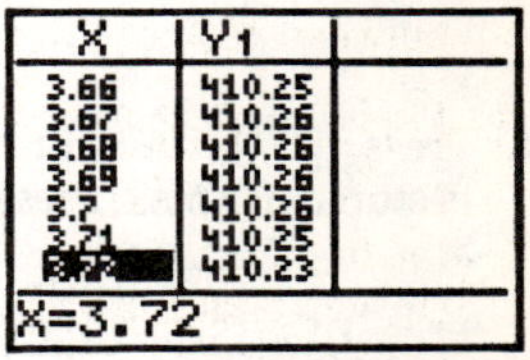

6. Press [▼] or [▲] to move the cursor to **3.67**. Press [▶] to move the cursor into the **Y1** column.

   The value of **Y1** at **X=3.67** is displayed on the bottom line in full precision as **410.261226**.

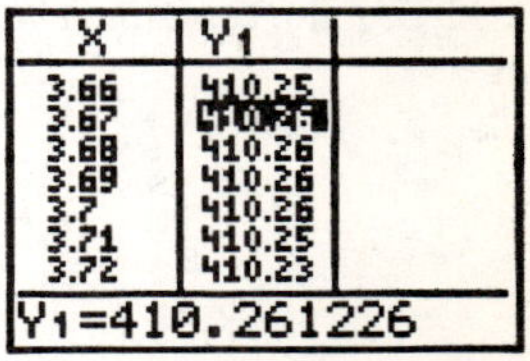

7. Press [▼] to display the other maximum.

   The value of **Y1** at **X=3.68** in full precision is **410.264064**, at **X=3.69** is **410.262318** and at **X=3.7** is **410.256**.

   The maximum volume of the box would occur at **3.68** if you could measure and cut the paper at .01-centimetre increments.

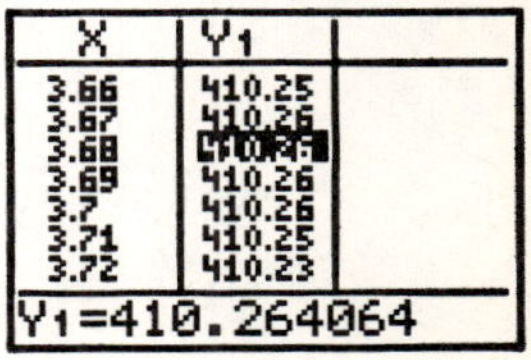

Setting the Viewing Window: Box with Lid

You also can use the graphing features of the TI-83 Plus to find the maximum value of a previously defined function. When the graph is activated, the viewing window defines the displayed portion of the coordinate plane. The values of the window variables determine the size of the viewing window.

1. Press [WINDOW] to display the window editor, where you can view and edit the values of the window variables.

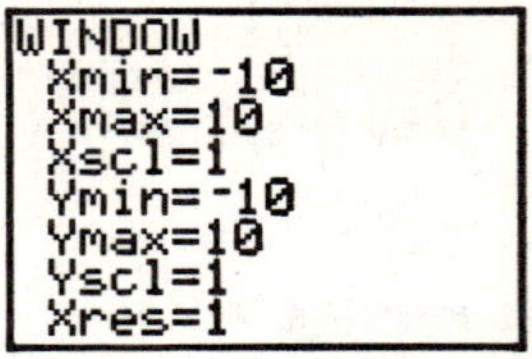

The standard window variables define the viewing window as shown. **Xmin**, **Xmax**, **Ymin**, and **Ymax** define the boundaries of the display. **Xscl** and **Yscl** define the distance between tick marks on the **X** and **Y** axes. **Xres** controls resolution.

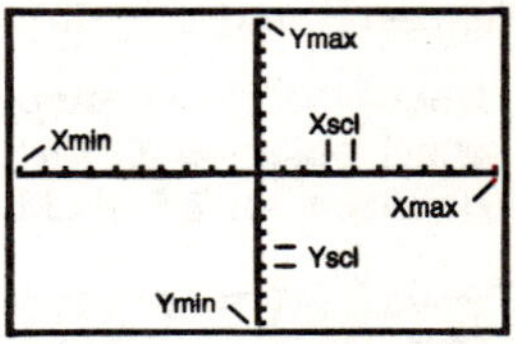

2. Press **0** [ENTER] to define **Xmin**.
3. Press **20** [÷] **2** to define **Xmax** using an expression.

4. Press [ENTER]. The expression is evaluated, and **10** is stored in **Xmax**. Press [ENTER] to accept **Xscl** as **1**.
5. Press **0** [ENTER] **500** [ENTER] **100** [ENTER] **1** [ENTER] to define the remaining window variables.

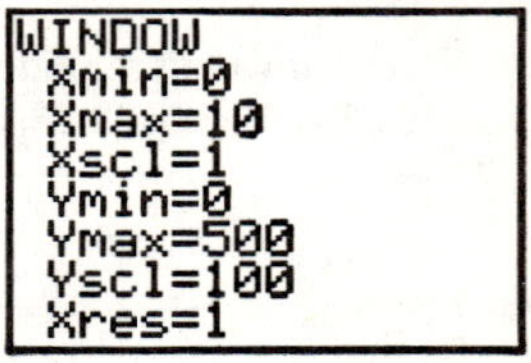

Now that you have defined the function to be graphed and the window in which to graph it, you can display and explore the graph. You can trace along a function using the TRACE feature.

1. Press [GRAPH] to graph the selected function in the viewing window.

   The graph of **Y1=(20−2X)(25/2−X)X** is displayed.

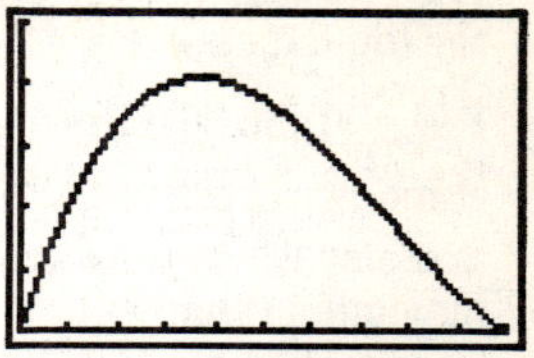

2. Press [▶] to activate the free-moving graph cursor.

   The **X** and **Y** coordinate values for the position of the graph cursor are displayed on the bottom line.

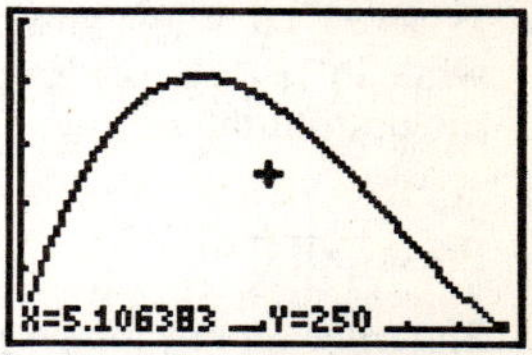

3. Press [◀], [▶], [▲], and [▼] to move the free-moving cursor to the apparent maximum of the function.

   As you move the cursor, the **X** and **Y** coordinate values are updated continually.

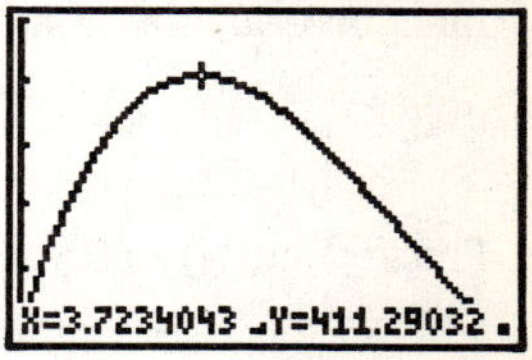

4. Press [TRACE]. The trace cursor is displayed on the **Y1** function.

   The function that you are tracing is displayed in the top-left corner.

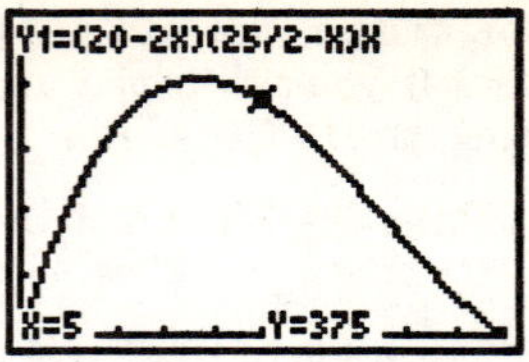

5. Press [◄] and [►] to trace along **Y1**, one **X** dot at a time, evaluating **Y1** at each **X**.

   You also can enter your estimate for the maximum value of **X**.

6. Press **3** [.] **8**. When you press a number key while in TRACE, the **X=** prompt is displayed in the bottom-left corner.

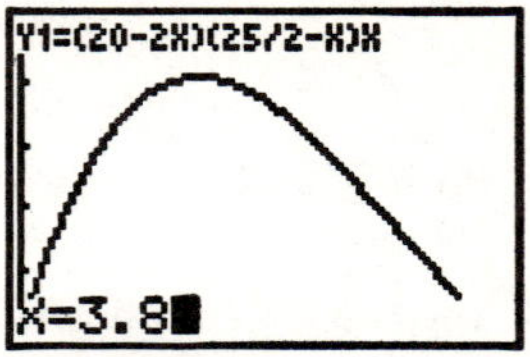

7. Press [ENTER].

   The trace cursor jumps to the point on the **Y1** function evaluated at **X=3.8**.

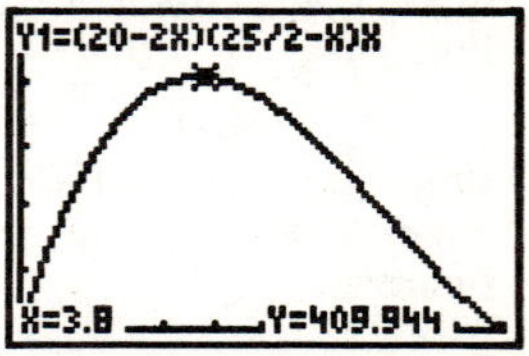

8. Press [◄] and [►] until you are on the maximum **Y** value.

   This is the maximum of **Y1(X)** for the **X** pixel values. The actual, precise maximum may lie between pixel values.

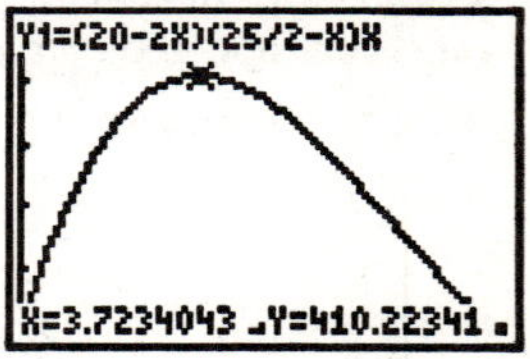

To help identify maximums, minimums, roots, and intersections of functions, you can magnify the viewing window at a specific location using the ZOOM instructions.

1. Press [ZOOM] to display the ZOOM menu.

   This menu is a typical TI-83 Plus menu. To select an item, you can either press the number or letter next to the item, or you can press [▼] until the item number or letter is highlighted, and then press [ENTER].

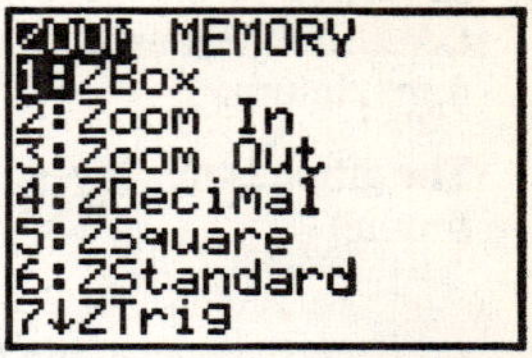

2. Press **2** to select **2:Zoom In**.

   The graph is displayed again. The cursor has changed to indicate that you are using a ZOOM instruction.

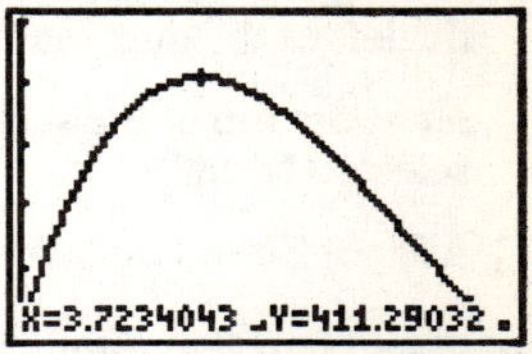

3. With the cursor near the maximum value of the function (as in step 8 on page 14), press [ENTER].

   The new viewing window is displayed. Both **Xmax−Xmin** and **Ymax−Ymin** have been adjusted by factors of 4, the default values for the zoom factors.

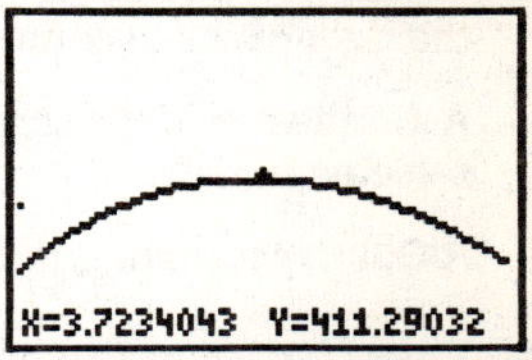

4. Press [WINDOW] to display the new window settings.

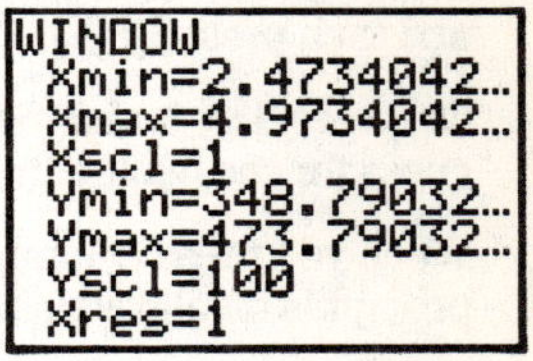

You can use a CALCULATE menu operation to calculate a local maximum of a function.

1. Press [2nd] [CALC] (above [TRACE]) to display the CALCULATE menu. Press **4** to select **4:maximum.**

   The graph is displayed again with a **Left Bound?** prompt.

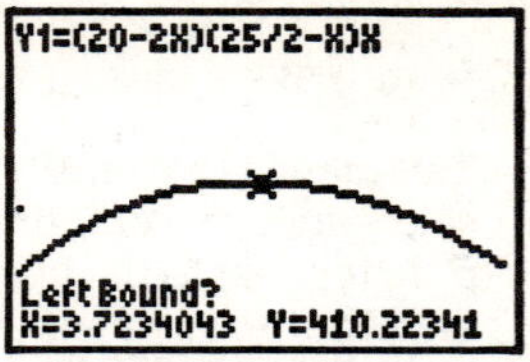

2. Press [◂] to trace along the curve to a point to the left of the maximum, and then press [ENTER].

   A ▸ at the top of the screen indicates the selected bound.

   A **Right Bound?** prompt is displayed.

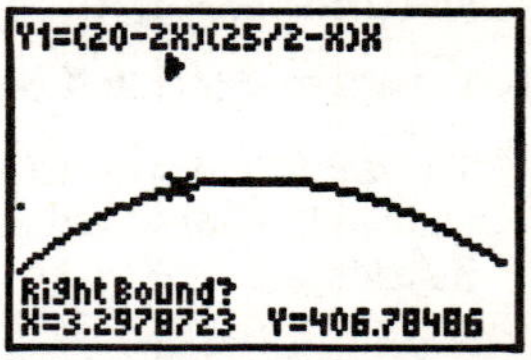

3. Press [▸] to trace along the curve to a point to the right of the maximum, and then press [ENTER].

   A ◂ at the top of the screen indicates the selected bound.

   A **Guess?** prompt is displayed.

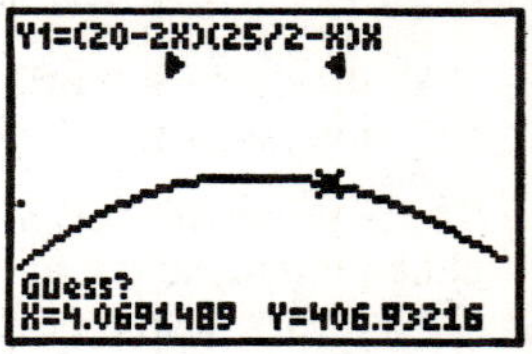

4. Press [◂] to trace to a point near the maximum, and then press [ENTER].

   Or, press **3** [.] **8**, and then press [ENTER] to enter a guess for the maximum.

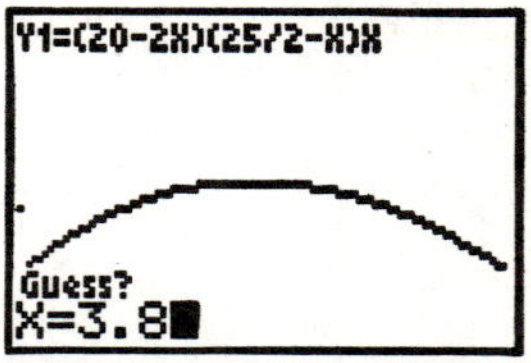

   When you press a number key in TRACE, the **X=** prompt is displayed in the bottom-left corner.

   Notice how the values for the calculated maximum compare with the maximums found with the free-moving cursor, the trace cursor, and the table.

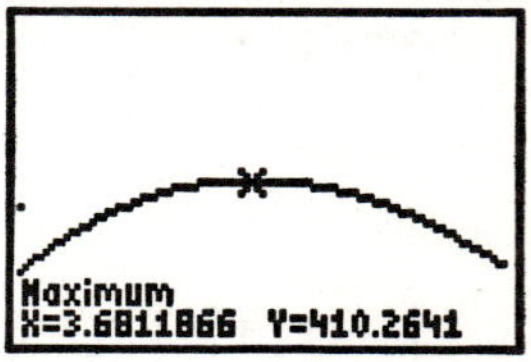

   Note: In steps 2 and 3 above, you can enter values directly for Left Bound and Right Bound, in the same way as described in step 4.

Getting Started has introduced you to basic TI-83 Plus operation. This guidebook describes in detail the features you used in Getting Started. It also covers the other features and capabilities of the TI-83 Plus.

Graphing

You can store, graph, and analyze up to 10 functions (Chapter 3), up to six parametric functions (Chapter 4), up to six polar functions (Chapter 5), and up to three sequences (Chapter 6). You can use DRAW instructions to annotate graphs (Chapter 8).

Sequences

You can generate sequences and graph them over time. Or, you can graph them as web plots or as phase plots (Chapter 6).

Tables

You can create function evaluation tables to analyze many functions simultaneously (Chapter 7).

Split Screen

You can split the screen horizontally to display both a graph and a related editor (such as the Y= editor), the table, the stat list editor, or the home screen. Also, you can split the screen vertically to display a graph and its table simultaneously (Chapter 9).

Matrices

You can enter and save up to 10 matrices and perform standard matrix operations on them (Chapter 10).

Lists

You can enter and save as many lists as memory allows for use in statistical analyses. You can attach formulas to lists for automatic computation. You can use lists to evaluate expressions at multiple values simultaneously and to graph a family of curves (Chapter 11).

Statistics

You can perform one- and two-variable, list-based statistical analyses, including logistic and sine regression analysis. You can plot the data as a histogram, xyLine, scatter plot, modified or regular box-and-whisker plot, or normal probability plot. You can define and store up to three stat plot definitions (Chapter 12).

Inferential Statistics

You can perform 16 hypothesis tests and confidence intervals and 15 distribution functions. You can display hypothesis test results graphically or numerically (Chapter 13).

Applications

You can use such applications as the Finance, the Calculator-Based Laboratory™ (CBL 2™ and CBL™), or the Calculator-Based Ranger™ (CBR™). With the Finance application you can use time-value-of-money (TVM) functions to analyze financial instruments such as annuities, loans, mortgages, leases, and savings. You can analyze the value of money over equal time periods using cash flow functions. You can amortize loans with the amortization functions. With the CBL/CBR applications and CBL 2/CBL or CBR (optional) accessories, you can use a variety of probes to collect real world data. (Chapter 14).

CATALOG

The CATALOG is a convenient, alphabetical list of all functions and instructions on the TI-83 Plus. You can paste any function or instruction from the CATALOG to the current cursor location (Chapter 15).

Programming

You can enter and store programs that include extensive control and input/output instructions (Chapter 16).

Archiving

Archiving allows you to store data, programs, or other variables to user data archive where they cannot be edited or deleted inadvertently. Archiving also allows you to free up RAM for variables that may require additional memory.

Archived variables are indicated by asterisks (*) to the left of the variable names (Chapter 16).

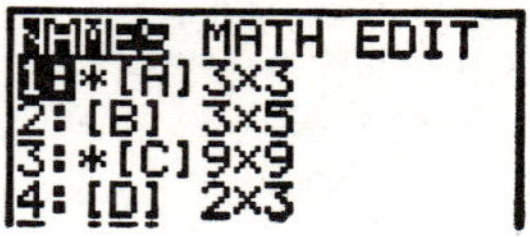

Communication Link

The TI-83 Plus has a port to connect and communicate with another TI-83 Plus, a TI-83 Plus Silver Edition, a TI-83, a TI-82, a TI-73, a Calculator-Based Laboratory (CBL 2/CBL) System, or a Calculator-Based Ranger (CBR) System. A unit-to-unit link cable is included with the TI-83 Plus for this purpose (Chapter 19). With a TI-GRAPH LINK™ (sold separately), you can also link the TI-83 Plus to a personal computer. As future software upgrades become available on the TI web site, you can download the software to your PC and then use the TI-GRAPH LINK™ to upgrade your TI-83 Plus.

1 Operating the TI-83 Plus

Contents

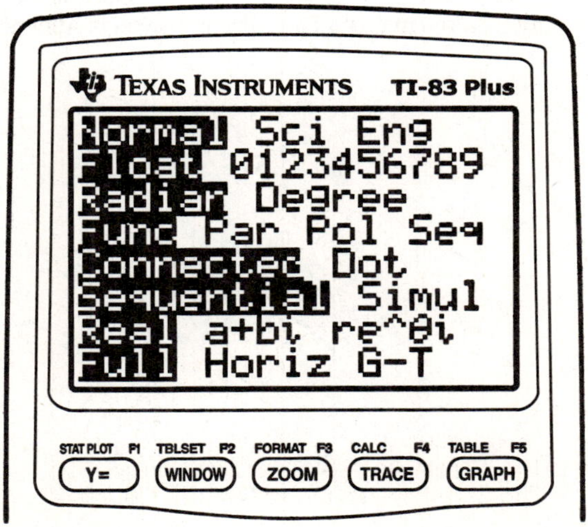

Turning On the Calculator

To turn on the TI-83 Plus, press [ON].

- If you previously had turned off the calculator by pressing [2nd] [OFF], the TI-83 Plus displays the home screen as it was when you last used it and clears any error.
- If Automatic Power Down™ (APD™) had previously turned off the calculator, the TI-83 Plus will return exactly as you left it, including the display, cursor, and any error.
- If the TI-83 Plus is turned off and you connect it to another calculator or personal computer, the TI-83 Plus will "wake up"when you complete the connection.
- If the TI-83 Plus is turned off and connected to another calculator or personal computer, any communication activity will "wake up" the TI-83 Plus.

To prolong the life of the batteries, APD turns off the TI-83 Plus automatically after about five minutes without any activity.

Turning Off the Calculator

To turn off the TI-83 Plus manually, press [2nd] [OFF].

- All settings and memory contents are retained by Constant Memory™.
- Any error condition is cleared.

Batteries

The TI-83 Plus uses four AAA alkaline batteries and has a user-replaceable backup lithium battery (CR1616 or CR1620). To replace batteries without losing any information stored in memory, follow the steps in Appendix B.

Adjusting the Display Contrast

You can adjust the display contrast to suit your viewing angle and lighting conditions. As you change the contrast setting, a number from **0** (lightest) to **9** (darkest) in the top-right corner indicates the current level. You may not be able to see the number if contrast is too light or too dark.

Note: The TI-83 Plus has 40 contrast settings, so each number **0** through **9** represents four settings.

The TI-83 Plus retains the contrast setting in memory when it is turned off.

To adjust the contrast, follow these steps.

1. Press and release the [2nd] key.
2. Press and hold [▼] or [▲], which are below and above the contrast symbol (yellow, half-shaded circle).
   - [▼] lightens the screen.
   - [▲] darkens the screen.

Note: If you adjust the contrast setting to **0**, the display may become completely blank. To restore the screen, press and release [2nd], and then press and hold [▲] until the display reappears.

When to Replace Batteries

When the batteries are low, a low-battery message is displayed when you:

- Turn on the calculator.
- Download a new application.
- Attempt to upgrade to new software.

To replace the batteries without losing any information in memory, follow the steps in Appendix B.

Generally, the calculator will continue to operate for one or two weeks after the low-battery message is first displayed. After this period, the TI-83 Plus will turn off automatically and the unit will not operate. Batteries must be replaced. All memory should be retained.

Note: The operating period following the first low-battery message could be longer than two weeks if you use the calculator infrequently.

Types of Displays

The TI-83 Plus displays both text and graphs. Chapter 3 describes graphs. Chapter 9 describes how the TI-83 Plus can display a horizontally or vertically split screen to show graphs and text simultaneously.

Home Screen

The home screen is the primary screen of the TI-83 Plus. On this screen, enter instructions to execute and expressions to evaluate. The answers are displayed on the same screen.

Displaying Entries and Answers

When text is displayed, the TI-83 Plus screen can display a maximum of 8 lines with a maximum of 16 characters per line. If all lines of the display are full, text scrolls off the top of the display. If an expression on the home screen, the Y= editor (Chapter 3), or the program editor (Chapter 16) is longer than one line, it wraps to the beginning of the next line. In numeric editors such as the window screen (Chapter 3), a long expression scrolls to the right and left.

When an entry is executed on the home screen, the answer is displayed on the right side of the next line.

```
log(2)              ← Entry
      .3010299957   ← Answer
```

The mode settings control the way the TI-83 Plus interprets expressions and displays answers (page 1-9).

If an answer, such as a list or matrix, is too long to display entirely on one line, an ellipsis (...) is displayed to the right or left. Press [▶] and [◀] to display the answer.

```
L1                  ← Entry
{25.12 874.2 36...  ← Answer
```

Returning to the Home Screen

To return to the home screen from any other screen, press [2nd] [QUIT].

Busy Indicator

When the TI-83 Plus is calculating or graphing, a vertical moving line is displayed as a busy indicator in the top-right corner of the screen. When you pause a graph or a program, the busy indicator becomes a vertical moving dotted line.

Display Cursors

In most cases, the appearance of the cursor indicates what will happen when you press the next key or select the next menu item to be pasted as a character.

Cursor	Appearance	Effect of Next Keystroke
Entry	Solid rectangle ■	A character is entered at the cursor; any existing character is overwritten
Insert	Underline __	A character is inserted in front of the cursor location
Second	Reverse arrow	A 2nd character (yellow on the keyboard) is entered or a 2nd operation is executed
Alpha	Reverse A	An alpha character (green on the keyboard) is entered or SOLVE is executed
Full	Checkerboard rectangle	No entry; the maximum characters are entered at a prompt or memory is full

If you press [ALPHA] during an insertion, the cursor becomes an underlined A (A). If you press [2nd] during an insertion, the underlined cursor becomes an underlined ↑ (↑).

Graphs and editors sometimes display additional cursors, which are described in other chapters.

What Is an Expression?

An expression is a group of numbers, variables, functions and their arguments, or a combination of these elements. An expression evaluates to a single answer. On the TI-83 Plus, you enter an expression in the same order as you would write it on paper. For example, πR^2 is an expression.

You can use an expression on the home screen to calculate an answer. In most places where a value is required, you can use an expression to enter a value.

```
(1/3)²
      .1111111111
```

```
WINDOW
 Xmin=-10
 Xmax=2π
```

Entering an Expression

To create an expression, you enter numbers, variables, and functions from the keyboard and menus. An expression is completed when you press [ENTER], regardless of the cursor location. The entire expression is evaluated according to Equation Operating System (EOS™) rules (page 1-23), and the answer is displayed.

Most TI-83 Plus functions and operations are symbols comprising several characters. You must enter the symbol from the keyboard or a menu; do not spell it out. For example, to calculate the log of 45, you must press [LOG] **45**. Do not enter the letters **L**, **O**, and **G**. If you enter **LOG**, the TI-83 Plus interprets the entry as implied multiplication of the variables **L**, **O**, and **G**.

Calculate 3.76 ÷ (⁻7.9 + √5) + 2 log 45.

3 [.] **76** [÷] [(] [(-)] **7** [.] **9** [+] [2nd] [√] **5** [)] [)] [+] **2** [LOG] **45** [)] [ENTER]

```
3.76/(-7.9+√(5))
+2log(45)
      2.642575252
```

Multiple Entries on a Line

To enter two or more expressions or instructions on a line, separate them with colons ([ALPHA] [:]). All instructions are stored together in last entry (ENTRY; page 1-17).

```
5→A:2→B:A/B
              2.5
```

Entering a Number in Scientific Notation

To enter a number in scientific notation, follow these steps.

1. Enter the part of the number that precedes the exponent. This value can be an expression.
2. Press [2nd] [EE]. **E** is pasted to the cursor location.
3. If the exponent is negative, press [(-)], and then enter the exponent, which can be one or two digits.

```
(19/2)E-2
                .095
```

When you enter a number in scientific notation, the TI-83 Plus does not automatically display answers in scientific or engineering notation. The mode settings (page 1-9) and the size of the number determine the display format.

Functions

A function returns a value. For example, ÷, ⁻, +, √(, and **log(** are the functions in the example on page 1-6. In general, the first letter of each function is lowercase on the TI-83 Plus. Most functions take at least one argument, as indicated by an open parenthesis (**(**) following the name. For example, **sin(** requires one argument, **sin(***value***)**.

Instructions

An instruction initiates an action. For example, **ClrDraw** is an instruction that clears any drawn elements from a graph. Instructions cannot be used in expressions. In general, the first letter of each instruction name is uppercase. Some instructions take more than one argument, as indicated by an open parenthesis (**(**) at the end of the name. For example, **Circle(** requires three arguments, **Circle(***X*,*Y*,*radius***)**.

Interrupting a Calculation

To interrupt a calculation or graph in progress, which is indicated by the busy indicator, press [ON].

When you interrupt a calculation, a menu is displayed.

- To return to the home screen, select **1:Quit**.
- To go to the location of the interruption, select **2:Goto**.

When you interrupt a graph, a partial graph is displayed.

- To return to the home screen, press [CLEAR] or any nongraphing key.
- To restart graphing, press a graphing key or select a graphing instruction.

Keystrokes	Result
▶ or ◀	Moves the cursor within an expression; these keys repeat.
▲ or ▼	Moves the cursor from line to line within an expression that occupies more than one line; these keys repeat. On the top line of an expression on the home screen, ▲ moves the cursor to the beginning of the expression. On the bottom line of an expression on the home screen, ▼ moves the cursor to the end of the expression.
2nd ◀	Moves the cursor to the beginning of an expression.
2nd ▶	Moves the cursor to the end of an expression.
ENTER	Evaluates an expression or executes an instruction.
CLEAR	On a line with text on the home screen, clears the current line. On a blank line on the home screen, clears everything on the home screen. In an editor, clears the expression or value where the cursor is located; it does not store a zero.
DEL	Deletes a character at the cursor; this key repeats.
2nd [INS]	Changes the cursor to an underline (__); inserts characters in front of the underline cursor; to end insertion, press 2nd [INS] or press ◀, ▲, ▶, or ▼.
2nd	Changes the cursor to ⬆; the next keystroke performs a 2nd operation (an operation in yellow above a key and to the left); to cancel 2nd, press 2nd again.
ALPHA	Changes the cursor to Ⓐ; the next keystroke pastes an alpha character (a character in green above a key and to the right) or executes SOLVE (Chapters 10 and 11); to cancel ALPHA, press ALPHA or press ◀, ▲, ▶, or ▼.
2nd [A-LOCK]	Changes the cursor to Ⓐ; sets alpha-lock; subsequent keystrokes (on an alpha key) paste alpha characters; to cancel alpha-lock, press ALPHA. If you are prompted to enter a name such as for a group or a program, alpha-lock is set automatically.
X,T,Θ,*n*	Pastes an **X** in **Func** mode, a **T** in **Par** mode, a θ in **Pol** mode, or an ***n*** in **Seq** mode with one keystroke.

Checking Mode Settings

Mode settings control how the TI-83 Plus displays and interprets numbers and graphs. Mode settings are retained by the Constant Memory feature when the TI-83 Plus is turned off. All numbers, including elements of matrices and lists, are displayed according to the current mode settings.

To display the mode settings, press [MODE]. The current settings are highlighted. Defaults are highlighted below. The following pages describe the mode settings in detail.

Normal Sci Eng	Numeric notation
Float 0123456789	Number of decimal places
Radian Degree	Unit of angle measure
Func Par Pol Seq	Type of graphing
Connected Dot	Whether to connect graph points
Sequential Simul	Whether to plot simultaneously
Real a+bi re^θi	Real, rectangular complex, or polar complex
Full Horiz G-T	Full screen, two split-screen modes

Changing Mode Settings

To change mode settings, follow these steps.

1. Press [▼] or [▲] to move the cursor to the line of the setting that you want to change.
2. Press [▶] or [◀] to move the cursor to the setting you want.
3. Press [ENTER].

Setting a Mode from a Program

You can set a mode from a program by entering the name of the mode as an instruction; for example, **Func** or **Float**. From a blank program command line, select the mode setting from the mode screen; the instruction is pasted to the cursor location.

```
PROGRAM:TEST
:Func■
```

Normal, Sci, Eng

Notation modes only affect the way an answer is displayed on the home screen. Numeric answers can be displayed with up to 10 digits and a two-digit exponent. You can enter a number in any format.

Normal notation mode is the usual way we express numbers, with digits to the left and right of the decimal, as in **12345.67**.

Sci (scientific) notation mode expresses numbers in two parts. The significant digits display with one digit to the left of the decimal. The appropriate power of 10 displays to the right of **E**, as in **1.234567E4**.

Eng (engineering) notation mode is similar to scientific notation. However, the number can have one, two, or three digits before the decimal; and the power-of-10 exponent is a multiple of three, as in **12.34567E3**.

Note: If you select **Normal** notation, but the answer cannot display in 10 digits (or the absolute value is less than .001), the TI-83 Plus expresses the answer in scientific notation.

Float, 0123456789

Float (floating) decimal mode displays up to 10 digits, plus the sign and decimal.

0123456789 (fixed) decimal mode specifies the number of digits (**0** through **9**) to display to the right of the decimal. Place the cursor on the desired number of decimal digits, and then press [ENTER].

The decimal setting applies to **Normal**, **Sci**, and **Eng** notation modes.

The decimal setting applies to these numbers:

- An answer displayed on the home screen
- Coordinates on a graph (Chapters 3, 4, 5, and 6)
- The **Tangent(** DRAW instruction equation of the line, **x**, and **dy/dx** values (Chapter 8)
- Results of CALCULATE operations (Chapters 3, 4, 5, and 6)
- The regression equation stored after the execution of a regression model (Chapter 12)

Radian, Degree

Angle modes control how the TI-83 Plus interprets angle values in trigonometric functions and polar/rectangular conversions.

Radian mode interprets angle values as radians. Answers display in radians.

Degree mode interprets angle values as degrees. Answers display in degrees.

Func, Par, Pol, Seq

Graphing modes define the graphing parameters. Chapters 3, 4, 5, and 6 describe these modes in detail.

Func (function) graphing mode plots functions, where **Y** is a function of **X** (Chapter 3).

Par (parametric) graphing mode plots relations, where **X** and **Y** are functions of **T** (Chapter 4).

Pol (polar) graphing mode plots functions, where **r** is a function of θ (Chapter 5).

Seq (sequence) graphing mode plots sequences (Chapter 6).

Connected, Dot

Connected plotting mode draws a line connecting each point calculated for the selected functions.

Dot plotting mode plots only the calculated points of the selected functions.

Sequential, Simul

Sequential graphing-order mode evaluates and plots one function completely before the next function is evaluated and plotted.

Simul (simultaneous) graphing-order mode evaluates and plots all selected functions for a single value of **X** and then evaluates and plots them for the next value of **X**.

Note: Regardless of which graphing mode is selected, the TI-83 Plus will sequentially graph all stat plots before it graphs any functions.

Real, a+b*i*, re^θ*i*

Real mode does not display complex results unless complex numbers are entered as input.

Two complex modes display complex results.

- **a+b*i*** (rectangular complex mode) displays complex numbers in the form a+b*i*.
- **re^θ*i*** (polar complex mode) displays complex numbers in the form re^θ*i*.

Full, Horiz, G-T

Full screen mode uses the entire screen to display a graph or edit screen.

Each split-screen mode displays two screens simultaneously.

- **Horiz** (horizontal) mode displays the current graph on the top half of the screen; it displays the home screen or an editor on the bottom half (Chapter 9).
- **G-T** (graph-table) mode displays the current graph on the left half of the screen; it displays the table screen on the right half (Chapter 9).

Variables and Defined Items

On the TI-83 Plus you can enter and use several types of data, including real and complex numbers, matrices, lists, functions, stat plots, graph databases, graph pictures, and strings.

The TI-83 Plus uses assigned names for variables and other items saved in memory. For lists, you also can create your own five-character names.

Variable Type	Names
Real numbers	**A**, **B**, . . . , **Z**
Complex numbers	**A**, **B**, . . . , **Z**
Matrices	[A], [B], [C], . . . , [J]
Lists	**L1**, **L2**, **L3**, **L4**, **L5**, **L6**, and user-defined names
Functions	**Y1**, **Y2**, . . . , **Y9**, **Y0**
Parametric equations	**X1T** and **Y1T**, . . . , **X6T** and **Y6T**
Polar functions	**r1**, **r2**, **r3**, **r4**, **r5**, **r6**
Sequence functions	**u**, **v**, **w**
Stat plots	**Plot1**, **Plot2**, **Plot3**
Graph databases	**GDB1**, **GDB2**, . . . , **GDB9**, **GDB0**
Graph pictures	**Pic1**, **Pic2**, . . . , **Pic9**, **Pic0**
Strings	**Str1**, **Str2**, . . . , **Str9**, **Str0**
Apps	Applications
AppVars	Application variables
Groups	Grouped variables
System variables	**Xmin**, **Xmax**, and others

Notes about Variables

- You can create as many list names as memory will allow (Chapter 11).
- Programs have user-defined names and share memory with variables (Chapter 16).
- From the home screen or from a program, you can store to matrices (Chapter 10), lists (Chapter 11), strings (Chapter 15), system variables such as **Xmax** (Chapter 1), **TblStart** (Chapter 7), and all Y= functions (Chapters 3, 4, 5, and 6).
- From an editor, you can store to matrices, lists, and Y= functions (Chapter 3).
- From the home screen, a program, or an editor, you can store a value to a matrix element or a list element.
- You can use DRAW STO menu items to store and recall graph databases and pictures (Chapter 8).
- Although most variables can be archived, system variables including **r**, **t**, x, y, and θ cannot be archived (Chapter 18)
- **Apps** are independent applications.which are stored in Flash ROM. **AppVars** is a variable holder used to store variables created by independent applications. You cannot edit or change variables in **AppVars** unless you do so through the application which created them.

Storing Values in a Variable

Values are stored to and recalled from memory using variable names. When an expression containing the name of a variable is evaluated, the value of the variable at that time is used.

To store a value to a variable from the home screen or a program using the [STO▶] key, begin on a blank line and follow these steps.

1. Enter the value you want to store. The value can be an expression.
2. Press [STO▶]. → is copied to the cursor location.
3. Press [ALPHA] and then the letter of the variable to which you want to store the value.
4. Press [ENTER]. If you entered an expression, it is evaluated. The value is stored to the variable.

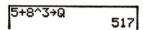

```
5+8^3→Q
                517
```

Displaying a Variable Value

To display the value of a variable, enter the name on a blank line on the home screen, and then press [ENTER].

```
Q
                517
```

Archiving Variables

You can archive data, programs, or other variables in a section of memory called user data archive where they cannot be edited or deleted inadvertently. Archived variables are indicated by asterisks (*) to the left of the variable names. Archived variables cannot be edited or executed. They can only be seen and unarchived. For example, if you archive list L1, you will see that L1 exists in memory but if you select it and paste the name L1 to the home screen, you won't be able to see its contents or edit it until they are unarchived.

Using Recall (RCL)

To recall and copy variable contents to the current cursor location, follow these steps. To leave RCL, press [CLEAR].

1. Press [2nd] [RCL]. **Rcl** and the edit cursor are displayed on the bottom line of the screen.
2. Enter the name of the variable in any of five ways.
   - Press [ALPHA] and then the letter of the variable.
   - Press [2nd] [LIST], and then select the name of the list, or press [2nd] [L*n*].
   - Press [2nd] [MATRX], and then select the name of the matrix.
   - Press [VARS] to display the VARS menu or [VARS] [▸] to display the VARS Y-VARS menu; then select the type and then the name of the variable or function.
   - Press [PRGM] [◂], and then select the name of the program (in the program editor only).

   The variable name you selected is displayed on the bottom line and the cursor disappears.

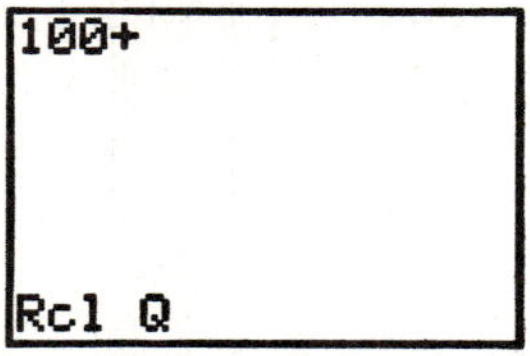

3. Press [ENTER]. The variable contents are inserted where the cursor was located before you began these steps.

Note: You can edit the characters pasted to the expression without affecting the value in memory.

Using ENTRY (Last Entry)

When you press [ENTER] on the home screen to evaluate an expression or execute an instruction, the expression or instruction is placed in a storage area called ENTRY (last entry). When you turn off the TI-83 Plus, ENTRY is retained in memory.

To recall ENTRY, press [2nd] [ENTRY]. The last entry is pasted to the current cursor location, where you can edit and execute it. On the home screen or in an editor, the current line is cleared and the last entry is pasted to the line.

Because the TI-83 Plus updates ENTRY only when you press [ENTER], you can recall the previous entry even if you have begun to enter the next expression.

Keystrokes	Screen
5 [+] **7** [ENTER] [2nd] [ENTRY]	5+7 12 5+7■

Accessing a Previous Entry

The TI-83 Plus retains as many previous entries as possible in ENTRY, up to a capacity of 128 bytes. To scroll those entries, press [2nd] [ENTRY] repeatedly. If a single entry is more than 128 bytes, it is retained for ENTRY, but it cannot be placed in the ENTRY storage area.

Keystrokes	Screen
1 [STO▸] [ALPHA] **A** [ENTER] **2** [STO▸] [ALPHA] **B** [ENTER] [2nd] [ENTRY]	1→A 1 2→B 2 2→B■

If you press [2nd] [ENTRY] after displaying the oldest stored entry, the newest stored entry is displayed again, then the next-newest entry, and so on.

Keystrokes	Screen
[2nd] [ENTRY]	1→A 1 2→B 2 1→A■

Reexecuting the Previous Entry

After you have pasted the last entry to the home screen and edited it (if you chose to edit it), you can execute the entry. To execute the last entry, press [ENTER].

To reexecute the displayed entry, press [ENTER] again. Each reexecution displays an answer on the right side of the next line; the entry itself is not redisplayed.

0 [STO▸] [ALPHA] **N**
[ENTER]
[ALPHA] **N** [+] **1** [STO▸] [ALPHA] **N** [ALPHA] [:] [ALPHA] **N**
[x^2] [ENTER]
[ENTER]
[ENTER]

```
0→N
                0
N+1→N:N²
                1
                4
                9
```

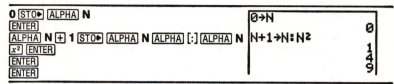

Multiple Entry Values on a Line

To store to ENTRY two or more expressions or instructions, separate each expression or instruction with a colon, then press [ENTER]. All expressions and instructions separated by colons are stored in ENTRY.

When you press [2nd] [ENTRY], all the expressions and instructions separated by colons are pasted to the current cursor location. You can edit any of the entries, and then execute all of them when you press [ENTER].

For the equation $A=\pi r^2$, use trial and error to find the radius of a circle that covers 200 square centimeters. Use 8 as your first guess.

8 [STO▸] [ALPHA] **R** [ALPHA] [:] [2nd] [π] [ALPHA] **R** [x^2]
[ENTER]
[2nd] [ENTRY]

```
8→R:πR²
      201.0619298
8→R:πR²█
```

[2nd] [◂] **7** [2nd] [INS] [.] **95**
[ENTER]

```
8→R:πR²
      201.0619298
7.95→R:πR²
      198.5565097
```

Continue until the answer is as accurate as you want.

Clearing ENTRY

Clear Entries (Chapter 18) clears all data that the TI-83 Plus is holding in the ENTRY storage area.

Using Ans in an Expression

When an expression is evaluated successfully from the home screen or from a program, the TI-83 Plus stores the answer to a storage area called **Ans** (last answer). **Ans** may be a real or complex number, a list, a matrix, or a string. When you turn off the TI-83 Plus, the value in **Ans** is retained in memory.

You can use the variable **Ans** to represent the last answer in most places. Press [2nd] [ANS] to copy the variable name **Ans** to the cursor location. When the expression is evaluated, the TI-83 Plus uses the value of **Ans** in the calculation.

Calculate the area of a garden plot 1.7 meters by 4.2 meters. Then calculate the yield per square meter if the plot produces a total of 147 tomatoes.

1 [.] 7 [×] 4 [.] 2
[ENTER]
147 [÷] [2nd] [ANS]
[ENTER]

```
1.7*4.2
                7.14
147/Ans
         20.58823529
```

Continuing an Expression

You can use **Ans** as the first entry in the next expression without entering the value again or pressing [2nd] [ANS]. On a blank line on the home screen, enter the function. The TI-83 Plus pastes the variable name **Ans** to the screen, then the function.

5 [÷] 2
[ENTER]
[×] 9 [.] 9
[ENTER]

```
5/2
                 2.5
Ans*9.9
               24.75
```

Storing Answers

To store an answer, store **Ans** to a variable before you evaluate another expression.

Calculate the area of a circle of radius 5 meters. Next, calculate the volume of a cylinder of radius 5 meters and height 3.3 meters, and then store the result in the variable V.

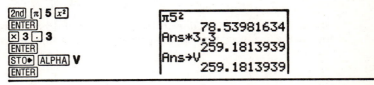

[2nd] [π] 5 [x^2]
[ENTER]
[×] 3 [.] 3
[ENTER]
[STO▸] [ALPHA] V
[ENTER]

```
π5²
         78.53981634
Ans*3.3
         259.1813939
Ans→V
         259.1813939
```

Using a TI-83 Plus Menu

You can access most TI-83 Plus operations using menus. When you press a key or key combination to display a menu, one or more menu names appear on the top line of the screen.

- The menu name on the left side of the top line is highlighted. Up to seven items in that menu are displayed, beginning with item **1**, which also is highlighted.
- A number or letter identifies each menu item's place in the menu. The order is **1** through **9**, then **0**, then **A**, **B**, **C**, and so on. The LIST NAMES, PRGM EXEC, and PRGM EDIT menus only label items **1** through **9** and **0**.
- When the menu continues beyond the displayed items, a down arrow (↓) replaces the colon next to the last displayed item.
- When a menu item ends in an ellipsis (...), the item displays a secondary menu or editor when you select it.
- When an asterisk (*) appears to the left of a menu item, that item is stored in user data archive (Chapter 18).

```
RAM FREE    19849
ARC FREE   147345
 *[A]          83
  [B]         146
▸*[C]          20
  [D]          65
```

To display any other menu listed on the top line, press [▶] or [◀] until that menu name is highlighted. The cursor location within the initial menu is irrelevant. The menu is displayed with the cursor on the first item.

Note: The Menu Map in Appendix A shows each menu, each operation under each menu, and the key or key combination you press to display each menu.

Scrolling a Menu

To scroll down the menu items, press [▼]. To scroll up the menu items, press [▲].

To page down six menu items at a time, press [ALPHA] [▼]. To page up six menu items at a time, press [ALPHA] [▲]. The green arrows on the calculator, between [▼] and [▲], are the page-down and page-up symbols.

To wrap to the last menu item directly from the first menu item, press [▲]. To wrap to the first menu item directly from the last menu item, press [▼].

Selecting an Item from a Menu

You can select an item from a menu in either of two ways.

- Press the number or letter of the item you want to select. The cursor can be anywhere on the menu, and the item you select need not be displayed on the screen.
- Press [▼] or [▲] to move the cursor to the item you want, and then press [ENTER].

After you select an item from a menu, the TI-83 Plus typically displays the previous screen.

Note: On the LIST NAMES, PRGM EXEC, and PRGM EDIT menus, only items **1** through **9** and **0** are labeled in such a way that you can select them by pressing the appropriate number key. To move the cursor to the first item beginning with any alpha character or θ, press the key combination for that alpha character or θ. If no items begin with that character, the cursor moves beyond it to the next item.

Calculate $^3\sqrt{27}$.

[MATH] [▼] [▼] [▼] [ENTER] **27** [)] [ENTER]	³√(27) 3

Leaving a Menu without Making a Selection

You can leave a menu without making a selection in any of four ways.

- Press [2nd] [QUIT] to return to the home screen.
- Press [CLEAR] to return to the previous screen.
- Press a key or key combination for a different menu, such as [MATH] or [2nd] [LIST].
- Press a key or key combination for a different screen, such as [Y=] or [2nd] [TABLE].

VARS Menu

You can enter the names of functions and system variables in an expression or store to them directly.

To display the VARS menu, press [VARS]. All VARS menu items display secondary menus, which show the names of the system variables. **1:Window, 2:Zoom**, and **5:Statistics** each access more than one secondary menu.

VARS Y-VARS	
1: Window...	X/Y, T/θ, and U/V/W variables
2: Zoom...	ZX/ZY, ZT/Zθ, and ZU variables
3: GDB...	Graph database variables
4: Picture...	Picture variables
5: Statistics...	XY, Σ, EQ, TEST, and PTS variables
6: Table...	TABLE variables
7: String...	String variables

Selecting a Variable from the VARS Menu or VARS Y-VARS Menu

To display the VARS Y-VARS menu, press [VARS] [▸]. **1:Function, 2:Parametric**, and **3:Polar** display secondary menus of the Y= function variables.

VARS Y-VARS	
1: Function...	**Y*n*** functions
2: Parametric...	**X*n*T**, **Y*n*T** functions
3: Polar...	**r*n*** functions
4: On/Off...	Lets you select/deselect functions

Note: The sequence variables (**u, v, w**) are located on the keyboard as the second functions of [7], [8], and [9].

To select a variable from the VARS or VARS Y-VARS menu, follow these steps.

1. Display the VARS or VARS Y-VARS menu.
   - Press [VARS] to display the VARS menu.
   - Press [VARS] [▸] to display the VARS Y-VARS menu.
2. Select the type of variable, such as **2:Zoom** from the VARS menu or **3:Polar** from the VARS Y-VARS menu. A secondary menu is displayed.
3. If you selected **1:Window**, **2:Zoom**, or **5:Statistics** from the VARS menu, you can press [▸] or [◂] to display other secondary menus.
4. Select a variable name from the menu. It is pasted to the cursor location.

Order of Evaluation

The Equation Operating System (EOS™) defines the order in which functions in expressions are entered and evaluated on the TI-83 Plus. EOS lets you enter numbers and functions in a simple, straightforward sequence.

EOS evaluates the functions in an expression in this order.

Order Number	Function
1	Functions that precede the argument, such as √(, **sin(**, or **log(**
2	Functions that are entered after the argument, such as ^{2}, $^{-1}$, **!**, °, ^{r}, and conversions
3	Powers and roots, such as **2^5** or **5ˣ√32**
4	Permutations (**nPr**) and combinations (**nCr**)
5	Multiplication, implied multiplication, and division
6	Addition and subtraction
7	Relational functions, such as > or ≤
8	Logic operator **and**
9	Logic operators **or** and **xor**

Within a priority level, EOS evaluates functions from left to right.

Calculations within parentheses are evaluated first.

Implied Multiplication

The TI-83 Plus recognizes implied multiplication, so you need not press [×] to express multiplication in all cases. For example, the TI-83 Plus interprets **2π**, **4sin(46)**, **5(1+2)**, and **(2*5)7** as implied multiplication.

Note: TI-83 Plus implied multiplication rules,although like theTI-83, differ from those of the TI-82. For example, the TI-83 Plus evaluates **1/2X** as **(1/2)*X**, while the TI-82 evaluates **1/2X** as **1/(2*X)** (Chapter 2).

Parentheses

All calculations inside a pair of parentheses are completed first. For example, in the expression **4(1+2)**, EOS first evaluates the portion inside the parentheses, **1+2**, and then multiplies the answer, **3**, by **4**.

```
4*1+2
                6
4(1+2)
               12
```

You can omit the close parenthesis (**)**) at the end of an expression. All open parenthetical elements are closed automatically at the end of an expression. This is also true for open parenthetical elements that precede the store or display-conversion instructions.

Note: An open parenthesis following a list name, matrix name, or Y= function name does not indicate implied multiplication. It specifies elements in the list (Chapter 11) or matrix (Chapter 10) and specifies a value for which to solve the Y= function.

Negation

To enter a negative number, use the negation key. Press [(-)] and then enter the number. On the TI-83 Plus, negation is in the third level in the EOS hierarchy. Functions in the first level, such as squaring, are evaluated before negation.

For example, $^{-}\mathbf{X}^{2}$, evaluates to a negative number (or 0). Use parentheses to square a negative number.

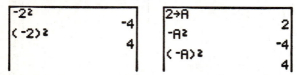

```
-2²
               -4
(-2)²
                4
```

```
2→A
                2
-A²
               -4
(-A)²
                4
```

Note: Use the [−] key for subtraction and the [(-)] key for negation. If you press [−] to enter a negative number, as in **9** [×] [−] **7**, or if you press [(-)] to indicate subtraction, as in **9** [(-)] **7**, an error occurs. If you press [ALPHA] **A** [(-)] [ALPHA] **B**, it is interpreted as implied multiplication (**A*⁻B**).

Diagnosing an Error

The TI-83 Plus detects errors while performing these tasks.

- Evaluating an expression
- Executing an instruction
- Plotting a graph
- Storing a value

When the TI-83 Plus detects an error, it returns an error message as a menu title, such as ERR:SYNTAX or ERR:DOMAIN. Appendix B describes each error type and possible reasons for the error.

- If you select **1:Quit** (or press [2nd] [QUIT] or [CLEAR]), then the home screen is displayed.
- If you select **2:Goto**, then the previous screen is displayed with the cursor at or near the error location.

Note: If a syntax error occurs in the contents of a Y= function during program execution, then the **Goto** option returns to the Y= editor, not to the program.

Correcting an Error

To correct an error, follow these steps.

1. Note the error type (**ERR:***error type*).
2. Select **2:Goto**, if it is available. The previous screen is displayed with the cursor at or near the error location.
3. Determine the error. If you cannot recognize the error, refer to Appendix B.
4. Correct the expression.

2 Math, Angle, and Test Operations

Contents

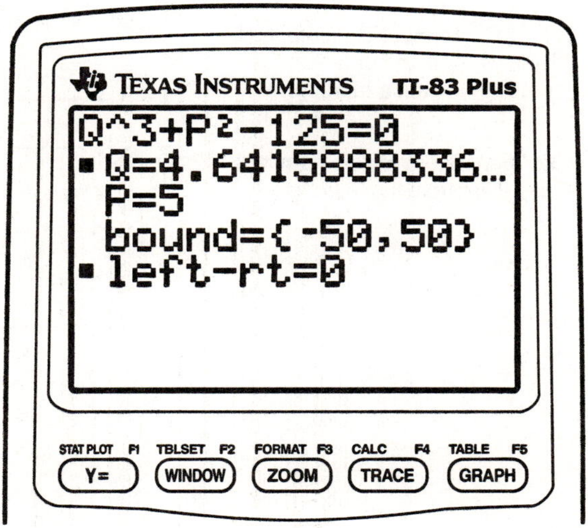

Getting Started is a fast-paced introduction. Read the chapter for details.

Suppose you want to model flipping a fair coin 10 times. You want to track how many of those 10 coin flips result in heads. You want to perform this simulation 40 times. With a fair coin, the probability of a coin flip resulting in heads is 0.5 and the probability of a coin flip resulting in tails is 0.5.

1. Begin on the home screen. Press [MATH] [◄] to display the MATH PRB menu. Press **7** to select **7:randBin(** (random Binomial). **randBin(** is pasted to the home screen. Press **10** to enter the number of coin flips. Press [,]. Press [.] **5** to enter the probability of heads. Press [,]. Press **40** to enter the number of simulations. Press [)].

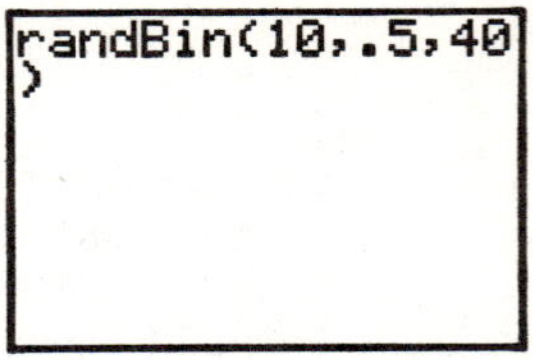

2. Press [ENTER] to evaluate the expression. A list of 40 elements is generated with the first 7 displayed. The list contains the count of heads resulting from each set of 10 coin flips. The list has 40 elements because this simulation was performed 40 times. In this example, the coin came up heads five times in the first set of 10 coin flips, five times in the second set of 10 coin flips, and so on.

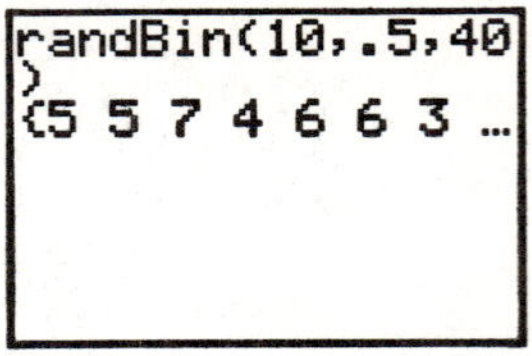

3. Press [►] or [◄] to view the additional counts in the list. Ellipses (...) indicate that the list continues beyond the screen.

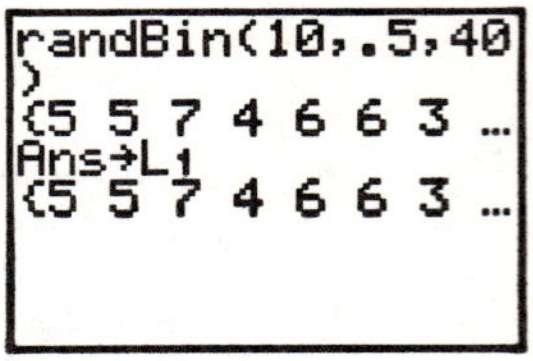

4. Press [STO►] [2nd] [L1] [ENTER] to store the data to the list name **L1**. You then can use the data for another activity, such as plotting a histogram (Chapter 12).

   Note: Since **randBin(** generates random numbers, your list elements may differ from those in the example.

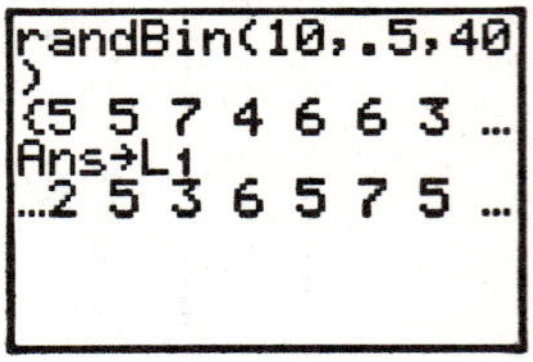

Using Lists with Math Operations

Math operations that are valid for lists return a list calculated element by element. If you use two lists in the same expression, they must be the same length.

```
{1,2}+{3,4}+5
          {9 11}
```

+ (Addition), – (Subtraction). * (Multiplication), / (Division)

You can use **+** (addition, [+]), – (subtraction, [–]), ***** (multiplication, [×]), and / (division, [÷]) with real and complex numbers, expressions, lists, and matrices. You cannot use / with matrices.

valueA+*valueB*   *valueA* – *valueB*
*valueA***valueB*   *valueA* / *valueB*

Trigonometric Functions

You can use the trigonometric (trig) functions (sine, [SIN]; cosine, [COS]; and tangent, [TAN]) with real numbers, expressions, and lists. The current angle mode setting affects interpretation. For example, **sin(30)** in **Radian** mode returns **⁻.9880316241**; in **Degree** mode it returns **.5**.

sin(*value***)**   **cos(***value***)**   **tan(***value***)**

You can use the inverse trig functions (arcsine, [2nd] [SIN^{-1}]; arccosine, [2nd] [COS^{-1}]; and arctangent, [2nd] [TAN^{-1}]) with real numbers, expressions, and lists. The current angle mode setting affects interpretation.

sin^{-1}(*value***)**   **cos^{-1}(***value***)**   **tan^{-1}(***value***)**

Note: The trig functions do not operate on complex numbers.

^ (Power), ^{2} (Square), √((Square Root)

You can use **^** (power, [^]), ^{2} (square, [x^2]), and **√(** (square root, [2nd] [√]) with real and complex numbers, expressions, lists, and matrices. You cannot use **√(** with matrices.

value^*power*   *value*^{2} **√(***value***)**

$^{-1}$ (Inverse)

You can use $^{-1}$ (inverse, [x^{-1}]) with real and complex numbers, expressions, lists, and matrices. The multiplicative inverse is equivalent to the reciprocal, $1/x$.

value^{-1}

```
5⁻¹
              .2
```

log(, 10^(, ln(

You can use **log(** (logarithm, [LOG]), **10^(** (power of 10, [2nd] [10^x]), and **ln(** (natural log, [LN]) with real or complex numbers, expressions, and lists.

log(*value***)** **10^(***power***)** **ln(***value***)**

e^((Exponential)

e^((exponential, [2nd] [e^x]) returns the constant **e** raised to a power. You can use **e^(** with real or complex numbers, expressions, and lists.

e^(*power***)**

```
e^(5)
         148.4131591
```

e (Constant)

e (constant, [2nd] [e]) is stored as a constant on the TI-83 Plus. Press [2nd] [e] to copy **e** to the cursor location. In calculations, the TI-83 Plus uses 2.718281828459 for **e**.

```
e
         2.718281828
```

⁻ (Negation)

⁻ (negation, [(-)]) returns the negative of *value*. You can use ⁻ with real or complex numbers, expressions, lists, and matrices.

⁻*value*

EOS rules (Chapter 1) determine when negation is evaluated. For example, ⁻$\mathbf{A^2}$ returns a negative number, because squaring is evaluated before negation. Use parentheses to square a negated number, as in **(⁻A)**$^{\mathbf{2}}$.

```
2→A:{⁻A²,(⁻A)²,⁻
2²,(⁻2)²}
         {⁻4 4 ⁻4 4}
```

Note: On the TI-83 Plus, the negation symbol (⁻) is shorter and higher than the subtraction sign (−), which is displayed when you press [−].

π (Pi)

π (Pi, [2nd] [π]) is stored as a constant in the TI-83 Plus. In calculations, the TI-83 Plus uses 3.1415926535898 for π.

```
π
         3.141592654
```

MATH Menu

To display the MATH menu, press [MATH].

MATH NUM CPX PRB	
1: ▸Frac	Displays the answer as a fraction.
2: ▸Dec	Displays the answer as a decimal.
3: ^{3}	Calculates the cube.
4: $^3\sqrt{}($	Calculates the cube root.
5: $^x\sqrt{}$	Calculates the xth root.
6: fMin(	Finds the minimum of a function.
7: fMax(	Finds the maximum of a function.
8: nDeriv(	Computes the numerical derivative.
9: fnInt(	Computes the function integral.
0: Solver...	Displays the equation solver.

▸Frac, ▸Dec

▸Frac (display as a fraction) displays an answer as its rational equivalent. You can use **▸Frac** with real or complex numbers, expressions, lists, and matrices. If the answer cannot be simplified or the resulting denominator is more than three digits, the decimal equivalent is returned. You can only use **▸Frac** following *value*.

value **▸Frac**

▸Dec (display as a decimal) displays an answer in decimal form. You can use **▸Dec** with real or complex numbers, expressions, lists, and matrices. You can only use **▸Dec** following *value*.

value **▸Dec**

```
1/2+1/3▸Frac
             5/6
Ans▸Dec
     .8333333333
```

^{3}(Cube), $^3\sqrt{}$((Cube Root)

^{3} (cube) returns the cube of *value*. You can use ^{3} with real or complex numbers, expressions, lists, and square matrices.

value^3

$^3\sqrt{}$((cube root) returns the cube root of *value*. You can use $^3\sqrt{}$(with real or complex numbers, expressions, and lists.

$^3\sqrt{}$(*value*)

```
{2,3,4,5}³
    {8 27 64 125}
³√(Ans)
        {2 3 4 5}
```

$^x\sqrt{}$ (Root)

$^x\sqrt{}$ (x^{th} root) returns the xth *root* of *value*. You can use $^x\sqrt{}$ with real or complex numbers, expressions, and lists.

x^{th}*root*$^x\sqrt{}$*value*

```
5ˣ√32
               2
```

fMin(, fMax(

fMin((function minimum) and **fMax(** (function maximum) return the value at which the local minimum or local maximum value of *expression* with respect to *variable* occurs, between *lower* and *upper* values for *variable*. **fMin(** and **fMax(** are not valid in *expression*. The accuracy is controlled by *tolerance* (if not specified, the default is 1E-5).

fMin(*expression*,*variable*,*lower*,*upper*[,*tolerance*]**)**
fMax(*expression*,*variable*,*lower*,*upper*[,*tolerance*]**)**

Note: In this guidebook, optional arguments and the commas that accompany them are enclosed in brackets ([ ]).

```
fMin(sin(A),A,-π
,π)
     -1.570797171
fMax(sin(A),A,-π
,π)
      1.570797171
```

nDeriv(

nDeriv((numerical derivative) returns an approximate derivative of *expression* with respect to *variable*, given the *value* at which to calculate the derivative and ε (if not specified, the default is 1E-3). **nDeriv(** is valid only for real numbers.

nDeriv(*expression*,*variable*,*value*[,ε]**)**

nDeriv(uses the symmetric difference quotient method, which approximates the numerical derivative value as the slope of the secant line through these points.

$$f'(x) = \frac{f(x+\varepsilon)-f(x-\varepsilon)}{2\varepsilon}$$

As ε becomes smaller, the approximation usually becomes more accurate.

```
nDeriv(A^3,A,5,.
01)
         75.0001
nDeriv(A^3,A,5,.
0001)
              75
```

You can use **nDeriv(** once in *expression*. Because of the method used to calculate **nDeriv(**, the TI-83 Plus can return a false derivative value at a nondifferentiable point.

fnInt(

fnInt((function integral) returns the numerical integral (Gauss-Kronrod method) of *expression* with respect to *variable*, given *lower* limit, *upper* limit, and a *tolerance* (if not specified, the default is 1E-5). **fnInt(** is valid only for real numbers.

fnInt(*expression*,*variable*,*lower*,*upper*[,*tolerance*]**)**

```
fnInt(A²,A,0,1)
     .3333333333
```

Tip: To speed the drawing of integration graphs (when **fnInt(** is used in a Y= equation), increase the value of the **Xres** window variable before you press [GRAPH].

Solver

Solver displays the equation solver, in which you can solve for any variable in an equation. The equation is assumed to be equal to zero. **Solver** is valid only for real numbers.

When you select **Solver**, one of two screens is displayed.

- The equation editor (see step 1 picture below) is displayed when the equation variable **eqn** is empty.
- The interactive solver editor (see step 3 picture on page 2-9) is displayed when an equation is stored in **eqn**.

Entering an Expression in the Equation Solver

To enter an expression in the equation solver, assuming that the variable **eqn** is empty, follow these steps.

1. Select **0:Solver** from the MATH menu to display the equation editor.

```
EQUATION SOLVER
eqn:0=█
```

2. Enter the expression in any of three ways.
   - Enter the expression directly into the equation solver.
   - Paste a Y= variable name from the VARS Y-VARS menu to the equation solver.
   - Press [2nd] [RCL], paste a Y= variable name from the VARS Y-VARS menu, and press [ENTER]. The expression is pasted to the equation solver.

   The expression is stored to the variable **eqn** as you enter it.

```
EQUATION SOLVER
eqn:0=Q^3+P²-125
█
```

Entering an Expression in the Equation Solver (continued)

3. Press [ENTER] or [▼]. The interactive solver editor is displayed.

```
Q^3+P²-125=0
 Q=0
 P=0
 bound={-1E99,1…
```

- The equation stored in **eqn** is set equal to zero and displayed on the top line.
- Variables in the equation are listed in the order in which they appear in the equation. Any values stored to the listed variables also are displayed.
- The default lower and upper bounds appear in the last line of the editor (**bound={-1E99,1E99}**).
- A ↓ is displayed in the first column of the bottom line if the editor continues beyond the screen.

Tip: To use the solver to solve an equation such as $\mathbf{K=.5MV^2}$, enter $\mathbf{eqn{:}0=K-.5MV^2}$ in the equation editor.

Entering and Editing Variable Values

When you enter or edit a value for a variable in the interactive solver editor, the new value is stored in memory to that variable.

You can enter an expression for a variable value. It is evaluated when you move to the next variable. Expressions must resolve to real numbers at each step during the iteration.

You can store equations to any VARS Y-VARS variables, such as **Y1** or **r6**, and then reference the variables in the equation. The interactive solver editor displays all variables of all Y= functions referenced in the equation.

```
\Y9■X²-4AC
\Y0=
```

```
EQUATION SOLVER
eqn:0=Y9+7
```

```
Y9+7=0
 X=0
 A=0
 C=0
 bound={-1E99,1…
```

Solving for a Variable in the Equation Solver

To solve for a variable using the equation solver after an equation has been stored to **eqn**, follow these steps.

1. Select **0:Solver** from the MATH menu to display the interactive solver editor, if not already displayed.

```
Q^3+P²-125=0
 Q=0
 P=0
 bound={-1E99,1…
```

2. Enter or edit the value of each known variable. All variables, except the unknown variable, must contain a value. To move the cursor to the next variable, press [ENTER] or [▼].

```
Q^3+P²-125=0
 Q=0
 P=5■
 bound={-1E99,1…
```

3. Enter an initial guess for the variable for which you are solving. This is optional, but it may help find the solution more quickly. Also, for equations with multiple roots, the TI-83 Plus will attempt to display the solution that is closest to your guess.

```
Q^3+P²-125=0
 Q=4■
 P=5
 bound={-1E99,1…
```

The default guess is calculated as $\frac{(upper+lower)}{2}$.

Solving for a Variable in the Equation Solver (continued)

4. Edit **bound={*lower*,*upper*}**. *lower* and *upper* are the bounds between which the TI-83 Plus searches for a solution. This is optional, but it may help find the solution more quickly. The default is **bound={⁻1E99,1E99}**.

5. Move the cursor to the variable for which you want to solve and press [ALPHA] [SOLVE] (above the [ENTER] key).

- The solution is displayed next to the variable for which you solved. A solid square in the first column marks the variable for which you solved and indicates that the equation is balanced. An ellipsis shows that the value continues beyond the screen.

  Note: When a number continues beyond the screen, be sure to press [▸] to scroll to the end of the number to see whether it ends with a negative or positive exponent. A very small number may appear to be a large number until you scroll right to see the exponent.

- The values of the variables are updated in memory.
- **left-rt=***diff* is displayed in the last line of the editor. *diff* is the difference between the left and right sides of the equation. A solid square in the first column next to **left-rt=** indicates that the equation has been evaluated at the new value of the variable for which you solved.

Editing an Equation Stored to eqn

To edit or replace an equation stored to **eqn** when the interactive equation solver is displayed, press [▲] until the equation editor is displayed. Then edit the equation.

Equations with Multiple Roots

Some equations have more than one solution. You can enter a new initial guess (page 2-10) or new bounds (page 2-11) to look for additional solutions.

Further Solutions

After you solve for a variable, you can continue to explore solutions from the interactive solver editor. Edit the values of one or more variables. When you edit any variable value, the solid squares next to the previous solution and **left-rt=***diff* disappear. Move the cursor to the variable for which you now want to solve and press [ALPHA] [SOLVE].

Controlling the Solution for Solver or solve(

The TI-83 Plus solves equations through an iterative process. To control that process, enter bounds that are relatively close to the solution and enter an initial guess within those bounds. This will help to find a solution more quickly. Also, it will define which solution you want for equations with multiple solutions.

Using solve(on the Home Screen or from a Program

The function **solve(** is available only from CATALOG or from within a program. It returns a solution (root) of *expression* for *variable*, given an initial *guess*, and *lower* and *upper* bounds within which the solution is sought. The default for *lower* is ⁻1E99. The default for *upper* is 1E99. **solve(** is valid only for real numbers.

solve(*expression*,*variable*,*guess*[,{*lower*,*upper*}])

expression is assumed equal to zero. The value of *variable* will not be updated in memory. *guess* may be a value or a list of two values. Values must be stored for every variable in *expression*, except *variable*, before *expression* is evaluated. *lower* and *upper* must be entered in list format.

```
5→P
                5
solve(Q^3+P²-125
,Q,4,{-50,50})
      4.641588834
```

MATH NUM Menu

To display the MATH NUM menu, press [MATH] [▶].

MATH NUM CPX PRB	
1: abs(	Absolute value
2: round(	Round
3: iPart(	Integer part
4: fPart(	Fractional part
5: int(	Greatest integer
6: min(	Minimum value
7: max(	Maximum value
8: lcm(	Least common multiple
9: gcd(	Greatest common divisor

abs(

abs((absolute value) returns the absolute value of real or complex (modulus) numbers, expressions, lists, and matrices.

abs(*value***)**

```
abs(-256)
              256
abs({1.25,-5.67}
)
      {1.25 5.67}
```

Note: abs(is also available on the MATH CPX menu.

round(

round(returns a number, expression, list, or matrix rounded to *#decimals* (≤9). If *#decimals* is omitted, *value* is rounded to the digits that are displayed, up to 10 digits.

round(*value*[,*#decimals*]**)**

```
round(π,4)
           3.1416
```

```
123456789012→C
   1.23456789E11
C-round(C)
              12
123456789012-123
456789000
              12
```

iPart(, fPart(

iPart((integer part) returns the integer part or parts of real or complex numbers, expressions, lists, and matrices.

iPart(*value***)**

fPart((fractional part) returns the fractional part or parts of real or complex numbers, expressions, lists, and matrices.

fPart(*value***)**

```
iPart(-23.45)
              -23
fPart(-23.45)
             -.45
```

int(

int((greatest integer) returns the largest integer ≤ real or complex numbers, expressions, lists, and matrices.

int(*value***)**

```
int(-23.45)
              -24
```

Note: For a given *value*, the result of **int(** is the same as the result of **iPart(** for nonnegative numbers and negative integers, but one integer less than the result of **iPart(** for negative noninteger numbers.

min(, max(

min((minimum value) returns the smaller of *valueA* and *valueB* or the smallest element in *list*. If *listA* and *listB* are compared, **min(** returns a list of the smaller of each pair of elements. If *list* and *value* are compared, **min(** compares each element in *list* with *value*.

max((maximum value) returns the larger of *valueA* and *valueB* or the largest element in *list*. If *listA* and *listB* are compared, **max(** returns a list of the larger of each pair of elements. If *list* and *value* are compared, **max(** compares each element in *list* with *value*.

min(*valueA*,*valueB***)**
min(*list***)**
min(*listA*,*listB***)**
min(*list*,*value***)**

max(*valueA*,*valueB***)**
max(*list***)**
max(*listA*,*listB***)**
max(*list*,*value***)**

```
min(3,2+2)
               3
min({3,4,5},4)
         {3 4 4}
max({4,5,6})
               6
```

Note: min(and **max(** also are available on the LIST MATH menu.

lcm(, gcd(

lcm(returns the least common multiple of *valueA* and *valueB*, both of which must be nonnegative integers. When *listA* and *listB* are specified, **lcm(** returns a list of the lcm of each pair of elements. If *list* and *value* are specified, **lcm(** finds the lcm of each element in *list* and *value*.

gcd(returns the greatest common divisor of *valueA* and *valueB*, both of which must be nonnegative integers. When *listA* and *listB* are specified, **gcd(** returns a list of the gcd of each pair of elements. If *list* and *value* are specified, **gcd(** finds the gcd of each element in *list* and *value*.

lcm(*valueA*,*valueB***)**
lcm(*listA*,*listB***)**
lcm(*list*,*value***)**

gcd(*valueA*,*valueB***)**
gcd(*listA*,*listB***)**
gcd(*list*,*value***)**

```
lcm(2,5)
              10
gcd({48,66},{64,
122})
          {16 2}
```

Complex-Number Modes

The TI-83 Plus displays complex numbers in rectangular form and polar form. To select a complex-number mode, press [MODE], and then select either of the two modes.

- **a+b*i*** (rectangular-complex mode)
- **re^θ*i*** (polar-complex mode)

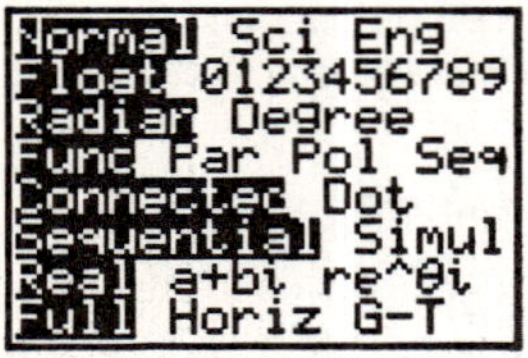

On the TI-83 Plus, complex numbers can be stored to variables. Also, complex numbers are valid list elements.

In **Real** mode, complex-number results return an error, unless you entered a complex number as input. For example, in **Real** mode **ln(⁻1)** returns an error; in **a+b*i*** mode **ln(⁻1)** returns an answer.

Real mode

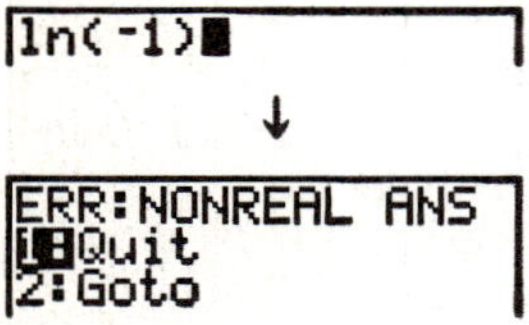

a+b*i* mode

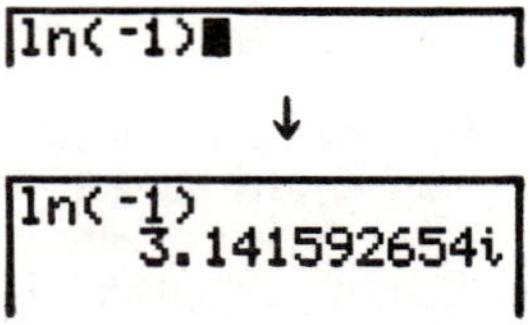

Entering Complex Numbers

Complex numbers are stored in rectangular form, but you can enter a complex number in rectangular form or polar form, regardless of the mode setting. The components of complex numbers can be real numbers or expressions that evaluate to real numbers; expressions are evaluated when the command is executed.

Note about Radian Versus Degree Mode

Radian mode is recommended for complex number calculations. Internally, the TI-83 Plus converts all entered trigonometric values to radians, but it does not convert values for exponential, logarithmic, or hyperbolic functions.

In degree mode, complex identities such as e^($i\theta$) = cos(θ) + i sin(θ) are not generally true because the values for cos and sin are converted to radians, while those for e^() are not. For example, e^(i45) = cos(45) + i sin(45) is treated internally as e^(i45) = cos(π/4) + i sin(π/4). Complex identities are always true in radian mode.

Interpreting Complex Results

Complex numbers in results, including list elements, are displayed in either rectangular or polar form, as specified by the mode setting or by a display conversion instruction (page 2-19). In the example below, polar-complex (**re^θ*i***) and **Radian** modes are set.

```
(2+i)−(1e^(π/4i)
)
1.325654296e^(...
```

Rectangular-Complex Mode

Rectangular-complex mode recognizes and displays a complex number in the form a+b***i***, where a is the real component, b is the imaginary component, and ***i*** is a constant equal to $\sqrt{-1}$.

```
ln(-1)
      3.141592654i
```

To enter a complex number in rectangular form, enter the value of *a* (*real component*), press [+] or [−], enter the value of *b* (*imaginary component*), and press [2nd] [***i***] (constant).

real component(+ or −)*imaginary component* ***i***

```
4+2i
             4+2i
```

Polar-Complex Mode

Polar-complex mode recognizes and displays a complex number in the form *re*^ ***θi***, where *r* is the magnitude, *e* is the base of the natural log, *θ* is the angle, and ***i*** is a constant equal to $\sqrt{-1}$.

```
ln(-1)
3.141592654e^(1...
```

To enter a complex number in polar form, enter the value of *r* (*magnitude*), press [2nd] [e^x] (exponential function), enter the value of *θ* (*angle*), press [2nd] [***i***] (constant), and then press [)].

*magnitude***e^(***anglei***)**

```
10e^(π/3i)
10e^(1.04719755...
```

MATH CPX Menu

To display the MATH CPX menu, press [MATH] [▸] [▸].

MATH NUM **CPX** PRB	
1:conj(	Returns the complex conjugate.
2:real(	Returns the real part.
3:imag(	Returns the imaginary part.
4:angle(	Returns the polar angle.
5:abs(	Returns the magnitude (modulus).
6:▸Rect	Displays the result in rectangular form.
7:▸Polar	Displays the result in polar form.

conj(

conj((conjugate) returns the complex conjugate of a complex number or list of complex numbers.

conj(*a*+*bi***)** returns *a*−*bi* in **a+b***i* mode.
conj(*r***e^(***θi***))** returns *r***e^(**-*θi***)** in **re^θ***i* mode.

```
conj(3+4i)
             3-4i
```

```
conj(3e^(4i))
3e^(2.283185307…
```

real(

real((real part) returns the real part of a complex number or list of complex numbers.

real(*a*+*bi***)** returns *a*.
real(*r***e^(***θi***))** returns *r*cos*(*θ*).

```
real(3+4i)
                3
```

```
real(3e^(4i))
     -1.960930863
```

imag(

imag((imaginary part) returns the imaginary (nonreal) part of a complex number or list of complex numbers.

imag(*a*+*bi***)** returns *b*.
imag(*r***e^(***θi***))** returns *r*sin*(*θ*).

```
imag(3+4i)
                4
```

```
imag(3e^(4i))
     -2.270407486
```

angle(

angle(returns the polar angle of a complex number or list of complex numbers, calculated as $\tan^{-1}$ (b/a), where b is the imaginary part and a is the real part. The calculation is adjusted by $+\pi$ in the second quadrant or $-\pi$ in the third quadrant.

angle($a+bi$**)** returns $tan^{-1}(b/a)$.
angle(r**e^(**θi**))** returns θ, where $^-\pi<\theta<\pi$.

```
angle(3+4i)
      .927295218
```

```
angle(3e^(4i))
    -2.283185307
```

abs(

abs((absolute value) returns the magnitude (modulus), $\sqrt{(real2+imag2)}$, of a complex number or list of complex numbers.

abs($a+bi$**)** returns $\sqrt{(a2+b2)}$.
abs(r**e^(**θi**))** returns r (magnitude).

```
abs(3+4i)
               5
```

```
abs(3e^(4i))
               3
```

▶Rect

▶Rect (display as rectangular) displays a complex result in rectangular form. It is valid only at the end of an expression. It is not valid if the result is real.

complex result▶**Rect** returns $a+bi$.

```
√(-2)▶Rect
     1.414213562i
```

▶Polar

▶Polar (display as polar) displays a complex result in polar form. It is valid only at the end of an expression. It is not valid if the result is real.

complex result▶**Polar** returns r**e^(**θi**)**.

```
√(-2)▶Polar
1.414213562e^(1…
```

MATH PRB Menu

To display the MATH PRB menu, press [MATH] [◄].

MATH NUM CPX PRB	
1: rand	Random-number generator
2: nPr	Number of permutations
3: nCr	Number of combinations
4: !	Factorial
5: randInt(	Random-integer generator
6: randNorm(	Random # from Normal distribution
7: randBin(	Random # from Binomial distribution

rand

rand (random number) generates and returns one or more random numbers > 0 and < 1. To generate a list of random-numbers, specify an integer > 1 for *numtrials* (number of trials). The default for *numtrials* is **1**.

rand[(*numtrials*)]

Tip: To generate random numbers beyond the range of 0 to 1, you can include **rand** in an expression. For example, **rand*5** generates a random number > 0 and < **5**.

With each **rand** execution, the TI-83 Plus generates the same random-number sequence for a given seed value. The TI-83 Plus factory-set seed value for **rand** is **0**. To generate a different random-number sequence, store any nonzero seed value to **rand**. To restore the factory-set seed value, store **0** to **rand** or reset the defaults (Chapter 18).

Note: The seed value also affects **randInt(**, **randNorm(**, and **randBin(** instructions (page 2-22).

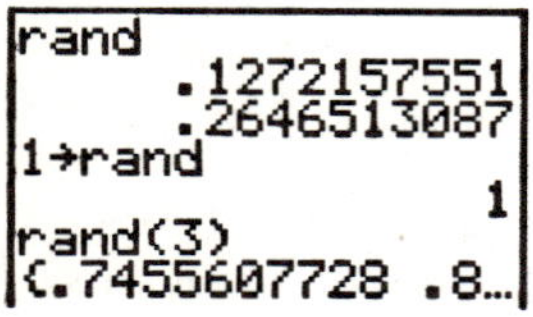

nPr, nCr

nPr (number of permutations) returns the number of permutations of *items* taken *number* at a time. *items* and *number* must be nonnegative integers. Both *items* and *number* can be lists.

items **nPr** *number*

nCr (number of combinations) returns the number of combinations of *items* taken *number* at a time. *items* and *number* must be nonnegative integers. Both *items* and *number* can be lists.

items **nCr** *number*

```
5 nPr 2
                  20
5 nCr 2
                  10
{2,3} nPr {2,2}
               {2 6}
```

! (Factorial)

! (factorial) returns the factorial of either an integer or a multiple of .5. For a list, it returns factorials for each integer or multiple of .5. *value* must be $\geq$⁻.5 and $\leq$69.

*value***!**

```
6!
                 720
{5,4,6}!
      {120 24 720}
```

Note: The factorial is computed recursively using the relationship $(n+1)! = n*n!$, until n is reduced to either 0 or ⁻1/2. At that point, the definition $0!=1$ or the definition $(-1/2)!=\sqrt{\pi}$ is used to complete the calculation. Hence:

$n!=n*(n-1)*(n-2)* \ldots *2*1$, if n is an integer ≥ 0
$n!= n*(n-1)*(n-2)* \ldots *1/2*\sqrt{\pi}$, if $n+1/2$ is an integer ≥ 0
n! is an error, if neither n nor $n+1/2$ is an integer ≥ 0.

(The variable n equals *value* in the syntax description above.)

randInt(

randInt((random integer) generates and displays a random integer within a range specified by *lower* and *upper* integer bounds. To generate a list of random numbers, specify an integer >1 for *numtrials* (number of trials); if not specified, the default is 1.

randInt(*lower*,*upper*[,*numtrials*]**)**

```
randInt(1,6)+ran
dInt(1,6)
               6
randInt(1,6,3)
        {2 1 5}
```

randNorm(

randNorm((random Normal) generates and displays a random real number from a specified Normal distribution. Each generated value could be any real number, but most will be within the interval [μ−3(σ), μ+3(σ)]. To generate a list of random numbers, specify an integer > 1 for *numtrials* (number of trials); if not specified, the default is 1.

randNorm(*μ*,*σ*[,*numtrials*]**)**

```
randNorm(0,1)
      .0772076175
randNorm(35,2,10
0)
{34.02701938 37…
```

randBin(

randBin((random Binomial) generates and displays a random integer from a specified Binomial distribution. *numtrials* (number of trials) must be ≥ 1. *prob* (probability of success) must be ≥ 0 and ≤ 1. To generate a list of random numbers, specify an integer > 1 for *numsimulations* (number of simulations); if not specified, the default is 1.

randBin(*numtrials*,*prob*[,*numsimulations*]**)**

```
randBin(5,.2)
               3
randBin(7,.4,10)
{3 3 2 5 1 2 2 …
```

Note: The seed value stored to **rand** also affects **randInt(**, **randNorm(**, and **randBin(** instructions (page 2-20).

ANGLE Menu

To display the ANGLE menu, press [2nd] [ANGLE]. The ANGLE menu displays angle indicators and instructions. The **Radian/Degree** mode setting affects the TI-83 Plus's interpretation of ANGLE menu entries.

ANGLE	
1: °	Degree notation
2: '	DMS minute notation
3: r	Radian notation
4: ▶DMS	Displays as degree/minute/second
5: R▶Pr(	Returns r, given X and Y
6: R▶Pθ(	Returns θ, given X and Y
7: P▶Rx(	Returns x, given R and θ
8: P▶Ry(	Returns y, given R and θ

DMS Entry Notation

DMS (degrees/minutes/seconds) entry notation comprises the degree symbol (°), the minute symbol ('), and the second symbol ("). *degrees* must be a real number; *minutes* and *seconds* must be real numbers ≥ 0.

degrees°minutes'seconds"

For example, enter for 30 degrees, 1 minute, 23 seconds. If the angle mode is not set to **Degree**, you must use ° so that the TI-83 Plus can interpret the argument as degrees, minutes, and seconds.

Degree mode

```
sin(30°1'23")
      .5003484441
```

Radian mode

```
sin(30°1'23")
     -.9842129995
sin(30°1'23"°)
      .5003484441
```

° (Degree)

° (degree) designates an angle or list of angles as degrees, regardless of the current angle mode setting. In **Radian** mode, you can use ° to convert degrees to radians.

value°
{*value1*,*value2*,*value3*,*value4*,...,*value n*}°

° also designates *degrees* (D) in DMS format.
' (minutes) designates *minutes* (M) in DMS format.
" (seconds) designates *seconds* (S) in DMS format.

Note: " is not on the ANGLE menu. To enter ", press [ALPHA] ["].

^{r} (Radians)

^{r} (radians) designates an angle or list of angles as radians, regardless of the current angle mode setting. In **Degree** mode, you can use ^{r} to convert radians to degrees.

value^r

Degree mode

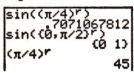

```
sin((π/4)r)
      .7071067812
sin({0,π/2}r)
            {0 1}
(π/4)r
               45
```

▶DMS

▶DMS (degree/minute/second) displays *answer* in DMS format (page 2-23). The mode setting must be **Degree** for *answer* to be interpreted as degrees, minutes, and seconds. **▶DMS** is valid only at the end of a line.

answer▶**DMS**

```
54°32'30"*2
      109.0833333
Ans▶DMS
         109°5'0"
```

R▶Pr(, R▶Pθ(, P▶Rx(, P▶Ry(

R▶Pr(converts rectangular coordinates to polar coordinates and returns **r**. **R▶Pθ(** converts rectangular coordinates to polar coordinates and returns **θ**. x and y can be lists.

R▶Pr(x,y**)**, **R▶Pθ(**x,y**)**

```
R▶Pr(-1,0)
                1
R▶Pθ(-1,0)
      3.141592654
```

Note: Radian mode is set.

P▶Rx(converts polar coordinates to rectangular coordinates and returns **x**. **P▶Ry(** converts polar coordinates to rectangular coordinates and returns **y**. r and θ can be lists.

P▶Rx(r,θ**)**, **P▶Ry(**r,θ**)**

```
P▶Rx(1,π)
               -1
P▶Ry(1,π)
                0
```

Note: Radian mode is set.

TEST Menu

To display the TEST menu, press [2nd] [TEST].

This operator...	Returns 1 (true) if...
TEST LOGIC	
1: =	Equal
2: ≠	Not equal to
3: >	Greater than
4: ≥	Greater than or equal to
5: <	Less than
6: ≤	Less than or equal to

=, ≠, >, ≥, <, ≤

Relational operators compare *valueA* and *valueB* and return **1** if the test is true or **0** if the test is false. *valueA* and *valueB* can be real numbers, expressions, or lists. For = and ≠ only, *valueA* and *valueB* also can be matrices or complex numbers. If *valueA* and *valueB* are matrices, both must have the same dimensions.

Relational operators are often used in programs to control program flow and in graphing to control the graph of a function over specific values.

valueA=valueB *valueA≠valueB*
valueA>valueB *valueA≥valueB*
valueA<valueB *valueA≤valueB*

```
25=26
                 0
{1,2,3}<3
           {1 1 0}
{1,2,3}≠{3,2,1}
           {1 0 1}
```

Using Tests

Relational operators are evaluated after mathematical functions according to EOS rules (Chapter 1).

- The expression **2+2=2+3** returns **0**. The TI-83 Plus performs the addition first because of EOS rules, and then it compares 4 to 5.
- The expression **2+(2=2)+3** returns **6**. The TI-83 Plus performs the relational test first because it is in parentheses, and then it adds 2, 1, and 3.

TEST LOGIC Menu

To display the TEST LOGIC menu, press [2nd] [TEST] [▶].

This operator...	Returns a 1 (true) if...
TEST LOGIC	
1: and	Both values are nonzero (true).
2: or	At least one value is nonzero (true).
3: xor	Only one value is zero (false).
4: not(	The value is zero (false).

Boolean Operators

Boolean operators are often used in programs to control program flow and in graphing to control the graph of the function over specific values. Values are interpreted as zero (false) or nonzero (true).

and, or, xor

and, **or**, and **xor** (exclusive or) return a value of **1** if an expression is true or **0** if an expression is false, according to the table below. *valueA* and *valueB* can be real numbers, expressions, or lists.

valueA **and** *valueB*
valueA **or** *valueB*
valueA **xor** *valueB*

valueA	valueB		and	or	xor
≠0	≠0	returns	**1**	**1**	**0**
≠0	0	returns	**0**	**1**	**1**
0	≠0	returns	**0**	**1**	**1**
0	0	returns	**0**	**0**	**0**

not(

not(returns **1** if *value* (which can be an expression) is **0**.

not(*value***)**

Using Boolean Operations

Boolean logic is often used with relational tests. In the following program, the instructions store **4** into **C**.

```
PROGRAM:BOOLEAN
:2→A:3→B
:If A=2 and B=3
:Then:4→C
:Else:5→C
:End
```

3 Function Graphing

Contents

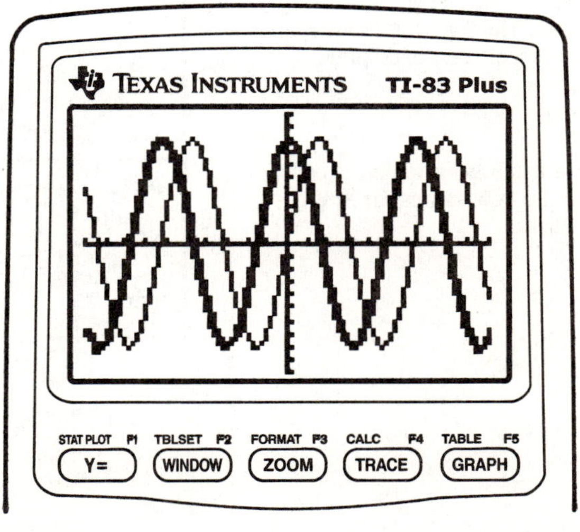

Getting Started is a fast-paced introduction. Read the chapter for details.

Graph a circle of radius 10, centered on the origin in the standard viewing window. To graph this circle, you must enter separate formulas for the upper and lower portions of the circle. Then use ZSquare (zoom square) to adjust the display and make the functions appear as a circle.

1. In **Func** mode, press [Y=] to display the Y= editor. Press [2nd] [√] **100** [−] [X,T,Θ,n] [x²] [)] [ENTER] to enter the expression Y=√(100−X²), which defines the top half of the circle.

   The expression Y=⁻√(100−X²) defines the bottom half of the circle. On the TI-83 Plus, you can define one function in terms of another. To define **Y2=⁻Y1**, press [(-)] to enter the negation sign. Press [VARS] [▶] to display the VARS Y-VARS menu. Then press [ENTER] to select **1:Function**. The FUNCTION secondary menu is displayed. Press **1** to select **1:Y1**.

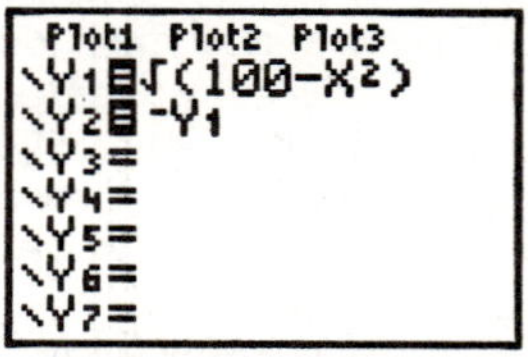

2. Press [ZOOM] **6** to select **6:ZStandard**. This is a quick way to reset the window variables to the standard values. It also graphs the functions; you do not need to press [GRAPH].

   Notice that the functions appear as an ellipse in the standard viewing window.

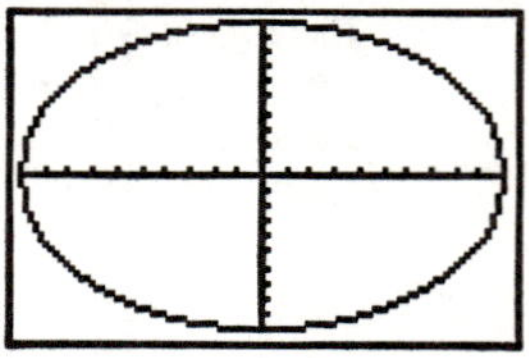

3. To adjust the display so that each pixel represents an equal width and height, press [ZOOM] **5** to select **5:ZSquare**. The functions are replotted and now appear as a circle on the display.

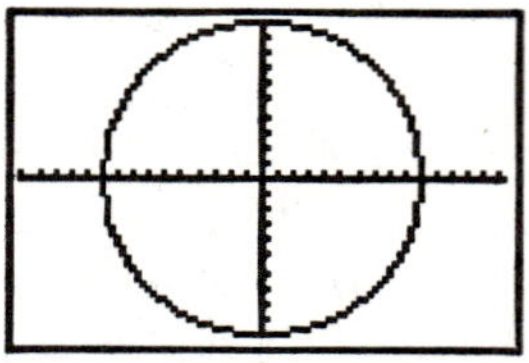

4. To see the **ZSquare** window variables, press [WINDOW] and notice the new values for **Xmin**, **Xmax**, **Ymin**, and **Ymax**.

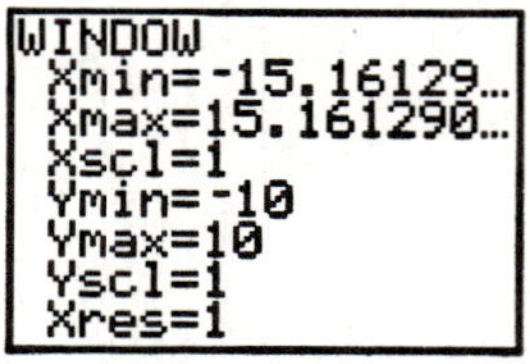

TI-83 Plus—Graphing Mode Similarities

Chapter 3 specifically describes function graphing, but the steps shown here are similar for each TI-83 Plus graphing mode. Chapters 4, 5, and 6 describe aspects that are unique to parametric graphing, polar graphing, and sequence graphing.

Defining a Graph

To define a graph in any graphing mode, follow these steps. Some steps are not always necessary.

1. Press [MODE] and set the appropriate graph mode (page 3-4).
2. Press [Y=] and enter, edit, or select one or more functions in the **Y=** editor (page 3-5 and 3-7).
3. Deselect stat plots, if necessary (page 3-7).
4. Set the graph style for each function (page 3-9).
5. Press [WINDOW] and define the viewing window variables (page 3-11).
6. Press [2nd] [FORMAT] and select the graph format settings (page 3-13).

Displaying and Exploring a Graph

After you have defined a graph, press [GRAPH] to display it. Explore the behavior of the function or functions using the TI-83 Plus tools described in this chapter.

Saving a Graph for Later Use

You can store the elements that define the current graph to any of 10 graph database variables (**GDB1** through **GDB9**, and **GDB0**; Chapter 8). To recreate the current graph later, simply recall the graph database to which you stored the original graph.

These types of information are stored in a **GDB**.

- **Y=** functions
- Graph style settings
- Window settings
- Format settings

You can store a picture of the current graph display to any of 10 graph picture variables (**Pic1** through **Pic9**, and **Pic0**; Chapter 8). Then you can superimpose one or more stored pictures onto the current graph.

Checking and Changing the Graphing Mode

To display the mode screen, press [MODE]. The default settings are highlighted below. To graph functions, you must select **Func** mode before you enter values for the window variables and before you enter the functions.

The TI-83 Plus has four graphing modes.

- **Func** (function graphing)
- **Par** (parametric graphing; Chapter 4)
- **Pol** (polar graphing; Chapter 5)
- **Seq** (sequence graphing; Chapter 6)

Other mode settings affect graphing results. Chapter 1 describes each mode setting.

- **Float** or **0123456789** (fixed) decimal mode affects displayed graph coordinates.
- **Radian** or **Degree** angle mode affects interpretation of some functions.
- **Connected** or **Dot** plotting mode affects plotting of selected functions.
- **Sequential** or **Simul** graphing-order mode affects function plotting when more than one function is selected.

Setting Modes from a Program

To set the graphing mode and other modes from a program, begin on a blank line in the program editor and follow these steps.

1. Press [MODE] to display the mode settings.
2. Press [▼], [▶], [◀], and [▲] to place the cursor on the mode that you want to select.
3. Press [ENTER] to paste the mode name to the cursor location.

The mode is changed when the program is executed.

Displaying Functions in the Y= Editor

To display the Y= editor, press [Y=]. You can store up to 10 functions to the function variables **Y1** through **Y9**, and **Y0**. You can graph one or more defined functions at once. In this example, functions **Y1** and **Y2** are defined and selected.

Defining or Editing a Function

To define or edit a function, follow these steps.

1. Press [Y=] to display the Y= editor.
2. Press [▼] to move the cursor to the function you want to define or edit. To erase a function, press [CLEAR].
3. Enter or edit the expression to define the function.
   - You may use functions and variables (including matrices and lists) in the expression. When the expression evaluates to a nonreal number, the value is not plotted; no error is returned.
   - The independent variable in the function is **X**. **Func** mode defines [X,T,Θ,*n*] as **X**. To enter **X**, press [X,T,Θ,*n*] or press [ALPHA] [X].
   - When you enter the first character, the **=** is highlighted, indicating that the function is selected.

   As you enter the expression, it is stored to the variable **Y*n*** as a user-defined function in the Y= editor.
4. Press [ENTER] or [▼] to move the cursor to the next function.

Defining a Function from the Home Screen or a Program

To define a function from the home screen or a program, begin on a blank line and follow these steps.

1. Press [ALPHA] ["], enter the expression, and then press [ALPHA] ["] again.
2. Press [STO▸].
3. Press [VARS] [▸] **1** to select **1:Function** from the VARS Y-VARS menu.
4. Select the function name, which pastes the name to the cursor location on the home screen or program editor.
5. Press [ENTER] to complete the instruction.

"*expression*"→Y*n*

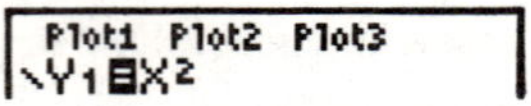

When the instruction is executed, the TI-83 Plus stores the expression to the designated variable **Y*n***, selects the function, and displays the message **Done**.

Evaluating Y= Functions in Expressions

You can calculate the value of a Y= function **Y*n*** at a specified *value* of **X**. A list of *values* returns a list.

Y*n*(*value*)
Y*n*({*value1*,*value2*,*value3*, . . .,*value n*})

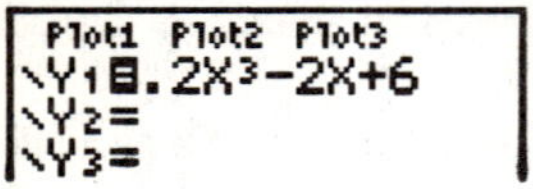

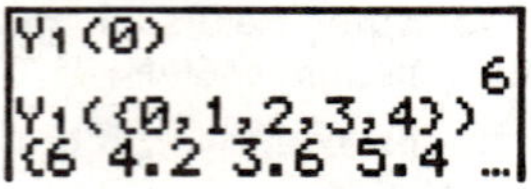

Selecting and Deselecting a Function

You can select and deselect (turn on and turn off) a function in the Y= editor. A function is selected when the = sign is highlighted. The TI-83 Plus graphs only the selected functions. You can select any or all functions **Y1** through **Y9**, and **Y0**.

To select or deselect a function in the Y= editor, follow these steps.

1. Press [Y=] to display the Y= editor.
2. Move the cursor to the function you want to select or deselect.
3. Press [◄] to place the cursor on the function's = sign.
4. Press [ENTER] to change the selection status.

When you enter or edit a function, it is selected automatically. When you clear a function, it is deselected.

Turning On or Turning Off a Stat Plot in the Y= Editor

To view and change the on/off status of a stat plot in the Y= editor, use **Plot1 Plot2 Plot3** (the top line of the Y= editor). When a plot is on, its name is highlighted on this line.

To change the on/off status of a stat plot from the Y= editor, press [▲] and [►] to place the cursor on **Plot1**, **Plot2**, or **Plot3**, and then press [ENTER].

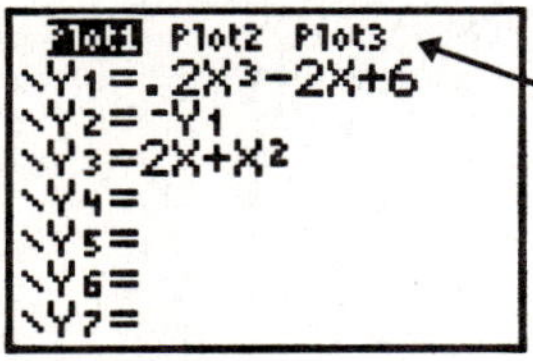

Plot1 is turned on.
Plot2 and **Plot3** are turned off.

Selecting and Deselecting Functions from the Home Screen or a Program

To select or deselect a function from the home screen or a program, begin on a blank line and follow these steps.

1. Press [VARS] [▶] to display the VARS Y-VARS menu.
2. Select **4:On/Off** to display the ON/OFF secondary menu.
3. Select **1:FnOn** to turn on one or more functions or **2:FnOff** to turn off one or more functions. The instruction you select is copied to the cursor location.
4. Enter the number (**1** through **9**, or **0**; not the variable **Y***n*) of each function you want to turn on or turn off.
   - If you enter two or more numbers, separate them with commas.
   - To turn on or turn off all functions, do not enter a number after **FnOn** or **FnOff**.

   FnOn[*function#,function#,*. . .*,function n*]
   FnOff[*function#,function#,*. . .*,function n*]

5. Press [ENTER]. When the instruction is executed, the status of each function in the current mode is set and **Done** is displayed.

For example, in **Func** mode, **FnOff :FnOn 1,3** turns off all functions in the Y= editor, and then turns on **Y1** and **Y3**.

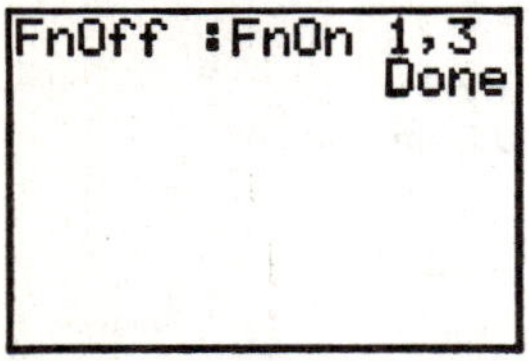

Graph Style Icons in the Y= Editor

This table describes the graph styles available for function graphing. Use the styles to visually differentiate functions to be graphed together. For example, you can set **Y1** as a solid line, **Y2** as a dotted line, and **Y3** as a thick line.

Icon	Style	Description
\	Line	A solid line connects plotted points; this is the default in **Connected** mode
◥	Thick	A thick solid line connects plotted points
◥	Above	Shading covers the area above the graph
◣	Below	Shading covers the area below the graph
-0	Path	A circular cursor traces the leading edge of the graph and draws a path
0	Animate	A circular cursor traces the leading edge of the graph without drawing a path
⋱	Dot	A small dot represents each plotted point; this is the default in **Dot** mode

Note: Some graph styles are not available in all graphing modes. Chapters 4, 5, and 6 list the styles for **Par**, **Pol**, and **Seq** modes.

Setting the Graph Style

To set the graph style for a function, follow these steps.

1. Press [Y=] to display the Y= editor.
2. Press [▼] and [▲] to move the cursor to the function.
3. Press [◄] [◄] to move the cursor left, past the = sign, to the graph style icon in the first column. The insert cursor is displayed. (Steps 2 and 3 are interchangeable.)
4. Press [ENTER] repeatedly to rotate through the graph styles. The seven styles rotate in the same order in which they are listed in the table above.
5. Press [►], [▲], or [▼] when you have selected a style.

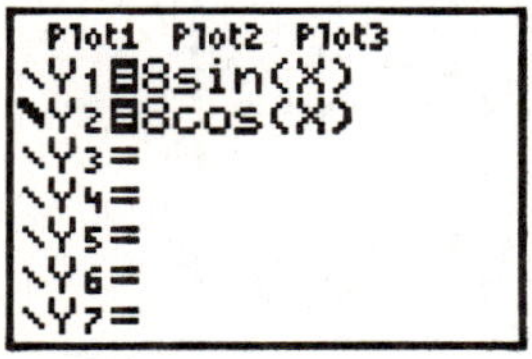

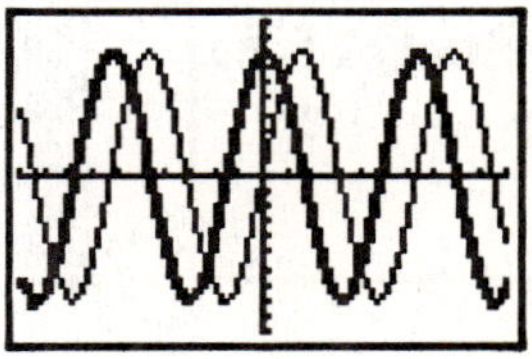

Shading Above and Below

When you select ◥ or ◣ for two or more functions, the TI-83 Plus rotates through four shading patterns.

- Vertical lines shade the first function with a ◥ or ◣ graph style.
- Horizontal lines shade the second.
- Negatively sloping diagonal lines shade the third.
- Positively sloping diagonal lines shade the fourth.
- The rotation returns to vertical lines for the fifth ◥ or ◣ function, repeating the order described above.

When shaded areas intersect, the patterns overlap.

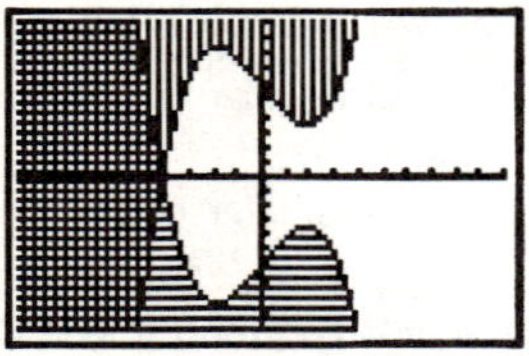

Note: When ◥ or ◣ is selected for a Y= function that graphs a family of curves, such as **Y1={1,2,3}X**, the four shading patterns rotate for each member of the family of curves.

Setting a Graph Style from a Program

To set the graph style from a program, select **H:GraphStyle(** from the PRGM CTL menu. To display this menu, press [PRGM] while in the program editor. *function#* is the number of the Y= function name in the current graphing mode. *graphstyle#* is an integer from **1** to **7** that corresponds to the graph style, as shown below.

1 = ╲ (line)	**2** = ╲ (thick)	**3** = ◥ (above)	
4 = ◣ (below)	**5** = -0 (path)	**6** = 0 (animate)	**7** = ⋱ (dot)

GraphStyle(*function#*,*graphstyle#***)**

For example, when this program is executed in **Func** mode, **GraphStyle(1,3)** sets **Y1** to ◥ (above).

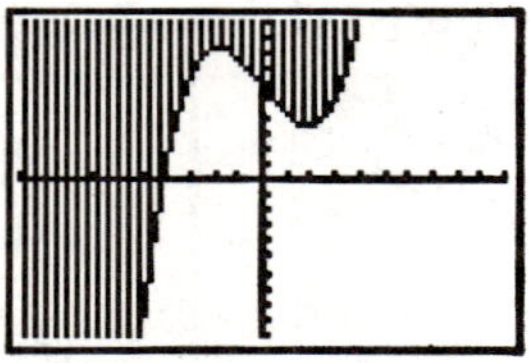

The TI-83 Plus Viewing Window

The viewing window is the portion of the coordinate plane defined by **Xmin**, **Xmax**, **Ymin**, and **Ymax**. **Xscl** (X scale) defines the distance between tick marks on the x-axis. **Yscl** (Y scale) defines the distance between tick marks on the y-axis. To turn off tick marks, set **Xscl=0** and **Yscl=0**.

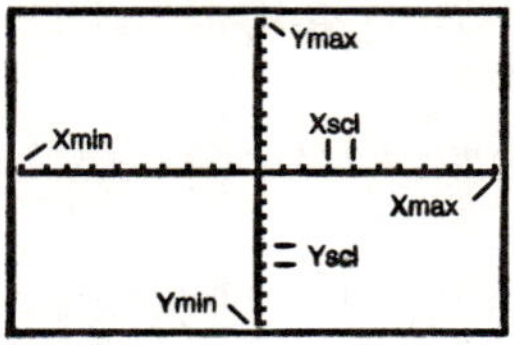

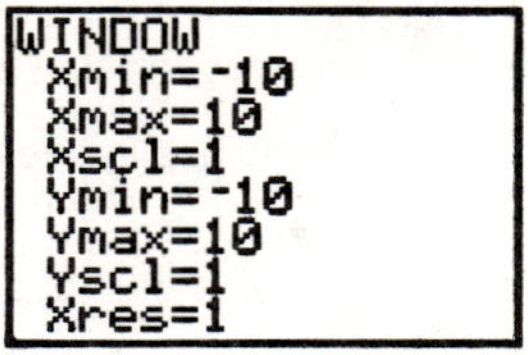

Displaying the Window Variables

To display the current window variable values, press [WINDOW]. The window editor above and to the right shows the default values in **Func** graphing mode and **Radian** angle mode. The window variables differ from one graphing mode to another.

Xres sets pixel resolution (**1** through **8**) for function graphs only. The default is **1**.

- At **Xres=1**, functions are evaluated and graphed at each pixel on the x-axis.
- At **Xres=8**, functions are evaluated and graphed at every eighth pixel along the x-axis.

Tip: Small **Xres** values improve graph resolution but may cause the TI-83 Plus to draw graphs more slowly.

Changing a Window Variable Value

To change a window variable value from the window editor, follow these steps.

1. Press [▼] or [▲] to move the cursor to the window variable you want to change.
2. Edit the value, which can be an expression.
   - Enter a new value, which clears the original value.
   - Move the cursor to a specific digit, and then edit it.
3. Press [ENTER], [▼], or [▲]. If you entered an expression, the TI-83 Plus evaluates it. The new value is stored.

Note: Xmin<Xmax and **Ymin<Ymax** must be true in order to graph.

Storing to a Window Variable from the Home Screen or a Program

To store a value, which can be an expression, to a window variable, begin on a blank line and follow these steps.

1. Enter the value you want to store.
2. Press [STO▸].
3. Press [VARS] to display the VARS menu.
4. Select **1:Window** to display the **Func** window variables (X/Y secondary menu).
   - Press [▸] to display the **Par** and **Pol** window variables (T/θ secondary menu).
   - Press [▸] [▸] to display the **Seq** window variables (U/V/W secondary menu).
5. Select the window variable to which you want to store a value. The name of the variable is pasted to the current cursor location.
6. Press [ENTER] to complete the instruction.

When the instruction is executed, the TI-83 Plus stores the value to the window variable and displays the value.

```
14→Xmax
                    14
```

ΔX and ΔY

The variables **ΔX** and **ΔY** (items **8** and **9** on the VARS (**1:Window**) X/Y secondary menu) define the distance from the center of one pixel to the center of any adjacent pixel on a graph (graphing accuracy). **ΔX** and **ΔY** are calculated from **Xmin, Xmax, Ymin,** and **Ymax** when you display a graph.

$$\Delta X = \frac{(Xmax - Xmin)}{94} \qquad \Delta Y = \frac{(Ymax - Ymin)}{62}$$

You can store values to **ΔX** and **ΔY**. If you do, **Xmax** and **Ymax** are calculated from **ΔX, Xmin, ΔY,** and **Ymin.**

Displaying the Format Settings

To display the format settings, press [2nd] [FORMAT]. The default settings are highlighted below.

RectGC	PolarGC	Sets cursor coordinates.
CoordOn	CoordOff	Sets coordinates display on or off.
GridOff	GridOn	Sets grid off or on.
AxesOn	AxesOff	Sets axes on or off.
LabelOff	LabelOn	Sets axes label off or on.
ExprOn	ExprOff	Sets expression display on or off.

Format settings define a graph's appearance on the display. Format settings apply to all graphing modes. **Seq** graphing mode has an additional mode setting (Chapter 6).

Changing a Format Setting

To change a format setting, follow these steps.

1. Press [▼], [▶], [▲], and [◀] as necessary to move the cursor to the setting you want to select.
2. Press [ENTER] to select the highlighted setting.

RectGC, PolarGC

RectGC (rectangular graphing coordinates) displays the cursor location as rectangular coordinates **X** and **Y**.

PolarGC (polar graphing coordinates) displays the cursor location as polar coordinates **R** and θ.

The **RectGC/PolarGC** setting determines which variables are updated when you plot the graph, move the free-moving cursor, or trace.

- **RectGC** updates **X** and **Y**; if **CoordOn** format is selected, **X** and **Y** are displayed.
- **PolarGC** updates **X**, **Y**, **R**, and θ; if **CoordOn** format is selected, **R** and θ are displayed.

CoordOn, CoordOff

CoordOn (coordinates on) displays the cursor coordinates at the bottom of the graph. If **ExprOff** format is selected, the function number is displayed in the top-right corner.

CoordOff (coordinates off) does not display the function number or coordinates.

GridOff, GridOn

Grid points cover the viewing window in rows that correspond to the tick marks (page 3-11) on each axis.

GridOff does not display grid points.

GridOn displays grid points.

AxesOn, AxesOff

AxesOn displays the axes.

AxesOff does not display the axes.

This overrides the **LabelOff/LabelOn** format setting.

LabelOff, LabelOn

LabelOff and **LabelOn** determine whether to display labels for the axes (**X** and **Y**), if **AxesOn** format is also selected.

ExprOn, ExprOff

ExprOn and **ExprOff** determine whether to display the **Y=** expression when the trace cursor is active. This format setting also applies to stat plots.

When **ExprOn** is selected, the expression is displayed in the top-left corner of the graph screen.

When **ExprOff** and **CoordOn** both are selected, the number in the top-right corner specifies which function is being traced.

Displaying a New Graph

To display the graph of the selected function or functions, press [GRAPH]. TRACE, ZOOM instructions, and CALC operations display the graph automatically. As the TI-83 Plus plots the graph, the busy indicator is on. As the graph is plotted, **X** and **Y** are updated.

Pausing or Stopping a Graph

While plotting a graph, you can pause or stop graphing.

- Press [ENTER] to pause; then press [ENTER] to resume.
- Press [ON] to stop; then press [GRAPH] to redraw.

Smart Graph

Smart Graph is a TI-83 Plus feature that redisplays the last graph immediately when you press [GRAPH], but only if all graphing factors that would cause replotting have remained the same since the graph was last displayed.

If you performed any of these actions since the graph was last displayed, the TI-83 Plus will replot the graph based on new values when you press [GRAPH].

- Changed a mode setting that affects graphs
- Changed a function in the current picture
- Selected or deselected a function or stat plot
- Changed the value of a variable in a selected function
- Changed a window variable or graph format setting
- Cleared drawings by selecting **ClrDraw**
- Changed a stat plot definition

Overlaying Functions on a Graph

On the TI-83 Plus, you can graph one or more new functions without replotting existing functions. For example, store **sin(X)** to **Y1** in the Y= editor and press [GRAPH]. Then store **cos(X)** to **Y2** and press [GRAPH] again. The function **Y2** is graphed on top of **Y1**, the original function.

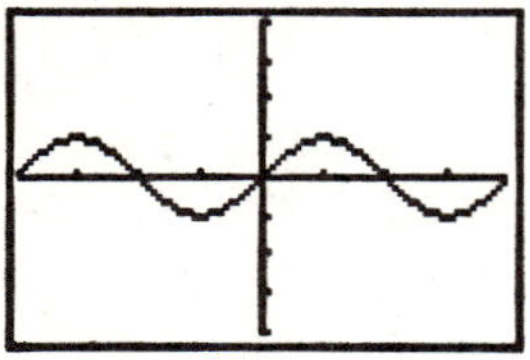

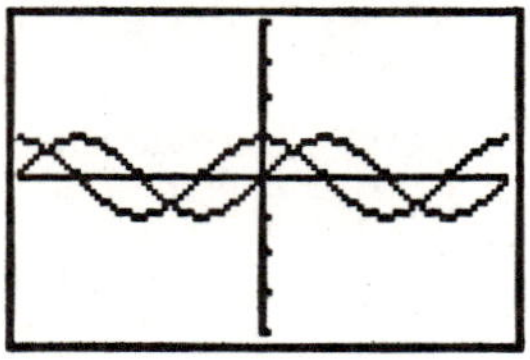

Graphing a Family of Curves

If you enter a list (Chapter 11) as an element in an expression, the TI-83 Plus plots the function for each value in the list, thereby graphing a family of curves. In **Simul** graphing-order mode, it graphs all functions sequentially for the first element in each list, and then for the second, and so on.

{2,4,6}sin(X) graphs three functions: **2 sin(X)**, **4 sin(X)**, and **6 sin(X)**.

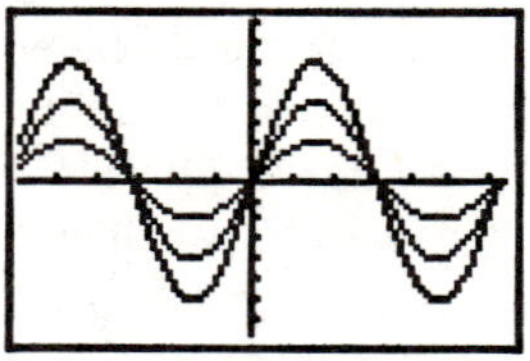

{2,4,6}sin({1,2,3}X) graphs **2 sin(X)**, **4 sin(2X)**, and **6 sin(3X)**.

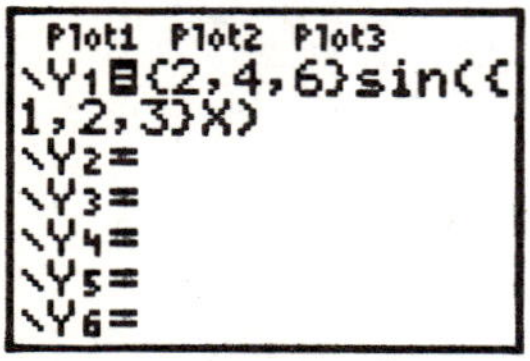

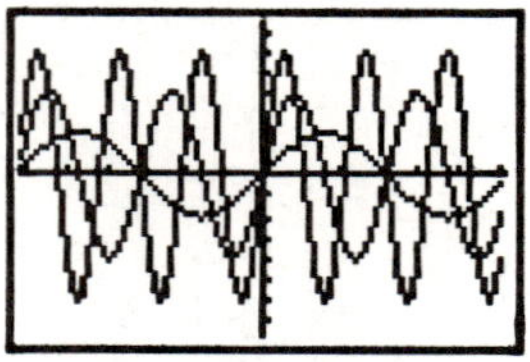

Note: When using more than one list, the lists must have the same dimensions.

Free-Moving Cursor

When a graph is displayed, press ◀, ▶, ▲, or ▼ to move the cursor around the graph. When you first display the graph, no cursor is visible. When you press ◀, ▶, ▲, or ▼, the cursor moves from the center of the viewing window.

As you move the cursor around the graph, the coordinate values of the cursor location are displayed at the bottom of the screen if **CoordOn** format is selected. The **Float/Fix** decimal mode setting determines the number of decimal digits displayed for the coordinate values.

To display the graph with no cursor and no coordinate values, press [CLEAR] or [ENTER]. When you press ◀, ▶, ▲, or ▼, the cursor moves from the same position.

Graphing Accuracy

The free-moving cursor moves from pixel to pixel on the screen. When you move the cursor to a pixel that appears to be on the function, the cursor may be near, but not actually on, the function. The coordinate value displayed at the bottom of the screen actually may not be a point on the function. To move the cursor along a function, use [TRACE] (page 3-18).

The coordinate values displayed as you move the cursor approximate actual math coordinates, *accurate to within the width and height of the pixel. As **Xmin**, **Xmax**, **Ymin**, and **Ymax** get closer together (as in a **Zoom In**) graphing accuracy increases, and the coordinate values more closely approximate the math coordinates.

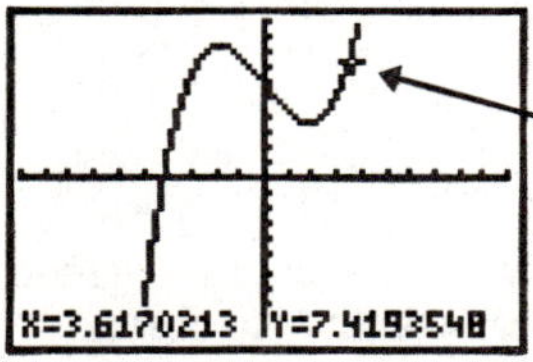

Free-moving cursor "on" the curve

Beginning a Trace

Use TRACE to move the cursor from one plotted point to the next along a function. To begin a trace, press [TRACE]. If the graph is not displayed already, press [TRACE] to display it. The trace cursor is on the first selected function in the Y= editor, at the middle **X** value on the screen. The cursor coordinates are displayed at the bottom of the screen if **CoordOn** format is selected. The Y= expression is displayed in the top-left corner of the screen, if **ExprOn** format is selected.

Moving the Trace Cursor

To move the TRACE cursor	do this:
To the previous or next plotted point,	press [◄] or [►].
Five plotted points on a function (**Xres** affects this),	press [2nd] [◄] or [2nd] [►].
To any valid **X** value on a function,	enter a value, and then press [ENTER].
From one function to another,	press [▲] or [▼].

When the trace cursor moves along a function, the **Y** value is calculated from the **X** value; that is, **Y=Y*n*(X)**. If the function is undefined at an **X** value, the **Y** value is blank.

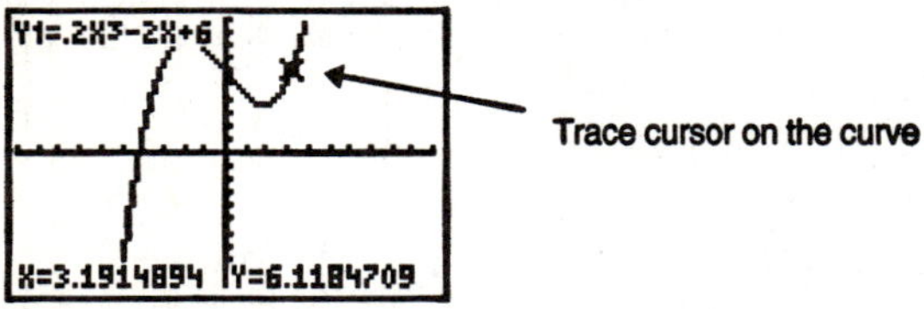

If you move the trace cursor beyond the top or bottom of the screen, the coordinate values at the bottom of the screen continue to change appropriately.

Moving the Trace Cursor from Function to Function

To move the trace cursor from function to function, press [▼] and [▲]. The cursor follows the order of the selected functions in the Y= editor. The trace cursor moves to each function at the same **X** value. If **ExprOn** format is selected, the expression is updated.

Moving the Trace Cursor to Any Valid X Value

To move the trace cursor to any valid **X** value on the current function, enter the value. When you enter the first digit, an **X=** prompt and the number you entered are displayed in the bottom-left corner of the screen. You can enter an expression at the **X=** prompt. The value must be valid for the current viewing window. When you have completed the entry, press [ENTER] to move the cursor.

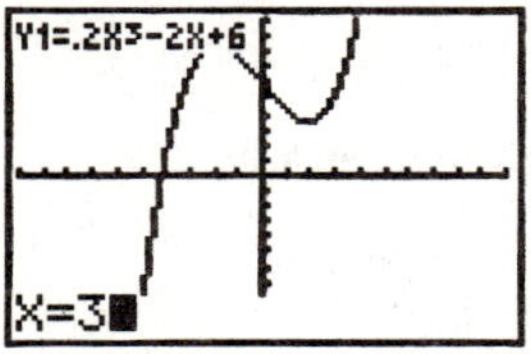

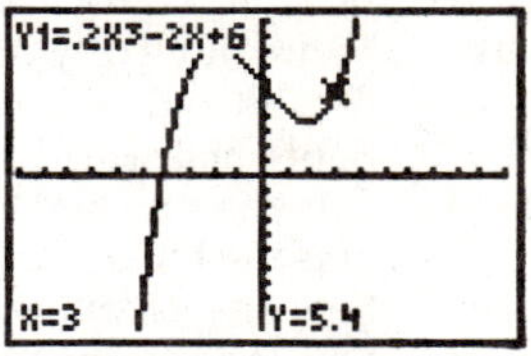

Note: This feature does not apply to stat plots.

Panning to the Left or Right

If you trace a function beyond the left or right side of the screen, the viewing window automatically pans to the left or right. **Xmin** and **Xmax** are updated to correspond to the new viewing window.

Quick Zoom

While tracing, you can press [ENTER] to adjust the viewing window so that the cursor location becomes the center of the new viewing window, even if the cursor is above or below the display. This allows panning up and down. After Quick Zoom, the cursor remains in TRACE.

Leaving and Returning to TRACE

When you leave and return to TRACE, the trace cursor is displayed in the same location it was in when you left TRACE, unless Smart Graph has replotted the graph (page 3-15).

Using TRACE in a Program

On a blank line in the program editor, press [TRACE]. The instruction **Trace** is pasted to the cursor location. When the instruction is encountered during program execution, the graph is displayed with the trace cursor on the first selected function. As you trace, the cursor coordinate values are updated. When you finish tracing the functions, press [ENTER] to resume program execution.

ZOOM Menu

To display the ZOOM menu, press [ZOOM]. You can adjust the viewing window of the graph quickly in several ways. All ZOOM instructions are accessible from programs.

ZOOM MEMORY	
1: ZBox	Draws a box to define the viewing window.
2: Zoom In	Magnifies the graph around the cursor.
3: Zoom Out	Views more of a graph around the cursor.
4: ZDecimal	Sets Δ**X** and Δ**Y** to 0.1.
5: ZSquare	Sets equal-size pixels on the **X** and **Y** axes.
6: ZStandard	Sets the standard window variables.
7: ZTrig	Sets the built-in trig window variables.
8: ZInteger	Sets integer values on the **X** and **Y** axes.
9: ZoomStat	Sets the values for current stat lists.
0: ZoomFit	Fits **YMin** and **YMax** between **XMin** and **XMax**.

Zoom Cursor

When you select **1:ZBox**, **2:Zoom In**, or **3:Zoom Out**, the cursor on the graph becomes the zoom cursor (+), a smaller version of the free-moving cursor (+).

ZBox

To define a new viewing window using **ZBox**, follow these steps.

1. Select **1:ZBox** from the ZOOM menu. The zoom cursor is displayed at the center of the screen.

2. Move the zoom cursor to any spot you want to define as a corner of the box, and then press [ENTER]. When you move the cursor away from the first defined corner, a small, square dot indicates the spot.

3. Press [◄], [▲], [►], or [▼]. As you move the cursor, the sides of the box lengthen or shorten proportionately on the screen.

   Note: To cancel **ZBox** before you press [ENTER], press [CLEAR].

4. When you have defined the box, press [ENTER] to replot the graph.

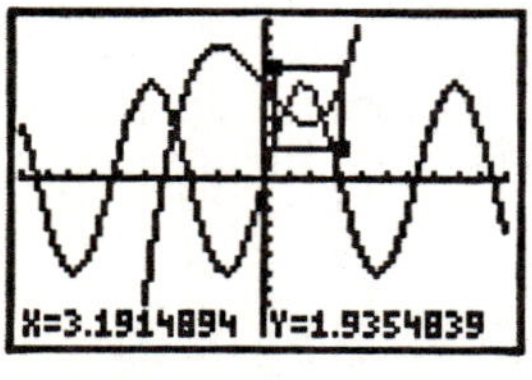

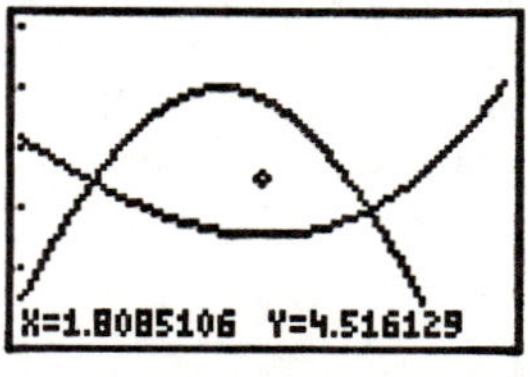

To use **ZBox** to define another box within the new graph, repeat steps 2 through 4. To cancel **ZBox**, press [CLEAR].

Zoom In, Zoom Out

Zoom In magnifies the part of the graph that surrounds the cursor location. **Zoom Out** displays a greater portion of the graph, centered on the cursor location. The **XFact** and **YFact** settings determine the extent of the zoom.

To zoom in on a graph, follow these steps.

1. Check **XFact** and **YFact** (page 3-24); change as needed.
2. Select **2:Zoom In** from the ZOOM menu. The zoom cursor is displayed.
3. Move the zoom cursor to the point that is to be the center of the new viewing window.
4. Press [ENTER]. The TI-83 Plus adjusts the viewing window by **XFact** and **YFact**; updates the window variables; and replots the selected functions, centered on the cursor location.
5. Zoom in on the graph again in either of two ways.
   - To zoom in at the same point, press [ENTER].
   - To zoom in at a new point, move the cursor to the point that you want as the center of the new viewing window, and then press [ENTER].

To zoom out on a graph, select **3:Zoom Out** and repeat steps 3 through 5.

To cancel **Zoom In** or **Zoom Out**, press [CLEAR].

ZDecimal

ZDecimal replots the functions immediately. It updates the window variables to preset values, as shown below. These values set ∆**X** and ∆**Y** equal to **0.1** and set the **X** and **Y** value of each pixel to one decimal place.

Xmin=⁻4.7	**Ymin=⁻3.1**
Xmax=4.7	**Ymax=3.1**
Xscl=1	**Yscl=1**

ZSquare

ZSquare replots the functions immediately. It redefines the viewing window based on the current values of the window variables. It adjusts in only one direction so that **∆X=∆Y**, which makes the graph of a circle look like a circle. **Xscl** and **Yscl** remain unchanged. The midpoint of the current graph (not the intersection of the axes) becomes the midpoint of the new graph.

ZStandard

ZStandard replots the functions immediately. It updates the window variables to the standard values shown below.

Xmin=⁻10	**Ymin=⁻10**	**Xres=1**
Xmax=10	**Ymax=10**	
Xscl=1	**Yscl=1**	

ZTrig

ZTrig replots the functions immediately. It updates the window variables to preset values that are appropriate for plotting trig functions. Those preset values in **Radian** mode are shown below.

Xmin=⁻(47/24)π	**Ymin=⁻4**
Xmax=(47/24)π	**Ymax=4**
Xscl=π/2	**Yscl=1**

ZInteger

ZInteger redefines the viewing window to the dimensions shown below. To use **ZInteger**, move the cursor to the point that you want to be the center of the new window, and then press [ENTER]; **ZInteger** replots the functions.

ΔX=1	**Xscl=10**
ΔY=1	**Yscl=10**

ZoomStat

ZoomStat redefines the viewing window so that all statistical data points are displayed. For regular and modified box plots, only **Xmin** and **Xmax** are adjusted.

ZoomFit

ZoomFit replots the functions immediately. **ZoomFit** recalculates **YMin** and **YMax** to include the minimum and maximum **Y** values of the selected functions between the current **XMin** and **XMax**. **XMin** and **XMax** are not changed.

ZOOM MEMORY Menu

To display the ZOOM MEMORY menu, press [ZOOM] [▶].

ZOOM MEMORY	
1:ZPrevious	Uses the previous viewing window.
2:ZoomSto	Stores the user-defined window.
3:ZoomRcl	Recalls the user-defined window.
4:SetFactors...	Changes **Zoom In** and **Zoom Out** factors.

ZPrevious

ZPrevious replots the graph using the window variables of the graph that was displayed before you executed the last ZOOM instruction.

ZoomSto

ZoomSto immediately stores the current viewing window. The graph is displayed, and the values of the current window variables are stored in the user-defined ZOOM variables **ZXmin, ZXmax, ZXscl, ZYmin, ZYmax, ZYscl,** and **ZXres.**

These variables apply to all graphing modes. For example, changing the value of **ZXmin** in **Func** mode also changes it in **Par** mode.

ZoomRcl

ZoomRcl graphs the selected functions in a user-defined viewing window. The user-defined viewing window is determined by the values stored with the **ZoomSto** instruction. The window variables are updated with the user-defined values, and the graph is plotted.

ZOOM FACTORS

The zoom factors, **XFact** and **YFact**, are positive numbers (not necessarily integers) greater than or equal to 1. They define the magnification or reduction factor used to **Zoom In** or **Zoom Out** around a point.

Checking XFact and YFact

To display the ZOOM FACTORS screen, where you can review the current values for **XFact** and **YFact**, select **4:SetFactors** from the ZOOM MEMORY menu. The values shown are the defaults.

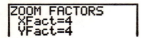

```
ZOOM FACTORS
 XFact=4
 YFact=4
```

Changing XFact and YFact

You can change **XFact** and **YFact** in either of two ways.

- Enter a new value. The original value is cleared automatically when you enter the first digit.
- Place the cursor on the digit you want to change, and then enter a value or press [DEL] to delete it.

Using ZOOM MEMORY Menu Items from the Home Screen or a Program

From the home screen or a program, you can store directly to any of the user-defined ZOOM variables.

```
-5→ZXmin:5→ZXmax
               5
```

From a program, you can select the **ZoomSto** and **ZoomRcl** instructions from the ZOOM MEMORY menu.

CALCULATE Menu

To display the CALCULATE menu, press [2nd] [CALC]. Use the items on this menu to analyze the current graph functions.

CALCULATE	
1: value	Calculates a function **Y** value for a given **X**.
2: zero	Finds a zero (x-intercept) of a function.
3: minimum	Finds a minimum of a function.
4: maximum	Finds a maximum of a function.
5: intersect	Finds an intersection of two functions.
6: dy/dx	Finds a numeric derivative of a function.
7: ∫f(x)dx	Finds a numeric integral of a function.

value

value evaluates one or more currently selected functions for a specified value of **X**.

Note: When a value is displayed for **X**, press [CLEAR] to clear the value. When no value is displayed, press [CLEAR] to cancel the **value** operation.

To evaluate a selected function at **X**, follow these steps.

1. Select **1:value** from the CALCULATE menu. The graph is displayed with **X=** in the bottom-left corner.
2. Enter a real value, which can be an expression, for **X** between **Xmin** and **Xmax**.
3. Press [ENTER].

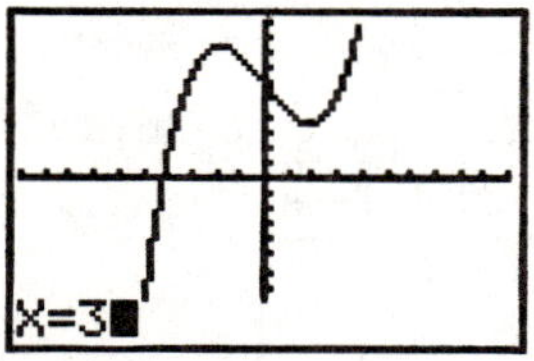

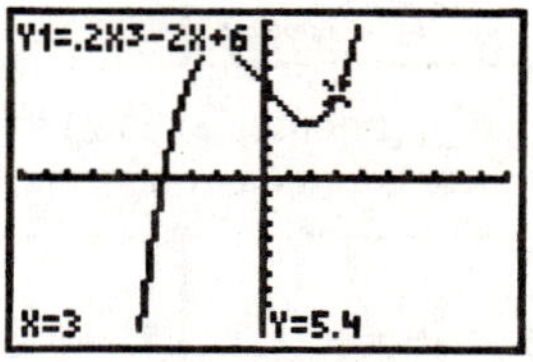

The cursor is on the first selected function in the Y= editor at the **X** value you entered, and the coordinates are displayed, even if **CoordOff** format is selected.

To move the cursor from function to function at the entered **X** value, press [▲] or [▼]. To restore the free-moving cursor, press [◄] or [►].

zero

zero finds a zero (x-intercept or root) of a function using **solve(**. Functions can have more than one x-intercept value; **zero** finds the zero closest to your guess.

The time **zero** spends to find the correct zero value depends on the accuracy of the values you specify for the left and right bounds and the accuracy of your guess.

To find a zero of a function, follow these steps.

1. Select **2:zero** from the CALCULATE menu. The current graph is displayed with **Left Bound?** in the bottom-left corner.
2. Press [▲] or [▼] to move the cursor onto the function for which you want to find a zero.
3. Press [◀] or [▶] (or enter a value) to select the x-value for the left bound of the interval, and then press [ENTER]. A ▶ indicator on the graph screen shows the left bound. **Right Bound?** is displayed in the bottom-left corner. Press [◀] or [▶] (or enter a value) to select the x-value for the right bound, and then press [ENTER]. A ◀ indicator on the graph screen shows the right bound. **Guess?** is then displayed in the bottom-left corner.

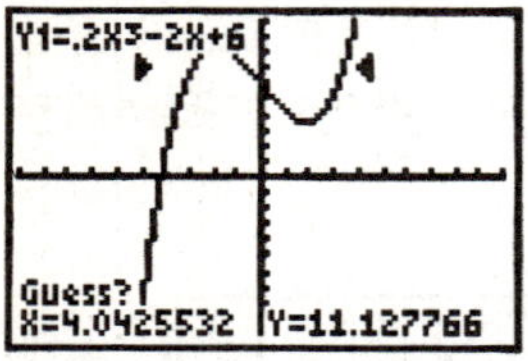

4. Press [◀] or [▶] (or enter a value) to select a point near the zero of the function, between the bounds, and then press [ENTER].

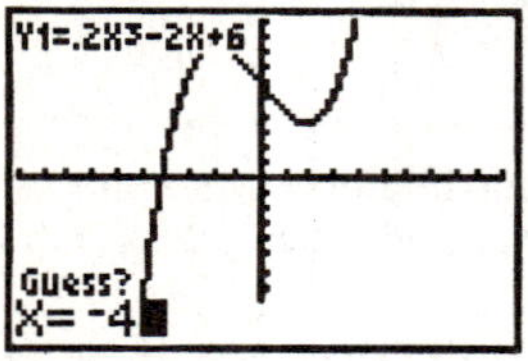

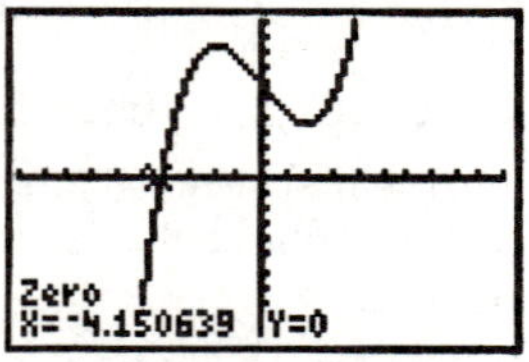

The cursor is on the solution and the coordinates are displayed, even if **CoordOff** format is selected. To move to the same x-value for other selected functions, press [▲] or [▼]. To restore the free-moving cursor, press [◀] or [▶].

minimum, maximum

minimum and **maximum** find a minimum or maximum of a function within a specified interval to a tolerance of 1E-5.

To find a minimum or maximum, follow these steps.

1. Select **3:minimum** or **4:maximum** from the CALCULATE menu. The current graph is displayed.
2. Select the function and set left bound, right bound, and guess as described for **zero** (steps 2 through 4; page 3-26).

The cursor is on the solution, and the coordinates are displayed, even if you have selected **CoordOff** format; **Minimum** or **Maximum** is displayed in the bottom-left corner.

To move to the same x-value for other selected functions, press [▲] or [▼]. To restore the free-moving cursor, press [◄] or [►].

intersect

intersect finds the coordinates of a point at which two or more functions intersect using **solve(**. The intersection must appear on the display to use **intersect**.

To find an intersection, follow these steps.

1. Select **5:intersect** from the CALCULATE menu. The current graph is displayed with **First curve?** in the bottom-left corner.

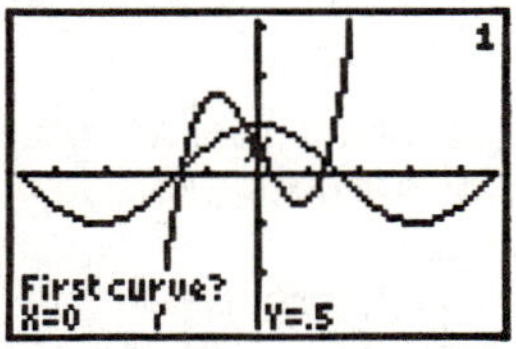

2. Press [▼] or [▲], if necessary, to move the cursor to the first function, and then press [ENTER]. **Second curve?** is displayed in the bottom-left corner.
3. Press [▼] or [▲], if necessary, to move the cursor to the second function, and then press [ENTER].
4. Press [►] or [◄] to move the cursor to the point that is your guess as to location of the intersection, and then press [ENTER].

The cursor is on the solution and the coordinates are displayed, even if **CoordOff** format is selected. **Intersection** is displayed in the bottom-left corner. To restore the free-moving cursor, press [◄], [▲], [►], or [▼].

dy/dx

dy/dx (numerical derivative) finds the numerical derivative (slope) of a function at a point, with ε=1E-3.

To find a function's slope at a point, follow these steps.

1. Select **6:dy/dx** from the CALCULATE menu. The current graph is displayed.
2. Press [▲] or [▼] to select the function for which you want to find the numerical derivative.
3. Press [◀] or [▶] (or enter a value) to select the **X** value at which to calculate the derivative, and then press [ENTER].

The cursor is on the solution and the numerical derivative is displayed.

To move to the same x-value for other selected functions, press [▲] or [▼]. To restore the free-moving cursor, press [◀] or [▶].

∫f(x)dx

∫f(x)dx (numerical integral) finds the numerical integral of a function in a specified interval. It uses the **fnInt(** function, with a tolerance of ε=1E-3.

To find the numerical derivative of a function, follow these steps.

1. Select **7:∫f(x)dx** from the CALCULATE menu. The current graph is displayed with **Lower Limit?** in the bottom-left corner.
2. Press [▲] or [▼] to move the cursor to the function for which you want to calculate the integral.
3. Set lower and upper limits as you would set left and right bounds for **zero** (step 3; page 3-26). The integral value is displayed, and the integrated area is shaded.

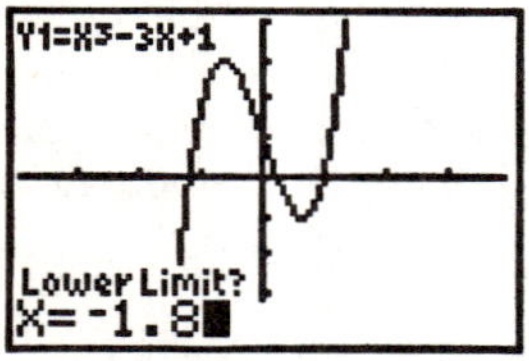

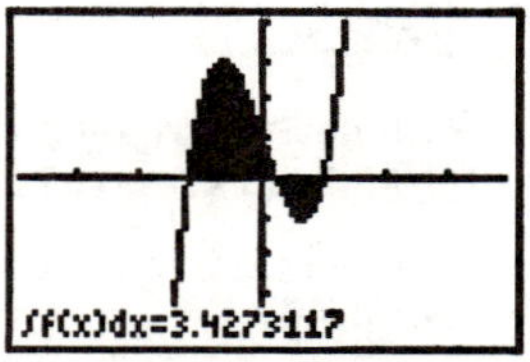

Note: The shaded area is a drawing. Use **ClrDraw** (Chapter 8) or any action that invokes Smart Graph to clear the shaded area.

4 Parametric Graphing

Contents

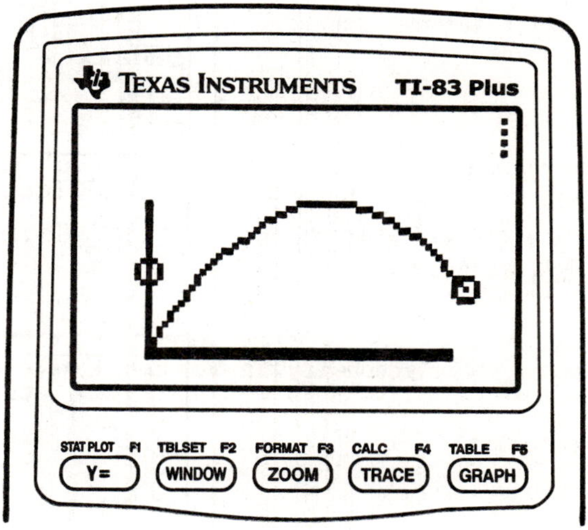

Getting Started is a fast-paced introduction. Read the chapter for details.

Graph the parametric equation that describes the path of a ball hit at an initial speed of 30 meters per second, at an initial angle of 25 degrees with the horizontal from ground level. How far does the ball travel? When does it hit the ground? How high does it go? Ignore all forces except gravity.

For initial velocity v_0 and angle θ, the position of the ball as a function of time has horizontal and vertical components.

Horizontal:	$X1(t)=tv_0\cos(\theta)$	**Vertical:**	$Y1(t)=tv_0\sin(\theta)-\frac{1}{2}gt^2$

The vertical and horizontal vectors of the ball's motion also will be graphed.

Vertical vector:	**X2(t)=0**	**Y2(t)=Y1(t)**
Horizontal vector:	**X3(t)=X1(t)**	**Y3(t)=0**
Gravity constant:	$g=9.8\ m/sec^2$	

1. Press [MODE]. Press ▼ ▼ ▼ ▶ [ENTER] to select **Par** mode. Press ▼ ▼ ▶ [ENTER] to select **Simul** for simultaneous graphing of all three parametric equations in this example.

2. Press [Y=]. Press **30** [X,T,Θ,n] [COS] **25** [2nd] [ANGLE] **1** (to select °) [)] [ENTER] to define **X1T** in terms of **T**.

3. Press **30** [X,T,Θ,n] [SIN] **25** [2nd] [ANGLE] **1** [)] [−] **9.8** [÷] **2** [X,T,Θ,n] [x²] [ENTER] to define **Y1T**.

The vertical component vector is defined by **X2T** and **Y2T**.

4. Press **0** [ENTER] to define **X2T**.

5. Press [VARS] ▶ to display the VARS Y-VARS menu. Press **2** to display the PARAMETRIC secondary menu. Press **2** [ENTER] to define **Y2T**.

The horizontal component vector is defined by **X3T** and **Y3T**.

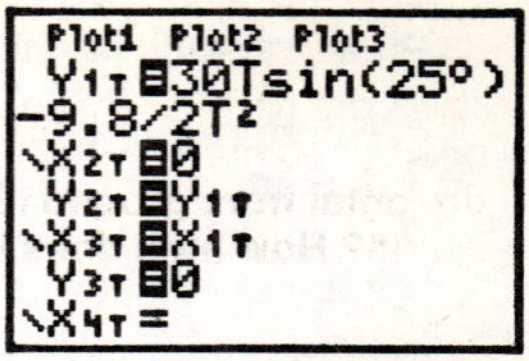

6. Press [VARS] [▶] **2**, and then press **1** [ENTER] to define **X3T**. Press **0** [ENTER] to define **Y3T**.

7. Press [◀] [◀] [▲] [ENTER] to change the graph style to ◥ for **X3T** and **Y3T**. Press [▲] [ENTER] [ENTER] to change the graph style to -0 for **X2T** and **Y2T**. Press [▲] [ENTER] [ENTER] to change the graph style to -0 for **X1T** and **Y1T**. (These keystrokes assume that all graph styles were set to \ originally.)

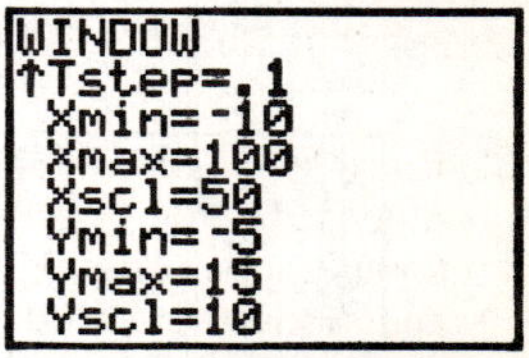

8. Press [WINDOW]. Enter these values for the window variables.

Tmin=0	**Xmin=-10**	**Ymin=-5**
Tmax=5	**Xmax=100**	**Ymax=15**
Tstep=.1	**Xscl=50**	**Yscl=10**

9. Press [2nd] [FORMAT] [▼] [▼] [▼] [▶] [ENTER] to set **AxesOff**, which turns off the axes.

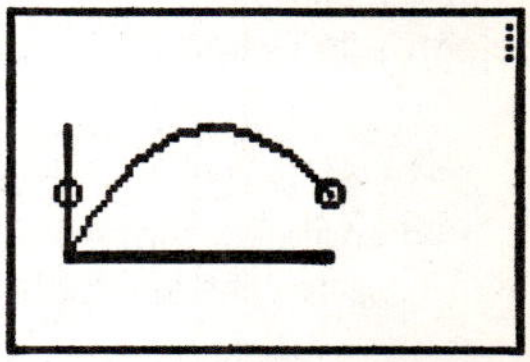

10. Press [GRAPH]. The plotting action simultaneously shows the ball in flight and the vertical and horizontal component vectors of the motion.

Tip: To simulate the ball flying through the air, set graph style to -0 (animate) for **X1T** and **Y1T**.

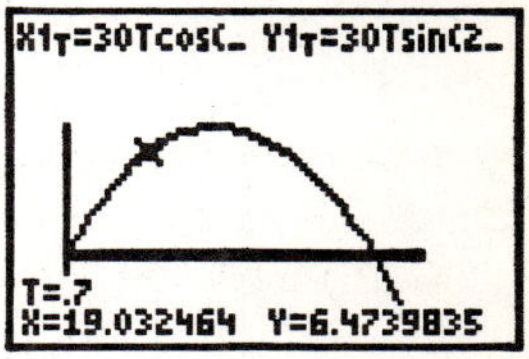

11. Press [TRACE] to obtain numerical results and answer the questions at the beginning of this section.

Tracing begins at **Tmin** on the first parametric equation (**X1T** and **Y1T**). As you press [▶] to trace the curve, the cursor follows the path of the ball over time. The values for **X** (distance), **Y** (height), and **T** (time) are displayed at the bottom of the screen.

TI-83 Plus Graphing Mode Similarities

The steps for defining a parametric graph are similar to the steps for defining a function graph. Chapter 4 assumes that you are familiar with Chapter 3: Function Graphing. Chapter 4 details aspects of parametric graphing that differ from function graphing.

Setting Parametric Graphing Mode

To display the mode screen, press [MODE]. To graph parametric equations, you must select **Par** graphing mode before you enter window variables and before you enter the components of parametric equations.

Displaying the Parametric Y= Editor

After selecting **Par** graphing mode, press [Y=] to display the parametric **Y=** editor.

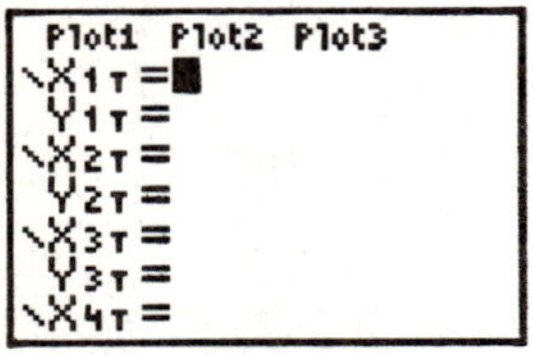

In this editor, you can display and enter both the **X** and **Y** components of up to six equations, **X1T** and **Y1T** through **X6T** and **Y6T**. Each is defined in terms of the independent variable **T**. A common application of parametric graphs is graphing equations over time.

Selecting a Graph Style

The icons to the left of **X1T** through **X6T** represent the graph style of each parametric equation (Chapter 3). The default in **Par** mode is ⟍ (line), which connects plotted points. Line, ⟍ (thick), -0 (path), 0 (animate), and ·. (dot) styles are available for parametric graphing.

Defining and Editing Parametric Equations

To define or edit a parametric equation, follow the steps in Chapter 3 for defining a function or editing a function. The independent variable in a parametric equation is **T**. In **Par** graphing mode, you can enter the parametric variable **T** in either of two ways.

- Press [X,T,Θ,n].
- Press [ALPHA] [T].

Two components, **X** and **Y**, define a single parametric equation. You must define both of them.

Selecting and Deselecting Parametric Equations

The TI-83 Plus graphs only the selected parametric equations. In the **Y=** editor, a parametric equation is selected when the **=** signs of both the **X** and **Y** components are highlighted. You may select any or all of the equations **X1T** and **Y1T** through **X6T** and **Y6T**.

To change the selection status, move the cursor onto the **=** sign of either the **X** or **Y** component and press [ENTER]. The status of both the **X** and **Y** components is changed.

Setting Window Variables

To display the window variable values, press [WINDOW]. These variables define the viewing window. The values below are defaults for **Par** graphing in **Radian** angle mode.

Tmin=0	Smallest **T** value to evaluate
Tmax=6.2831853...	Largest **T** value to evaluate (2π)
Tstep=.1308996...	**T** value increment (π/24)
Xmin=-10	Smallest **X** value to be displayed
Xmax=10	Largest **X** value to be displayed
Xscl=1	Spacing between the **X** tick marks
Ymin=-10	Smallest **Y** value to be displayed
Ymax=10	Largest **Y** value to be displayed
Yscl=1	Spacing between the **Y** tick marks

Note: To ensure that sufficient points are plotted, you may want to change the **T** window variables.

Setting the Graph Format

To display the current graph format settings, press [2nd] [FORMAT]. Chapter 3 describes the format settings in detail. The other graphing modes share these format settings; **Seq** graphing mode has an additional axes format setting.

Displaying a Graph

When you press [GRAPH], the TI-83 Plus plots the selected parametric equations. It evaluates the **X** and **Y** components for each value of **T** (from **Tmin** to **Tmax** in intervals of **Tstep**), and then plots each point defined by **X** and **Y**. The window variables define the viewing window.

As the graph is plotted, **X**, **Y**, and **T** are updated.

Smart Graph applies to parametric graphs (Chapter 3).

Window Variables and Y-VARS Menus

You can perform these actions from the home screen or a program.

- Access functions by using the name of the **X** or **Y** component of the equation as a variable.

```
X1T*.5
      94.70916375
```

- Store parametric equations.

```
"sin(T)"→X1T
             Done
"cos(T)"→Y1T
             Done
```

```
Plot1 Plot2 Plot3
\X1T=sin(T)
 Y1T=cos(T)
\X2T=
 Y2T=
```

- Select or deselect parametric equations.

```
FnOff 1
             Done
```

```
Plot1 Plot2 Plot3
\X1T=cos(T)
 Y1T=sin(T)
\X2T=
 Y2T=
```

- Store values directly to window variables.

```
360→Tmax
              360
```

Free-Moving Cursor

The free-moving cursor in **Par** graphing works the same as in **Func** graphing.

In **RectGC** format, moving the cursor updates the values of **X** and **Y**; if **CoordOn** format is selected, **X** and **Y** are displayed.

In **PolarGC** format, **X**, **Y**, **R**, and θ are updated; if **CoordOn** format is selected, **R** and θ are displayed.

TRACE

To activate TRACE, press [TRACE]. When TRACE is active, you can move the trace cursor along the graph of the equation one **Tstep** at a time. When you begin a trace, the trace cursor is on the first selected function at **Tmin.** If **ExprOn** is selected, then the function is displayed.

In **RectGC** format, TRACE updates and displays the values of **X, Y,** and **T** if **CoordOn** format is on.

In **PolarGC** format, **X**, **Y**, **R**, θ and **T** are updated; if **CoordOn** format is selected, **R**, θ, and **T** are displayed. The **X** and **Y** (or **R** and θ) values are calculated from **T**.

To move five plotted points at a time on a function, press [2nd] [◄] or [2nd] [►]. If you move the cursor beyond the top or bottom of the screen, the coordinate values at the bottom of the screen continue to change appropriately.

Quick Zoom is available in **Par** graphing; panning is not (Chapter 3).

Moving the Trace Cursor to Any Valid T Value

To move the trace cursor to any valid **T** value on the current function, enter the number. When you enter the first digit, a **T=** prompt and the number you entered are displayed in the bottom-left corner of the screen. You can enter an expression at the **T=** prompt. The value must be valid for the current viewing window. When you have completed the entry, press [ENTER] to move the cursor.

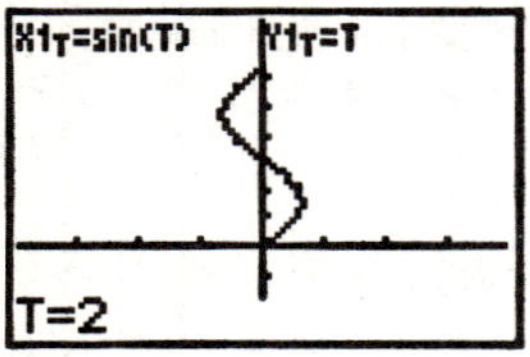

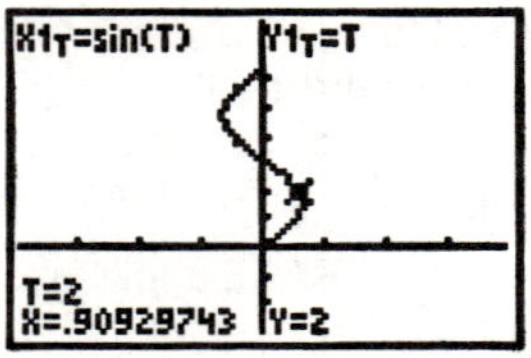

ZOOM

ZOOM operations in **Par** graphing work the same as in **Func** graphing. Only the **X** (**Xmin**, **Xmax**, and **Xscl**) and **Y** (**Ymin**, **Ymax**, and **Yscl**) window variables are affected.

The **T** window variables (**Tmin**, **Tmax**, and **Tstep**) are only affected when you select **ZStandard**. The VARS ZOOM secondary menu ZT/Zθ items **1:ZTmin**, **2:ZTmax**, and **3:ZTstep** are the zoom memory variables for **Par** graphing.

CALC

CALC operations in **Par** graphing work the same as in **Func** graphing. The CALCULATE menu items available in **Par** graphing are **1:value**, **2:dy/dx**, **3:dy/dt**, and **4:dx/dt**.

5 Polar Graphing

Contents

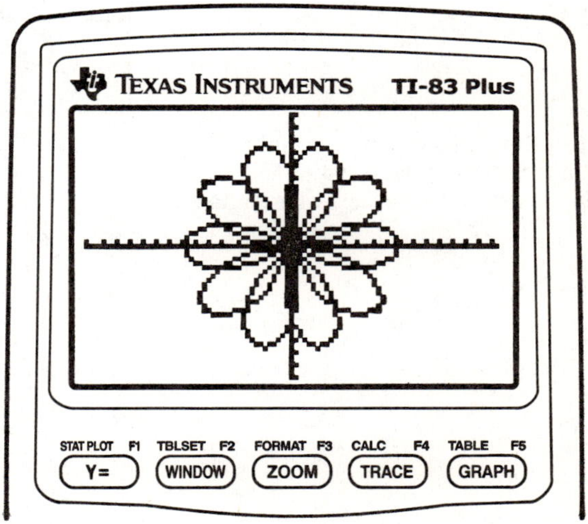

Getting Started is a fast-paced introduction. Read the chapter for details.

The polar equation R=Asin(Bθ) graphs a rose. Graph the rose for A=8 and B=2.5, and then explore the appearance of the rose for other values of A and B.

1. Press [MODE] to display the mode screen. Press [▼] [▼] [▼] [▶] [▶] [ENTER] to select **Pol** graphing mode. Select the defaults (the options on the left) for the other mode settings.

2. Press [Y=] to display the polar Y= editor. Press **8** [SIN] **2.5** [X,T,Θ,n] [)] [ENTER] to define **r1**.

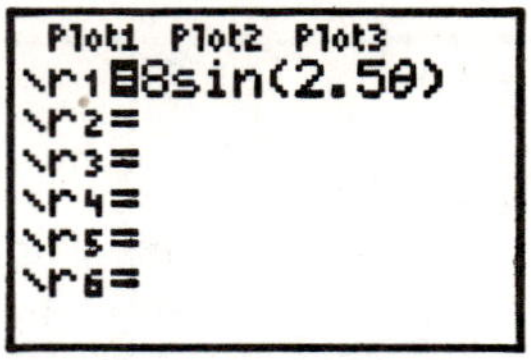

3. Press [ZOOM] **6** to select **6:ZStandard** and graph the equation in the standard viewing window. The graph shows only five petals of the rose, and the rose does not appear to be symmetrical. This is because the standard window sets **θmax=2π** and defines the window, rather than the pixels, as square.

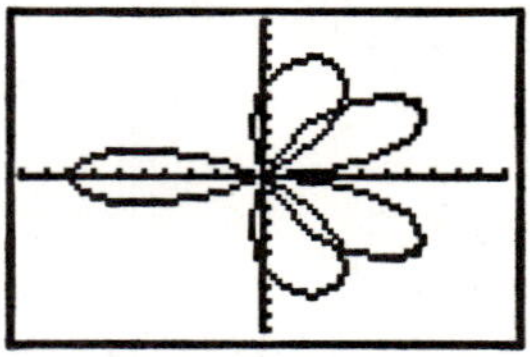

4. Press [WINDOW] to display the window variables. Press [▼] **4** [2nd] [π] to increase the value of θ**max** to 4π.

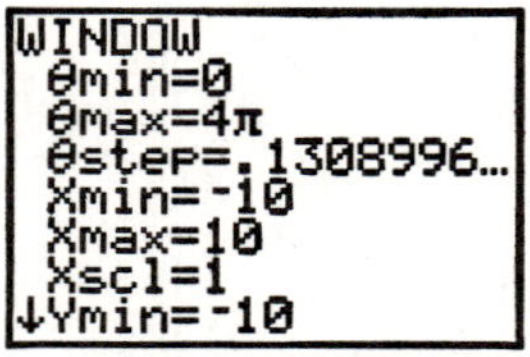

5. Press [ZOOM] **5** to select **5:ZSquare** and plot the graph.

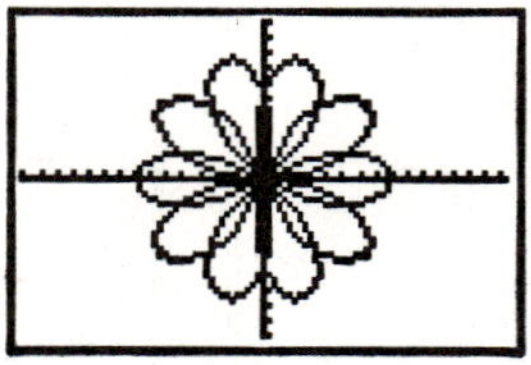

6. Repeat steps 2 through 5 with new values for the variables **A** and **B** in the polar equation **r1=Asin(Bθ)**. Observe how the new values affect the graph.

TI-83 Plus Graphing Mode Similarities

The steps for defining a polar graph are similar to the steps for defining a function graph. Chapter 5 assumes that you are familiar with Chapter 3: Function Graphing. Chapter 5 details aspects of polar graphing that differ from function graphing.

Setting Polar Graphing Mode

To display the mode screen, press [MODE]. To graph polar equations, you must select **Pol** graphing mode before you enter values for the window variables and before you enter polar equations.

Displaying the Polar Y= Editor

After selecting **Pol** graphing mode, press [Y=] to display the polar Y= editor.

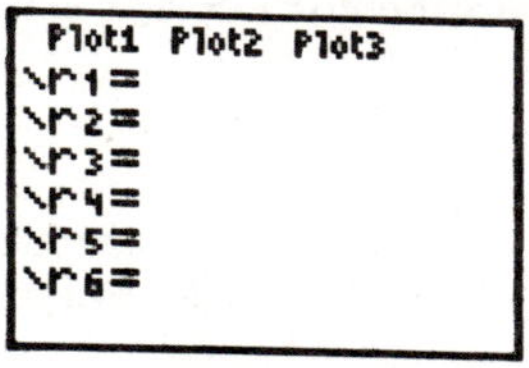

In this editor, you can enter and display up to six polar equations, **r1** through **r6**. Each is defined in terms of the independent variable θ (page 5-4).

Selecting Graph Styles

The icons to the left of **r1** through **r6** represent the graph style of each polar equation (Chapter 3). The default in **Pol** graphing mode is \ (line), which connects plotted points. Line, **** (thick), -0 (path), 0 (animate), and ∵ (dot) styles are available for polar graphing.

Defining and Editing Polar Equations

To define or edit a polar equation, follow the steps in Chapter 3 for defining a function or editing a function. The independent variable in a polar equation is θ. In **Pol** graphing mode, you can enter the polar variable θ in either of two ways.

- Press [X,T,Θ,n].
- Press [ALPHA] [θ].

Selecting and Deselecting Polar Equations

The TI-83 Plus graphs only the selected polar equations. In the Y= editor, a polar equation is selected when the = sign is highlighted. You may select any or all of the equations.

To change the selection status, move the cursor onto the = sign, and then press [ENTER].

Setting Window Variables

To display the window variable values, press [WINDOW]. These variables define the viewing window. The values below are defaults for **Pol** graphing in **Radian** angle mode.

θmin=0	Smallest θ value to evaluate
θmax=6.2831853...	Largest θ value to evaluate (2π)
θstep=.1308996...	Increment between θ values (π/24)
Xmin=-10	Smallest **X** value to be displayed
Xmax=10	Largest **X** value to be displayed
Xscl=1	Spacing between the **X** tick marks
Ymin=-10	Smallest **Y** value to be displayed
Ymax=10	Largest **Y** value to be displayed
Yscl=1	Spacing between the **Y** tick marks

Note: To ensure that sufficient points are plotted, you may want to change the θ window variables.

Setting the Graph Format

To display the current graph format settings, press [2nd] [FORMAT]. Chapter 3 describes the format settings in detail. The other graphing modes share these format settings.

Displaying a Graph

When you press [GRAPH], the TI-83 Plus plots the selected polar equations. It evaluates **R** for each value of θ (from **θmin** to **θmax** in intervals of **θstep**) and then plots each point. The window variables define the viewing window.

As the graph is plotted, **X**, **Y**, **R**, and θ are updated.

Smart Graph applies to polar graphs (Chapter 3).

Window Variables and Y-VARS Menus

You can perform these actions from the home screen or a program.

- Access functions by using the name of the equation as a variable.

- Store polar equations.

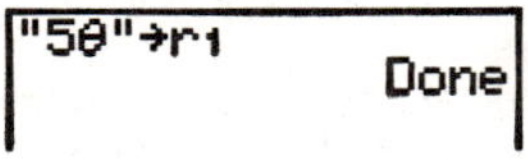

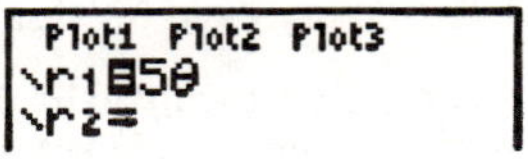

- Select or deselect polar equations.

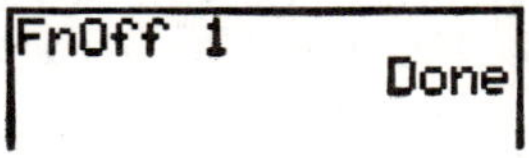

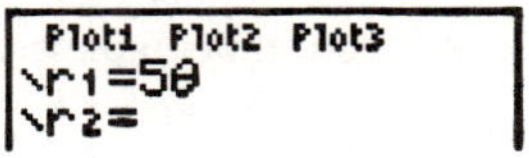

- Store values directly to window variables.

Free-Moving Cursor

The free-moving cursor in **Pol** graphing works the same as in **Func** graphing. In **RectGC** format, moving the cursor updates the values of **X** and **Y**; if **CoordOn** format is selected, **X** and **Y** are displayed. In **PolarGC** format, **X**, **Y**, **R**, and θ are updated; if **CoordOn** format is selected, **R** and θ are displayed.

TRACE

To activate TRACE, press [TRACE]. When TRACE is active, you can move the trace cursor along the graph of the equation one θ**step** at a time. When you begin a trace, the trace cursor is on the first selected function at θ**min.** If **ExprOn** format is selected, then the equation is displayed.

In **RectGC** format, TRACE updates the values of **X**, **Y**, and θ; if **CoordOn** format is selected, **X**, **Y**, and θ are displayed. In **PolarGC** format, TRACE updates **X**, **Y**, **R**, and θ; if **CoordOn** format is selected, **R** and θ are displayed.

To move five plotted points at a time on a function, press [2nd] [◄] or [2nd] [►]. If you move the trace cursor beyond the top or bottom of the screen, the coordinate values at the bottom of the screen continue to change appropriately.

Quick Zoom is available in **Pol** graphing mode; panning is not (Chapter 3).

Moving the Trace Cursor to Any Valid θ Value

To move the trace cursor to any valid θ value on the current function, enter the number. When you enter the first digit, a θ= prompt and the number you entered are displayed in the bottom-left corner of the screen. You can enter an expression at the θ= prompt. The value must be valid for the current viewing window. When you complete the entry, press [ENTER] to move the cursor.

ZOOM

ZOOM operations in **Pol** graphing work the same as in **Func** graphing. Only the **X** (**Xmin**, **Xmax**, and **Xscl**) and **Y** (**Ymin**, **Ymax**, and **Yscl**) window variables are affected.

The θ window variables (θ**min**, θ**max**, and θ**step**) are not affected, except when you select **ZStandard**. The VARS ZOOM secondary menu ZT/Zθ items **4:Zθmin**, **5:Zθmax**, and **6:Zθstep** are zoom memory variables for **Pol** graphing.

CALC

CALC operations in **Pol** graphing work the same as in **Func** graphing. The CALCULATE menu items available in **Pol** graphing are **1:value, 2:dy/dx**, and **3:dr/dθ.**

6 Sequence Graphing

Contents

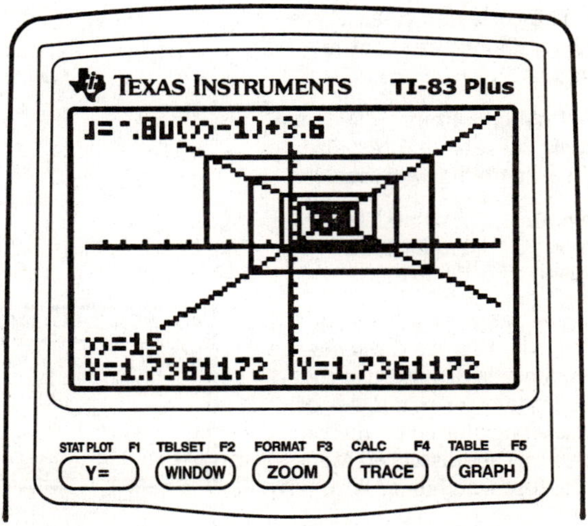

Getting Started is a fast-paced introduction. Read the chapter for details.

A small forest of 4,000 trees is under a new forestry plan. Each year 20 percent of the trees will be harvested and 1,000 new trees will be planted. Will the forest eventually disappear? Will the forest size stabilize? If so, in how many years and with how many trees?

1. Press [MODE]. Press [▼] [▼] [▼] [►] [►] [►] [ENTER] to select **Seq** graphing mode.

2. Press [2nd] [FORMAT] and select **Time** axes format and **ExprOn** format if necessary.

3. Press [Y=]. If the graph-style icon is not ⋱ (dot), press [◄] [◄], press [ENTER] until ⋱ is displayed, and then press [►] [►].

4. Press [MATH] [►] **3** to select **iPart(** (integer part) because only whole trees are harvested. After each annual harvest, 80 percent (.80) of the trees remain. Press [.] **8** [2nd] [u] [(] [X,T,Θ,n] [−] **1** [)] to define the number of trees after each harvest. Press [+] **1000** [)] to define the new trees. Press [▼] **4000** to define the number of trees at the beginning of the program.

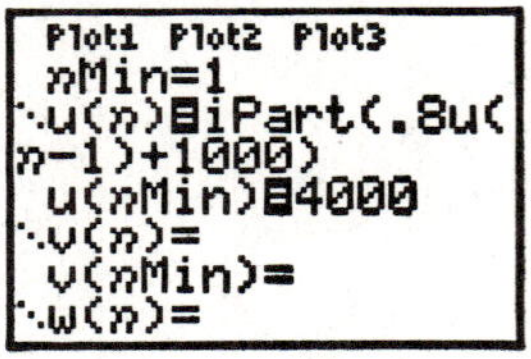

5. Press [WINDOW] **0** to set ***n*Min=0**. Press [▼] **50** to set ***n*Max=50**. ***n*Min** and ***n*Max** evaluate forest size over 50 years. Set the other window variables.

PlotStart=1	**Xmin=0**	**Ymin=0**
PlotStep=1	**Xmax=50**	**Ymax=6000**
	Xscl=10	**Yscl=1000**

6. Press [TRACE]. Tracing begins at ***n*Min** (the start of the forestry plan). Press [►] to trace the sequence year by year. The sequence is displayed at the top of the screen. The values for ***n*** (number of years), **X** (**X=*n***, because ***n*** is plotted on the x-axis), and **Y** (tree count) are displayed at the bottom. When will the forest stabilize? With how many trees?

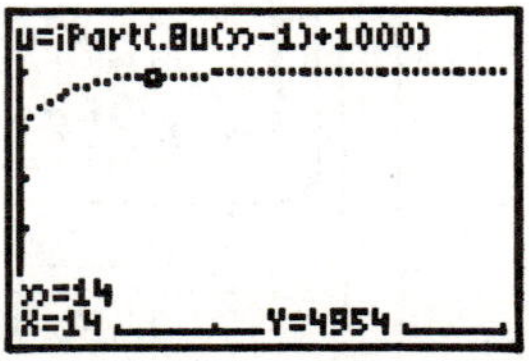

TI-83 Plus Graphing Mode Similarities

The steps for defining a sequence graph are similar to the steps for defining a function graph. Chapter 6 assumes that you are familiar with Chapter 3: Function Graphing. Chapter 6 details aspects of sequence graphing that differ from function graphing.

Setting Sequence Graphing Mode

To display the mode screen, press [MODE]. To graph sequence functions, you must select **Seq** graphing mode before you enter window variables and before you enter sequence functions.

Sequence graphs automatically plot in **Simul** mode, regardless of the current plotting-order mode setting.

TI-83 Plus Sequence Functions u, v, and w

The TI-83 Plus has three sequence functions that you can enter from the keyboard: **u**, **v**, and **w**. They are above the [7], [8], and [9] keys.

You can define sequence functions in terms of:

- The independent variable ***n***
- The previous term in the sequence function, such as **u(*n*−1)**
- The term that precedes the previous term in the sequence function, such as **u(*n*−2)**
- The previous term or the term that precedes the previous term in another sequence function, such as **u(*n*−1)** or **u(*n*−2)** referenced in the sequence **v(*n*)**.

Note: Statements in this chapter about **u(*n*)** are also true for **v(*n*)** and **w(*n*)**; statements about **u(*n*−1)** are also true for **v(*n*−1)** and **w(*n*−1)**; statements about **u(*n*−2)** are also true for **v(*n*−2)** and **w(*n*−2)**.

Displaying the Sequence Y= Editor

After selecting **Seq** mode, press [Y=] to display the sequence Y= editor.

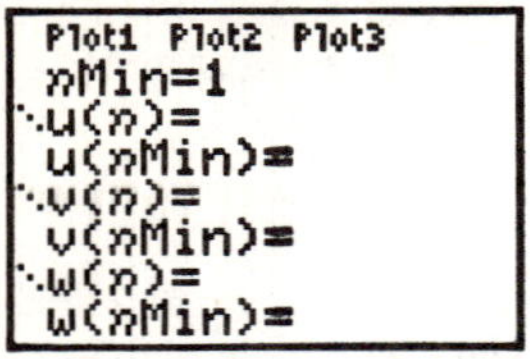

In this editor, you can display and enter sequences for **u(*n*)**, **v(*n*)**, and **w(*n*)**. Also, you can edit the value for ***n*Min**, which is the sequence window variable that defines the minimum ***n*** value to evaluate.

The sequence Y= editor displays the ***n*Min** value because of its relevance to **u(*n*Min)**, **v(*n*Min)**, and **w(*n*Min)**, which are the initial values for the sequence equations **u(*n*)**, **v(*n*)**, and **w(*n*)**, respectively.

***n*Min** in the Y= editor is the same as ***n*Min** in the window editor. If you enter a new value for ***n*Min** in one editor, the new value for ***n*Min** is updated in both editors.

Note: Use **u(*n*Min)**, **v(*n*Min)**, or **w(*n*Min)** only with a recursive sequence, which requires an initial value.

Selecting Graph Styles

The icons to the left of **u(*n*)**, **v(*n*)**, and **w(*n*)** represent the graph style of each sequence (Chapter 3). The default in **Seq** mode is ⋱ (dot), which shows discrete values. Dot, ⋱ (line), and ⋱ (thick) styles are available for sequence graphing. Graph styles are ignored in **Web** format.

Selecting and Deselecting Sequence Functions

The TI-83 Plus graphs only the selected sequence functions. In the Y= editor, a sequence function is selected when the = signs of both **u(*n*)=** and **u(*n*Min)=** are highlighted.

To change the selection status of a sequence function, move the cursor onto the = sign of the function name, and then press [ENTER]. The status is changed for both the sequence function **u(*n*)** and its initial value **u(*n*Min)**.

Defining and Editing a Sequence Function

To define or edit a sequence function, follow the steps in Chapter 3 for defining a function. The independent variable in a sequence is ***n***.

In **Seq** graphing mode, you can enter the sequence variable in either of two ways.

- Press [X,T,Θ,*n*].
- Press [2nd] [CATALOG] [N].

You can enter the function name from the keyboard.

- To enter the function name **u**, press [2nd] [u] (above [7]).
- To enter the function name **v**, press [2nd] [v] (above [8]).
- To enter the function name **w**, press [2nd] [w] (above [9]).

Generally, sequences are either nonrecursive or recursive. Sequences are evaluated only at consecutive integer values. ***n*** is always a series of consecutive integers, starting at zero or any positive integer.

Nonrecursive Sequences

In a nonrecursive sequence, the ***n***th term is a function of the independent variable ***n***. Each term is independent of all other terms.

For example, in the nonrecursive sequence below, you can calculate **u(5)** directly, without first calculating **u(1)** or any previous term.

The sequence equation above returns the sequence **2, 4, 6, 8, 10, . . .** for ***n*** = **1, 2, 3, 4, 5, ...**

Note: You may leave blank the initial value **u(nMin)** when calculating nonrecursive sequences.

Recursive Sequences

In a recursive sequence, the *n*th term in the sequence is defined in relation to the previous term or the term that precedes the previous term, represented by **u(*n*−1)** and **u(*n*−2)**. A recursive sequence may also be defined in relation to *n*, as in **u(*n*)=u(*n*−1)+*n*.**

For example, in the sequence below you cannot calculate **u(5)** without first calculating **u(1)**, **u(2)**, **u(3)**, and **u(4)**.

```
Plot1 Plot2 Plot3
 nMin=1
·.u(n)=2*u(n-1)
 u(nMin)=1
```

Using an initial value **u(*n*Min) = 1,** the sequence above returns **1, 2, 4, 8, 16, . . .**

Tip: On the TI-83 Plus, you must type each character of the terms. For example, to enter **u(*n*−1)**, press [2nd] [u] [(] [X,T,Θ,n] [−] [1] [)].

Recursive sequences require an initial value or values, since they reference undefined terms.

- If each term in the sequence is defined in relation to the previous term, as in **u(*n*−1)**, you must specify an initial value for the first term.

```
Plot1 Plot2 Plot3
 nMin=1
·.u(n)=.8u(n-1)+5
0
 u(nMin)=100
```

- If each term in the sequence is defined in relation to the term that precedes the previous term, as in **u(*n*−2)**, you must specify initial values for the first two terms. Enter the initial values as a list enclosed in braces ({ }) with commas separating the values.

```
Plot1 Plot2 Plot3
 nMin=1
·.u(n)=u(n-1)+u(n
-2)
 u(nMin)={1,0}
```

The value of the first term is 0 and the value of the second term is 1 for the sequence **u(*n*)**.

Setting Window Variables

To display the window variables, press [WINDOW]. These variables define the viewing window. The values below are defaults for **Seq** graphing in both **Radian** and **Degree** angle modes.

*n*Min=1	Smallest ***n*** value to evaluate
*n*Max=10	Largest ***n*** value to evaluate
PlotStart=1	First term number to be plotted
PlotStep=1	Incremental ***n*** value (for graphing only)
Xmin=-10	Smallest **X** value to be displayed
Xmax=10	Largest **X** value to be displayed
Xscl=1	Spacing between the **X** tick marks
Ymin=-10	Smallest **Y** value to be displayed
Ymax=10	Largest **Y** value to be displayed
Yscl=1	Spacing between the **Y** tick marks

***n*Min** must be an integer ≥ 0. ***n*Max, PlotStart**, and **PlotStep** must be integers ≥ 1.

n*Min** is the smallest ***n value to evaluate. ***n*Min** also is displayed in the sequence Y= editor. ***n*Max** is the largest ***n*** value to evaluate. Sequences are evaluated at **u(*n*Min)**, **u(*n*Min+1)**, **u(*n*Min+2)**, . . . , **u(*n*Max)**.

PlotStart is the first term to be plotted. **PlotStart=1** begins plotting on the first term in the sequence. If you want plotting to begin with the fifth term in a sequence, for example, set **PlotStart=5**. The first four terms are evaluated but are not plotted on the graph.

PlotStep is the incremental ***n*** value for graphing only. **PlotStep** does not affect sequence evaluation; it only designates which points are plotted on the graph. If you specify **PlotStep=2**, the sequence is evaluated at each consecutive integer, but it is plotted on the graph only at every other integer.

Setting the Graph Format

To display the current graph format settings, press [2nd] [FORMAT]. Chapter 3 describes the format settings in detail. The other graphing modes share these format settings. The axes setting on the top line of the screen is available only in **Seq** mode.

`Time Web uv vw uw`	Type of sequence plot (axes)
`RectGC PolarGC`	Rectangular or polar output
`CoordOn CoordOff`	Cursor coordinate display on/off
`GridOff GridOn`	Grid display off or on
`AxesOn AxesOff`	Axes display on or off
`LabelOff LabelOn`	Axes label display off or on
`ExprOn ExprOff`	Expression display on or off

Setting Axes Format

For sequence graphing, you can select from five axes formats. The table below shows the values that are plotted on the x-axis and y-axis for each axes setting.

Axes Setting	x-axis	y-axis
Time	***n***	**u(*n*), v(*n*), w(*n*)**
Web	**u(*n*−1), v(*n*−1), w(*n*−1)**	**u(*n*), v(*n*), w(*n*)**
uv	**u(*n*)**	**v(*n*)**
vw	**v(*n*)**	**w(*n*)**
uw	**u(*n*)**	**w(*n*)**

See pages 6-11 and 6-12 for more information on **Web** plots. See page 6-13 for more information on phase plots (**uv**, **vw**, and **uw** axes settings).

Displaying a Sequence Graph

To plot the selected sequence functions, press [GRAPH]. As a graph is plotted, the TI-83 Plus updates **X**, **Y**, and ***n***.

Smart Graph applies to sequence graphs (Chapter 3).

Free-Moving Cursor

The free-moving cursor in **Seq** graphing works the same as in **Func** graphing. In **RectGC** format, moving the cursor updates the values of **X** and **Y**; if **CoordOn** format is selected, **X** and **Y** are displayed. In **PolarGC** format, **X**, **Y**, **R**, and θ are updated; if **CoordOn** format is selected, **R** and θ are displayed.

TRACE

The axes format setting affects TRACE.

When **Time**, **uv**, **vw**, or **uw** axes format is selected, TRACE moves the cursor along the sequence one **PlotStep** increment at a time. To move five plotted points at once, press [2nd] [▶] or [2nd] [◀].

- When you begin a trace, the trace cursor is on the first selected sequence at the term number specified by **PlotStart**, even if it is outside the viewing window.
- Quick Zoom applies to all directions. To center the viewing window on the current cursor location after you have moved the trace cursor, press [ENTER]. The trace cursor returns to ***n*Min**.

In **Web** format, the trail of the cursor helps identify points with attracting and repelling behavior in the sequence. When you begin a trace, the cursor is on the x-axis at the initial value of the first selected function.

Tip: To move the cursor to a specified ***n*** during a trace, enter a value for ***n***, and press [ENTER]. For example, to quickly return the cursor to the beginning of the sequence, paste ***n*Min** to the ***n*=** prompt and press [ENTER].

Moving the Trace Cursor to Any Valid *n* Value

To move the trace cursor to any valid ***n*** value on the current function, enter the number. When you enter the first digit, an ***n*** = prompt and the number you entered are displayed in the bottom-left corner of the screen. You can enter an expression at the ***n*** = prompt. The value must be valid for the current viewing window. When you have completed the entry, press [ENTER] to move the cursor.

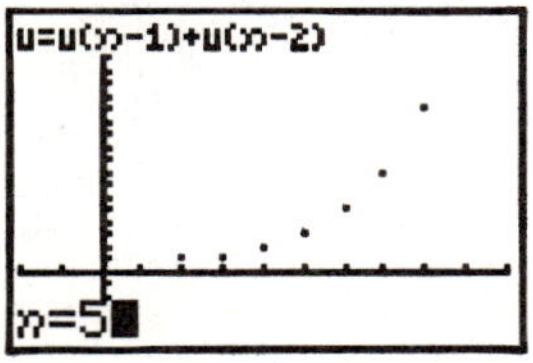

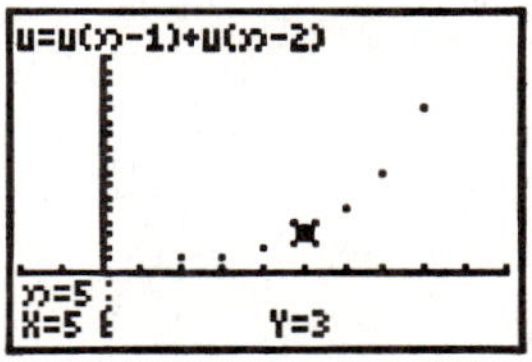

ZOOM

ZOOM operations in **Seq** graphing work the same as in **Func** graphing. Only the **X** (**Xmin**, **Xmax**, and **Xscl**) and **Y** (**Ymin, Ymax**, and **Yscl**) window variables are affected.

PlotStart, **PlotStep**, ***n*Min**, and ***n*Max** are only affected when you select **ZStandard**. The VARS Zoom secondary menu ZU items **1** through **7** are the ZOOM MEMORY variables for **Seq** graphing.

CALC

The only CALC operation available in **Seq** graphing is **value**.

- When **Time** axes format is selected, **value** displays **Y** (the **u(*n*)** value) for a specified ***n*** value.
- When **Web** axes format is selected, **value** draws the web and displays **Y** (the **u(*n*)** value) for a specified ***n*** value.
- When **uv**, **vw**, or **uw** axes format is selected, **value** displays **X** and **Y** according to the axes format setting. For example, for **uv** axes format, **X** represents **u(*n*)** and **Y** represents **v(*n*)**.

Evaluating u, v, and w

To enter the sequence names **u**, **v**, or **w**, press [2nd] [u], [v], or [w]. You can evaluate these names in any of three ways.

- Calculate the *n*th value in a sequence.
- Calculate a list of values in a sequence.
- Generate a sequence with **u(*nstart*,*nstop*[,*nstep*])**. *nstep* is optional; default is 1.

```
"n²"→u:u(3)
                 9
u({1,3,5,7,9})
  {1 9 25 49 81}
u(1,9,2)
  {1 9 25 49 81}
```

Graphing a Web Plot

To select **Web** axes format, press [2nd] [FORMAT] [▶] [ENTER]. A web plot graphs **u(*n*)** versus **u(*n*−1)**, which you can use to study long-term behavior (convergence, divergence, or oscillation) of a recursive sequence. You can see how the sequence may change behavior as its initial value changes.

Valid Functions for Web Plots

When **Web** axes format is selected, a sequence will not graph properly or will generate an error.

- It must be recursive with only one recursion level (**u(*n*−1)** but not **u(*n*−2)**).
- It cannot reference ***n*** directly.
- It cannot reference any defined sequence except itself.

Displaying the Graph Screen

In **Web** format, press [GRAPH] to display the graph screen. The TI-83 Plus:

- Draws a y=x reference line in **AxesOn** format.
- Plots the selected sequences with **u(*n*−1)** as the independent variable.

Note: A potential convergence point occurs whenever a sequence intersects the y=x reference line. However, the sequence may or may not actually converge at that point, depending on the sequence's initial value.

Drawing the Web

To activate the trace cursor, press [TRACE]. The screen displays the sequence and the current ***n***, **X**, and **Y** values (**X** represents **u(*n*−1)** and **Y** represents **u(*n*)**). Press [▶] repeatedly to draw the web step by step, starting at ***n*Min**. In **Web** format, the trace cursor follows this course.

1. It starts on the x-axis at the initial value **u(*n*Min)** (when **PlotStart=1**).
2. It moves vertically (up or down) to the sequence.
3. It moves horizontally to the y=x reference line.
4. It repeats this vertical and horizontal movement as you continue to press [▶].

Example: Convergence

1. Press [Y=] in **Seq** mode to display the sequence Y= editor. Make sure the graph style is set to ⠑ (dot), and then define ***n*Min**, **u(*n*)** and **u(*n*Min)** as shown below.

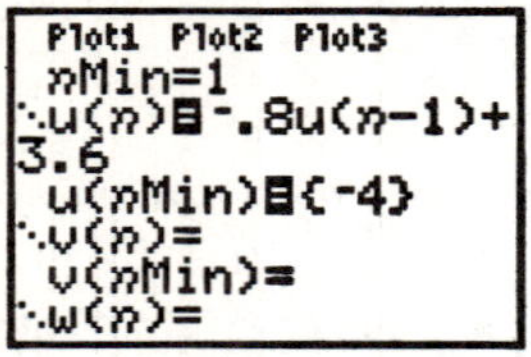

2. Press [2nd] [FORMAT] [ENTER] to set **Time** axes format.

3. Press [WINDOW] and set the variables as shown below.

***n*Min=1**	**Xmin=0**	**Ymin=⁻10**
***n*Max=25**	**Xmax=25**	**Ymax=10**
PlotStart=1	**Xscl=1**	**Yscl=1**
PlotStep=1		

4. Press [GRAPH] to graph the sequence.

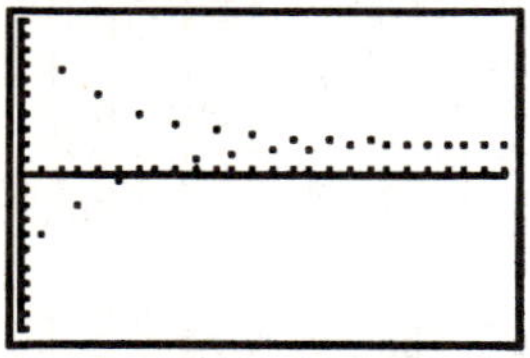

5. Press [2nd] [FORMAT] and select the **Web** axes setting.

6. Press [WINDOW] and change the variables below.
Xmin=⁻10 **Xmax=10**

7. Press [GRAPH] to graph the sequence.

8. Press [TRACE], and then press [▶] to draw the web. The displayed cursor coordinates ***n***, **X** (**u(*n*−1)**), and
Y (**u(*n*)**) change accordingly. When you press [▶], a new ***n*** value is displayed, and the trace cursor is on the sequence. When you press [▶] again, the ***n*** value remains the same, and the cursor moves to the **y=x** reference line. This pattern repeats as you trace the web.

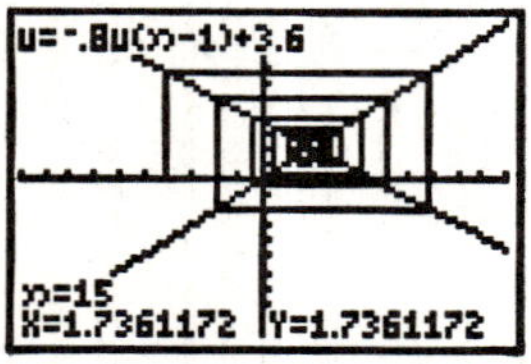

Graphing with uv, vw, and uw

The phase-plot axes settings **uv**, **vw**, and **uw** show relationships between two sequences. To select a phase-plot axes setting, press [2nd] [FORMAT], press [▶] until the cursor is on **uv**, **vw**, or **uw**, and then press [ENTER].

Axes Setting	x-axis	y-axis
uv	**u(*n*)**	**v(*n*)**
vw	**v(*n*)**	**w(*n*)**
uw	**u(*n*)**	**w(*n*)**

Example: Predator-Prey Model

Use the predator-prey model to determine the regional populations of a predator and its prey that would maintain population equilibrium for the two species.

This example uses the model to determine the equilibrium populations of wolves and rabbits, with initial populations of 200 rabbits (**u(*n*Min)**) and 50 wolves (**v(*n*Min)**).

These are the variables (given values are in parentheses):

R	=	number of rabbits	
M	=	rabbit population growth rate without wolves	(.05)
K	=	rabbit population death rate with wolves	(.001)
W	=	number of wolves	
G	=	wolf population growth rate with rabbits	(.0002)
D	=	wolf population death rate without rabbits	(.03)
n	=	time (in months)	
R_n	=	$R_{n-1}(1+M-KW_{n-1})$	
W_n	=	$W_{n-1}(1+GR_{n-1}-D)$	

1. Press [Y=] in **Seq** mode to display the sequence Y= editor. Define the sequences and initial values for R_n and W_n as shown below. Enter the sequence R_n as **u(*n*)** and enter the sequence W_n as **v(*n*)**.

Example: Predator-Prey Model (continued)

2. Press [2nd] [FORMAT] [ENTER] to select **Time** axes format.

3. Press [WINDOW] and set the variables as shown below.
   ***n*Min=0** **Xmin=0** **Ymin=0**
   ***n*Max=400** **Xmax=400** **Ymax=300**
   PlotStart=1 **Xscl=100** **Yscl=100**
   PlotStep=1

4. Press [GRAPH] to graph the sequence.

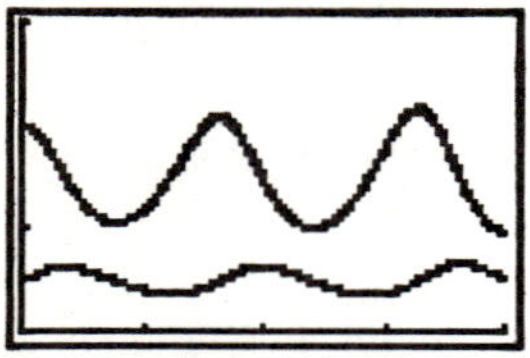

5. Press [TRACE] [▶] to individually trace the number of rabbits (**u(*n*)**) and wolves (**v(*n*)**) over time (***n***).

   Tip: Press a number, and then press [ENTER] to jump to a specific ***n*** value (month) while in TRACE.

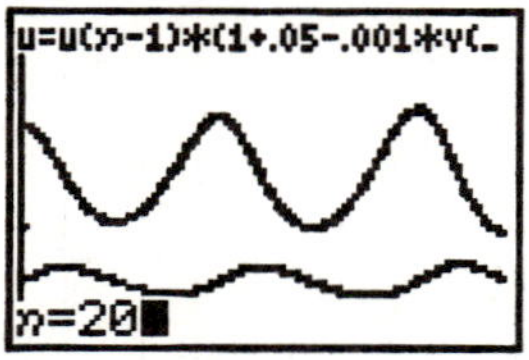

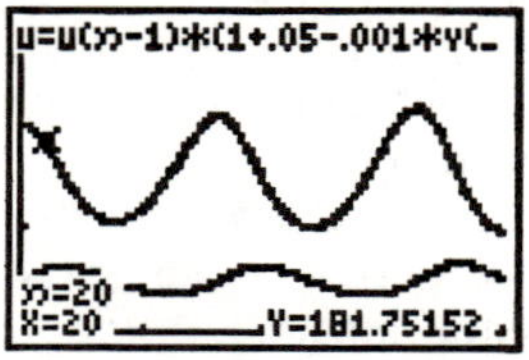

6. Press [2nd] [FORMAT] [▶] [▶] [ENTER] to select **uv** axes format.

7. Press [WINDOW] and change these variables as shown below.
   Xmin=84 **Ymin=25**
   Xmax=237 **Ymax=75**
   Xscl=50 **Yscl=10**

8. Press [TRACE]. Trace both the number of rabbits (**X**) and the number of wolves (**Y**) through 400 generations.

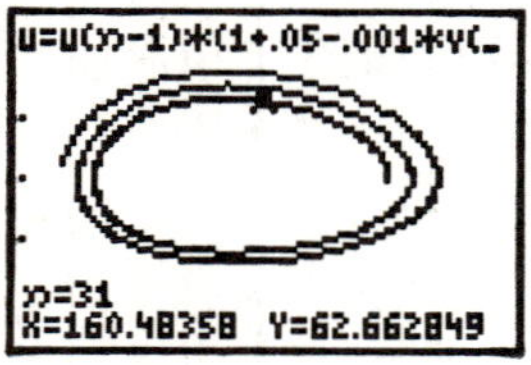

Note: When you press [TRACE], the equation for **u** is displayed in the top-left corner. Press [▲] or [▼] to see the equation for **v**.

Comparing TI-83 Plus and TI-82 Sequence Variables

Sequences and Window Variables

Refer to the table if you are familiar with the TI-82. It shows TI-83 Plus sequences and sequence window variables, as well as their TI-82 counterparts.

TI-83 Plus	TI-82
In the Y= editor:	
u(*n*)	**U*n***
u(*n*Min)	**U*n*Start** (window variable)
v(*n*)	**V*n***
v(*n*Min)	**V*n*Start** (window variable)
w(*n*)	not available
w(*n*Min)	not available
In the window editor:	
***n*Min**	***n*Start**
***n*Max**	***n*Max**
PlotStart	***n*Min**
PlotStep	not available

Sequence Keystroke Changes

Refer to the table if you are familiar with the TI-82. It compares TI-83 Plus sequence-name syntax and variable syntax with TI-82 sequence-name syntax and variable syntax.

TI-83 Plus / TI-82	On TI-83 Plus, press:	On TI-82, press:
n* / *n	[X,T,Θ,*n*]	[2nd] [*n*]
u(*n*) / U*n*	[2nd] [u] [(] [X,T,Θ,*n*] [)]	[2nd] [Y-VARS] [4] [1]
v(*n*) / V*n*	[2nd] [v] [(] [X,T,Θ,*n*] [)]	[2nd] [Y-VARS] [4] [2]
w(*n*)	[2nd] [w] [(] [X,T,Θ,*n*] [)]	not available
u(*n*−1) / U*n*−1	[2nd] [u] [(] [X,T,Θ,*n*] [−] [1] [)]	[2nd] [U_{n-1}]
v(*n*−1) / V*n*−1	[2nd] [v] [(] [X,T,Θ,*n*] [−] [1] [)]	[2nd] [V_{n-1}]
w(*n*−1)	[2nd] [w] [(] [X,T,Θ,*n*] [−] [1] [)]	not available

7 Tables

Contents

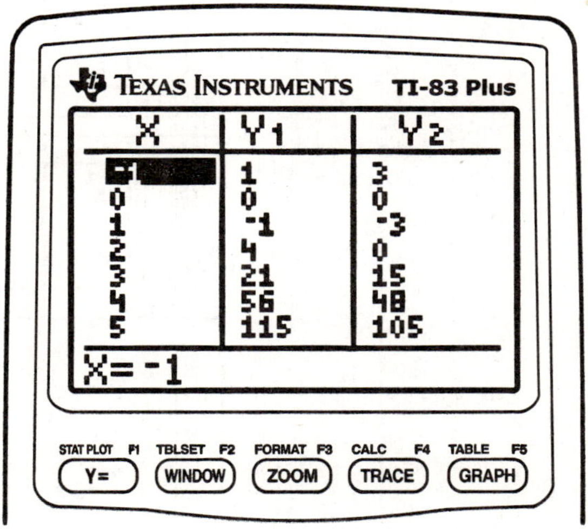

Getting Started is a fast-paced introduction. Read the chapter for details.

Evaluate the function Y = X³ – 2X at each integer between ⁻10 and 10. How many sign changes occur, and at what X values?

1. Press [MODE] [▼] [▼] [▼] [ENTER] to set **Func** graphing mode.

2. Press [Y=]. Press [X,T,Θ,n] [MATH] **3** to select ³. Then press [–] **2** [X,T,Θ,n] to enter the function **Y1=X³–2X**.

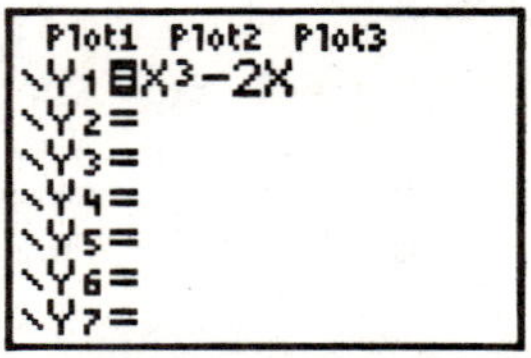

3. Press [2nd] [TBLSET] to display the TABLE SETUP screen. Press [(-)] **10** [ENTER] to set **TblStart=⁻10**. Press **1** [ENTER] to set **ΔTbl=1**.

   Press [ENTER] to select **Indpnt: Auto** (automatically generated independent values). Press [▼] [ENTER] to select **Depend: Auto** (automatically generated dependent values).

4. Press [2nd] [TABLE] to display the table screen.

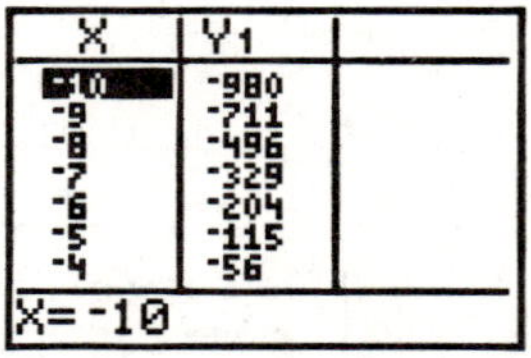

X
Y1
-10
-9
-8
-7
-6
-5
-4
-980
-711
-496
-329
-204
-115
-56
X=-10

5. Press [▼] until you see the sign changes in the value of **Y1**. How many sign changes occur, and at what **X** values?

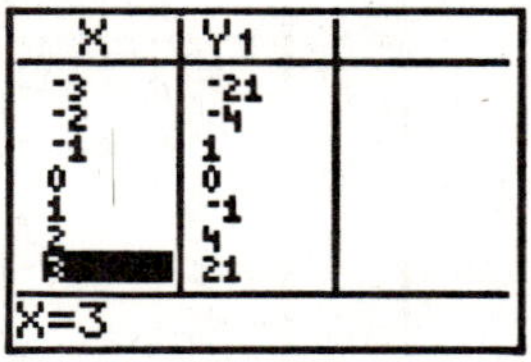

Setting Up the Table

TABLE SETUP Screen

To display the TABLE SETUP screen, press [2nd] [TBLSET].

TblStart, ΔTbl

TblStart (table start) defines the initial value for the independent variable. **TblStart** applies only when the independent variable is generated automatically (when **Indpnt: Auto** is selected).

ΔTbl (table step) defines the increment for the independent variable.

Note: In **Seq** mode, both **TblStart** and **ΔTbl** must be integers.

Indpnt: Auto, Indpnt: Ask, Depend: Auto, Depend: Ask

Selections	Table Characteristics
Indpnt: Auto **Depend: Auto**	Values are displayed automatically in both the independent-variable column and in all dependent-variable columns.
Indpnt: Ask **Depend: Auto**	The table is empty; when you enter a value for the independent variable, all corresponding dependent-variable values are calculated and displayed automatically.
Indpnt: Auto **Depend: Ask**	Values are displayed automatically for the independent variable; to generate a value for a dependent variable, move the cursor to that cell and press [ENTER].
Indpnt: Ask **Depend: Ask**	The table is empty; enter values for the independent variable; to generate a value for a dependent variable, move the cursor to that cell and press [ENTER].

Setting Up the Table from the Home Screen or a Program

To store a value to **TblStart**, **ΔTbl**, or **TblInput** from the home screen or a program, select the variable name from the VARS TABLE secondary menu. **TblInput** is a list of independent-variable values in the current table.

When you press [2nd] [TBLSET] in the program editor, you can select **IndpntAuto**, **IndpntAsk**, **DependAuto**, and **DependAsk**.

Defining Dependent Variables from the Y= Editor

In the Y= editor, enter the functions that define the dependent variables. Only functions that are selected in the Y= editor are displayed in the table. The current graphing mode is used. In **Par** mode, you must define both components of each parametric equation (Chapter 4).

Editing Dependent Variables from the Table Editor

To edit a selected Y= function from the table editor, follow these steps.

1. Press [2nd] [TABLE] to display the table, then press [▶] or [◀] to move the cursor to a dependent-variable column.
2. Press [▲] until the cursor is on the function name at the top of the column. The function is displayed on the bottom line.

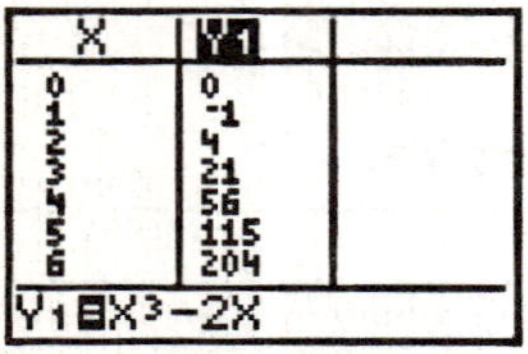

3. Press [ENTER]. The cursor moves to the bottom line. Edit the function.

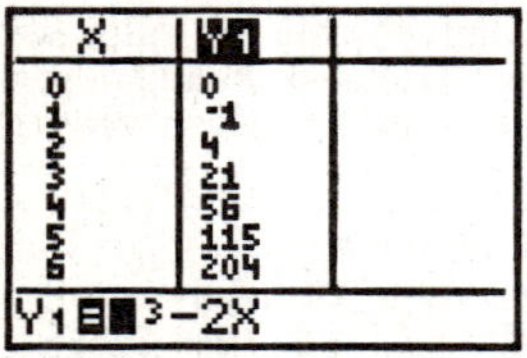

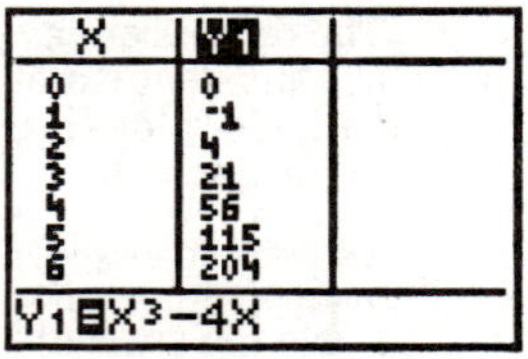

4. Press [ENTER] or [▼]. The new values are calculated. The table and the Y= function are updated automatically.

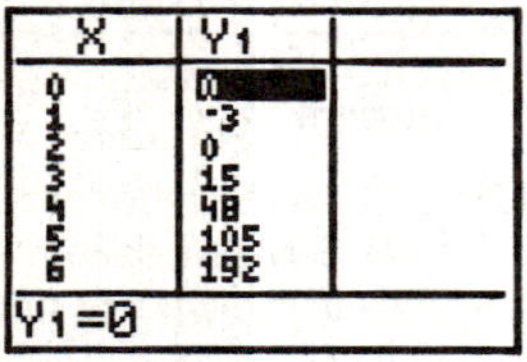

Note: You also can use this feature to view the function that defines a dependent variable without having to leave the table.

The Table

To display the table, press [2nd] [TABLE].

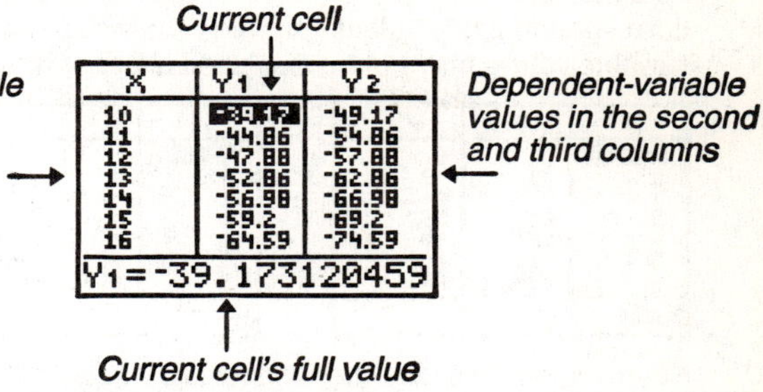

Note: The table abbreviates the values, if necessary.

Independent and Dependent Variables

The current graphing mode determines which independent and dependent variables are displayed in the table (Chapter 1). In the table above, for example, the independent variable **X** and the dependent variables **Y1** and **Y2** are displayed because **Func** graphing mode is set.

Graphing Mode	Independent Variable	Dependent Variable
Func (function)	**X**	**Y1** through **Y9**, and **Y0**
Par (parametric)	**T**	**X1T/Y1T** through **X6T/Y6T**
Pol (polar)	**θ**	**r1** through **r6**
Seq (sequence)	**n**	**u(*n*)**, **v(*n*)**, and **w(*n*)**

Clearing the Table from the Home Screen or a Program

From the home screen, select the **ClrTable** instruction from the CATALOG. To clear the table, press [ENTER].

From a program, select **9:ClrTable** from the PRGM I/O menu or from the CATALOG. The table is cleared upon execution. If **IndpntAsk** is selected, all independent and dependent variable values on the table are cleared. If **DependAsk** is selected, all dependent variable values on the table are cleared.

Scrolling Independent-Variable Values

If **Indpnt: Auto** is selected, you can press [▲] and [▼] in the independent-variable column to display more values. As you scroll the column, the corresponding dependent-variable values also are displayed. All dependent-variable values may not be displayed if **Depend: Ask** is selected.

X	Y1	Y2
0	0	0
1	-1	-3
2	4	0
3	21	15
4	56	48
5	115	105
6	204	192

X=0

X	Y1	Y2
-1	1	3
0	0	0
1	-1	-3
2	4	0
3	21	15
4	56	48
5	115	105

X= -1

Note: You can scroll back from the value entered for **TblStart**. As you scroll, **TblStart** is updated automatically to the value shown on the top line of the table. In the example above, **TblStart=0** and **ΔTbl=1** generates and displays values of **X=0**, . . . , **6**; but you can press [▲] to scroll back and display the table for **X=-1**, . . ., **5**.

Displaying Other Dependent Variables

If you have defined more than two dependent variables, the first two selected Y= functions are displayed initially. Press [▶] or [◀] to display dependent variables defined by other selected Y= functions. The independent variable always remains in the left column, except during a trace with **Par** graphing mode and **G-T** split-screen mode set.

X	Y2	Y3
-4	-4	-28
-3	-6	-18
-2	-6	-10
-1	-4	-4
0	0	0
1	6	2
2	14	2

Y3= -28

Tip: To simultaneously display on the table two dependent variables that are not defined as consecutive Y= functions, go to the Y= editor and deselect the Y= functions between the two you want to display. For example, to simultaneously display **Y4** and **Y7** on the table, go to the Y= editor and deselect **Y5** and **Y6**.

8 Draw Instructions

Contents

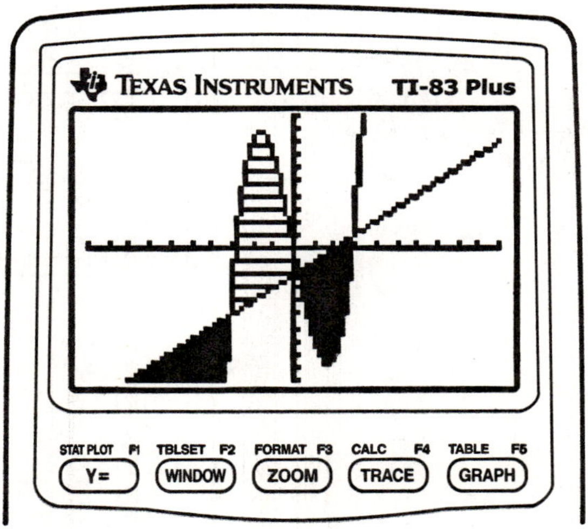

Getting Started is a fast-paced introduction. Read the chapter for details.

Suppose you want to find the equation of the tangent line at X = √2/2 for the function Y = sin(X).

Before you begin, select **Radian** and **Func** mode from the mode screen, if necessary.

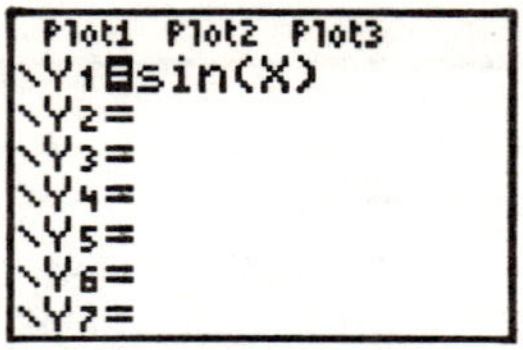

1. Press [Y=] to display the Y= editor. Press [SIN] [X,T,Θ,n] [)] to store **sin(X)** in **Y1**.

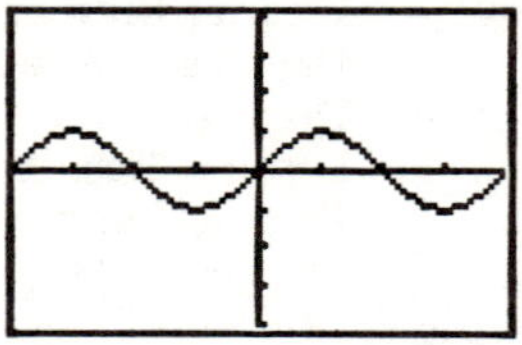

2. Press [ZOOM] **7** to select **7:ZTrig**, which graphs the equation in the Zoom Trig window.

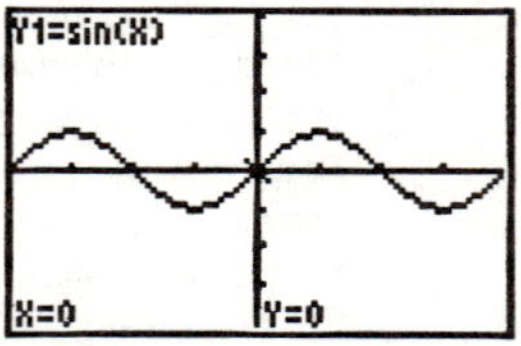

3. Press [2nd] [DRAW] **5** to select **5:Tangent(**. The tangent instruction is initiated.

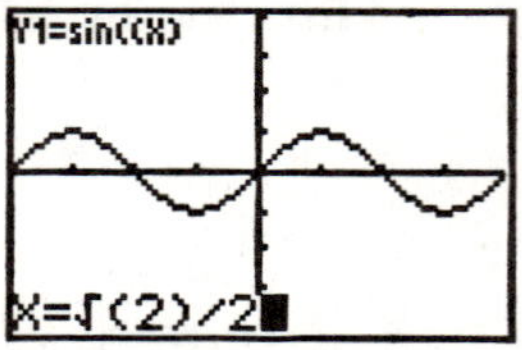

4. Press [2nd] [√] **2** [)] [÷] **2**.

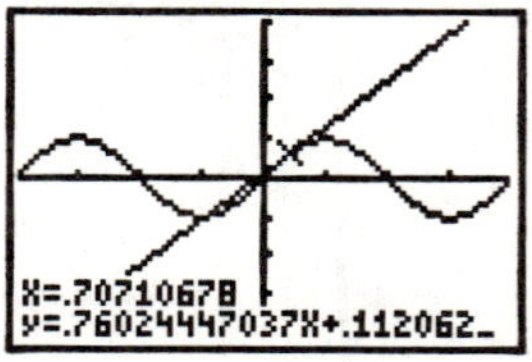

5. Press [ENTER]. The tangent line is drawn; the **X** value and the tangent-line equation are displayed on the graph.

DRAW Menu

To display the DRAW menu, press [2nd] [DRAW]. The TI-83 Plus's interpretation of these instructions depends on whether you accessed the menu from the home screen or the program editor or directly from a graph.

DRAW POINTS STO	
1: ClrDraw	Clears all drawn elements.
2: Line(	Draws a line segment between 2 points.
3: Horizontal	Draws a horizontal line.
4: Vertical	Draws a vertical line.
5: Tangent(	Draws a line segment tangent to a function.
6: DrawF	Draws a function.
7: Shade(	Shades an area between two functions.
8: DrawInv	Draws the inverse of a function.
9: Circle(	Draws a circle.
0: Text(	Draws text on a graph screen.
A: Pen	Activates the free-form drawing tool.

Before Drawing on a Graph

The DRAW instructions draw on top of graphs. Therefore, before you use the DRAW instructions, consider whether you want to perform one or more of the following actions.

- Change the mode settings on the mode screen.
- Change the format settings on the format screen.
- Enter or edit functions in the Y= editor.
- Select or deselect functions in the Y= editor.
- Change the window variable values.
- Turn stat plots on or off.
- Clear existing drawings with **ClrDraw** (page 8-4).

Note: If you draw on a graph and then perform any of the actions listed above, the graph is replotted without the drawings when you display the graph again.

Drawing on a Graph

You can use any DRAW menu instructions except **DrawInv** to draw on **Func**, **Par**, **Pol**, and **Seq** graphs. **DrawInv** is valid only in **Func** graphing. The coordinates for all DRAW instructions are the display's x-coordinate and y-coordinate values.

You can use most DRAW menu and DRAW POINTS menu instructions to draw directly on a graph, using the cursor to identify the coordinates. You also can execute these instructions from the home screen or from within a program. If a graph is not displayed when you select a DRAW menu instruction, the home screen is displayed.

Clearing Drawings

Clearing Drawings When a Graph Is Displayed

All points, lines, and shading drawn on a graph with DRAW instructions are temporary.

To clear drawings from the currently displayed graph, select **1:ClrDraw** from the DRAW menu. The current graph is replotted and displayed with no drawn elements.

Clearing Drawings from the Home Screen or a Program

To clear drawings on a graph from the home screen or a program, begin on a blank line on the home screen or in the program editor. Select **1:ClrDraw** from the DRAW menu. The instruction is copied to the cursor location. Press [ENTER].

When **ClrDraw** is executed, it clears all drawings from the current graph and displays the message **Done**. When you display the graph again, all drawn points, lines, circles, and shaded areas will be gone.

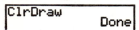

```
ClrDraw
                Done
```

Note: Before you clear drawings, you can store them with **StorePic** (page 8-17).

Drawing a Line Segment Directly on a Graph

To draw a line segment when a graph is displayed, follow these steps.

1. Select **2:Line(** from the DRAW menu.
2. Place the cursor on the point where you want the line segment to begin, and then press [ENTER].
3. Move the cursor to the point where you want the line segment to end. The line is displayed as you move the cursor. Press [ENTER].

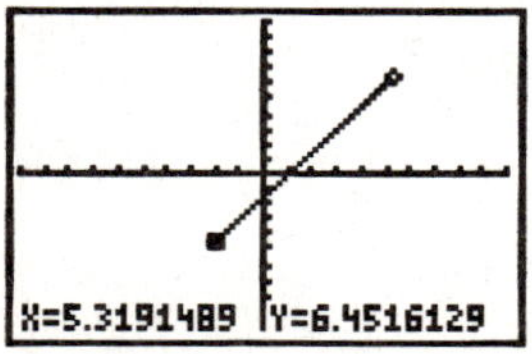

To continue drawing line segments, repeat steps 2 and 3. To cancel **Line(**, press [CLEAR].

Drawing a Line Segment from the Home Screen or a Program

Line(also draws a line segment between the coordinates (*X1*,*Y1*) and (*X2*,*Y2*). The values may be entered as expressions.

Line(*X1*,*Y1*,*X2*,*Y2***)**

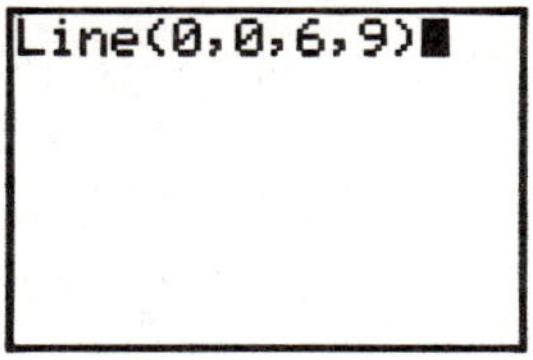

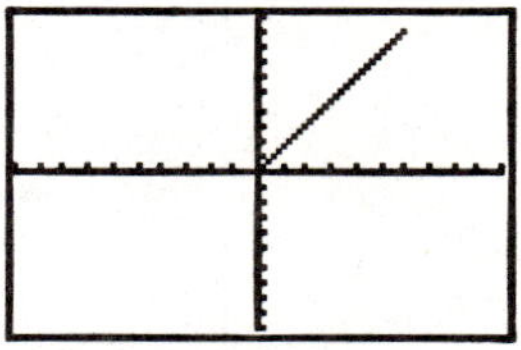

To erase a line segment, enter **Line(***X1*,*Y1*,*X2*,*Y2*,**0)**

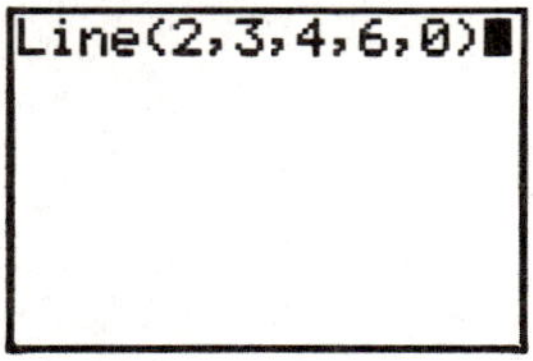

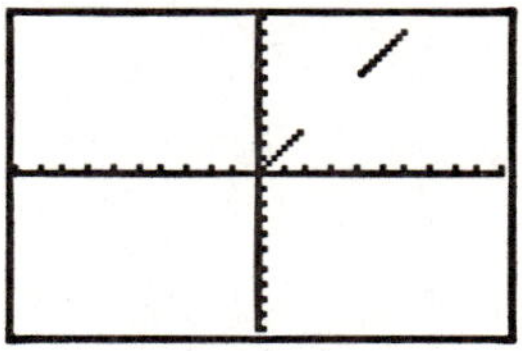

Drawing a Line Directly on a Graph

To draw a horizontal or vertical line when a graph is displayed, follow these steps.

1. Select **3:Horizontal** or **4:Vertical** from the DRAW menu. A line is displayed that moves as you move the cursor.
2. Place the cursor on the y-coordinate (for horizontal lines) or x-coordinate (for vertical lines) through which you want the drawn line to pass.
3. Press [ENTER] to draw the line on the graph.

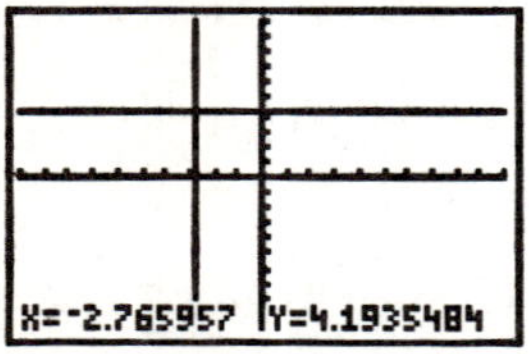

To continue drawing lines, repeat steps 2 and 3.

To cancel **Horizontal** or **Vertical**, press [CLEAR].

Drawing a Line from the Home Screen or a Program

Horizontal (horizontal line) draws a horizontal line at **Y**=y. y can be an expression but not a list.

Horizontal y

Vertical (vertical line) draws a vertical line at **X**=x. x can be an expression but not a list.

Vertical x

To instruct the TI-83 Plus to draw more than one horizontal or vertical line, separate each instruction with a colon (:).

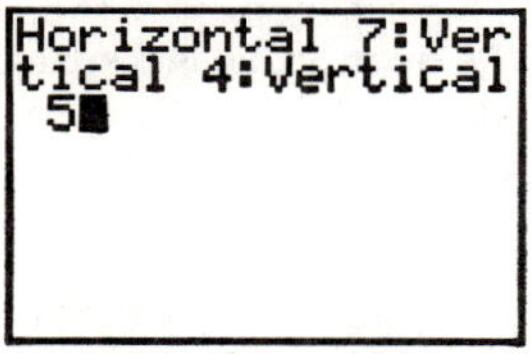

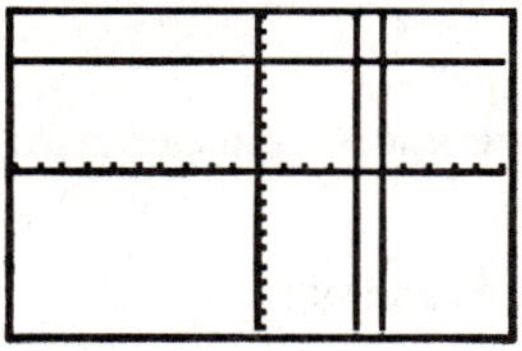

Drawing a Tangent Line Directly on a Graph

To draw a tangent line when a graph is displayed, follow these steps.

1. Select **5:Tangent(** from the DRAW menu.
2. Press [▼] and [▲] to move the cursor to the function for which you want to draw the tangent line. The current graph's **Y=** function is displayed in the top-left corner, if **ExprOn** is selected.
3. Press [▶] and [◀] or enter a number to select the point on the function at which you want to draw the tangent line.
4. Press [ENTER]. In **Func** mode, the **X** value at which the tangent line was drawn is displayed on the bottom of the screen, along with the equation of the tangent line. In all other modes, the **dy/dx** value is displayed.

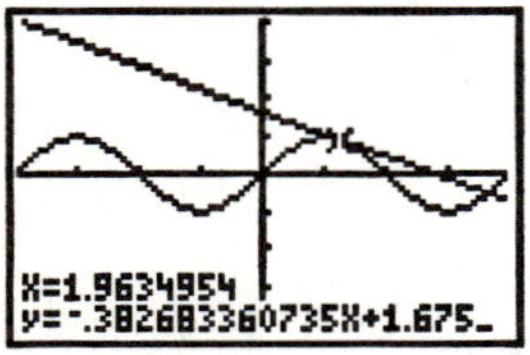

Tip: Change the fixed decimal setting on the mode screen if you want to see fewer digits displayed for **X** and the equation for **Y**.

Drawing a Tangent Line from the Home Screen or a Program

Tangent((tangent line) draws a line tangent to *expression* in terms of **X**, such as **Y1** or X^2, at point **X**=*value*. **X** can be an expression. *expression* is interpreted as being in **Func** mode.

Tangent(*expression*,*value***)**

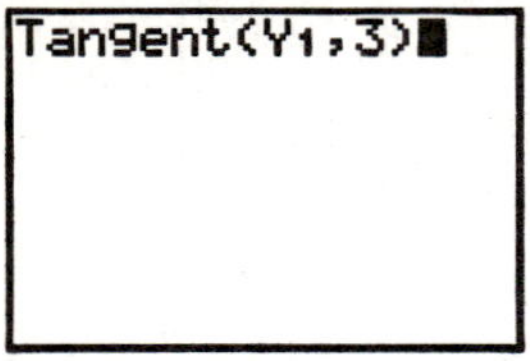

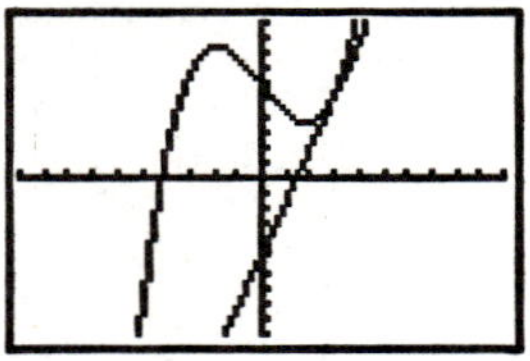

Drawing a Function

DrawF (draw function) draws *expression* as a function in terms of **X** on the current graph. When you select **6:DrawF** from the DRAW menu, the TI-83 Plus returns to the home screen or the program editor. **DrawF** is not interactive.

DrawF *expression*

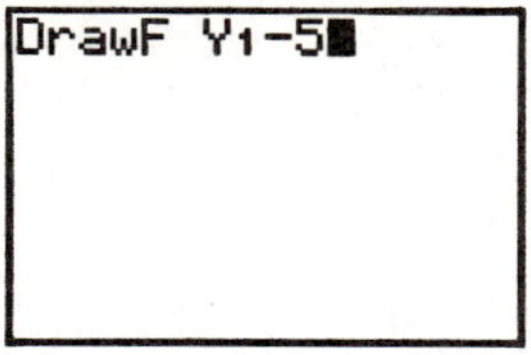

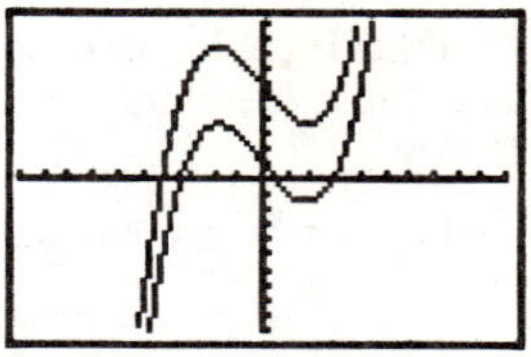

Note: You cannot use a list in *expression* to draw a family of curves.

Drawing an Inverse of a Function

DrawInv (draw inverse) draws the inverse of *expression* by plotting **X** values on the y-axis and **Y** values on the x-axis. When you select **8:DrawInv** from the DRAW menu, the TI-83 Plus returns to the home screen or the program editor. **DrawInv** is not interactive. **DrawInv** works in **Func** mode only.

DrawInv *expression*

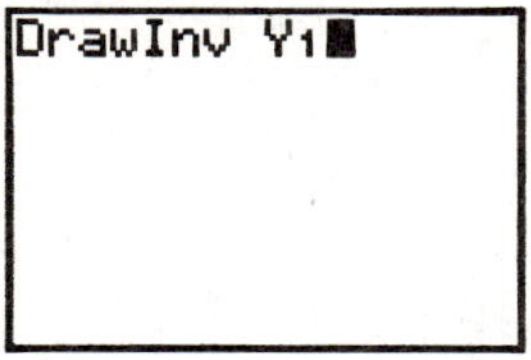

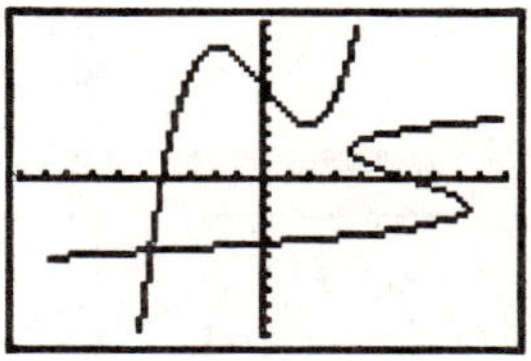

Note: You cannot use a list in *expression* to draw a family of curves.

Shading a Graph

To shade an area on a graph, select **7:Shade(** from the DRAW menu. The instruction is pasted to the home screen or to the program editor.

Shade(draws *lowerfunc* and *upperfunc* in terms of **X** on the current graph and shades the area that is specifically above *lowerfunc* and below *upperfunc*. Only the areas where *lowerfunc* < *upperfunc* are shaded.

Xleft and *Xright*, if included, specify left and right boundaries for the shading. *Xleft* and *Xright* must be numbers between **Xmin** and **Xmax**, which are the defaults.

pattern specifies one of four shading patterns.

pattern=**1**	vertical (default)
pattern=**2**	horizontal
pattern=**3**	negative—slope 45°
pattern=**4**	positive—slope 45°

patres specifies one of eight shading resolutions.

patres=**1**	shades every pixel (default)
patres=**2**	shades every second pixel
patres=**3**	shades every third pixel
patres=**4**	shades every fourth pixel
patres=**5**	shades every fifth pixel
patres=**6**	shades every sixth pixel
patres=**7**	shades every seventh pixel
patres=**8**	shades every eighth pixel

Shade(*lowerfunc*,*upperfunc*[,*Xleft*,*Xright*,*pattern*,*patres*]**)**

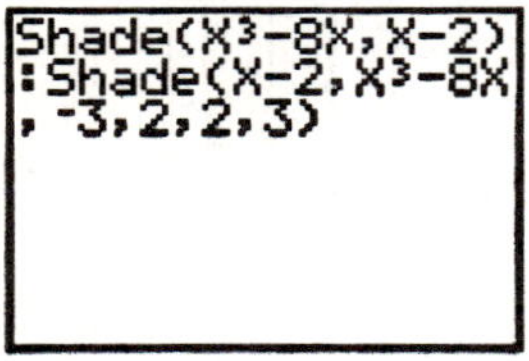

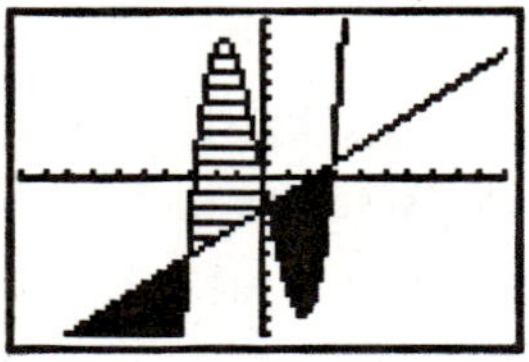

Drawing a Circle Directly on a Graph

To draw a circle directly on a displayed graph using the cursor, follow these steps.

1. Select **9:Circle(** from the DRAW menu.
2. Place the cursor at the center of the circle you want to draw. Press [ENTER].
3. Move the cursor to a point on the circumference. Press [ENTER] to draw the circle on the graph.

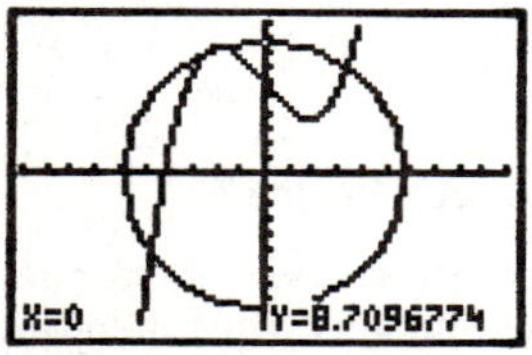

Note: This circle is displayed as circular, regardless of the window variable values, because you drew it directly on the display. When you use the **Circle(** instruction from the home screen or a program, the current window variables may distort the shape.

To continue drawing circles, repeat steps 2 and 3. To cancel **Circle(**, press [CLEAR].

Drawing a Circle from the Home Screen or a Program

Circle(draws a circle with center (*X*,*Y*) and *radius*. These values can be expressions.

Circle(*X*,*Y*,*radius***)**

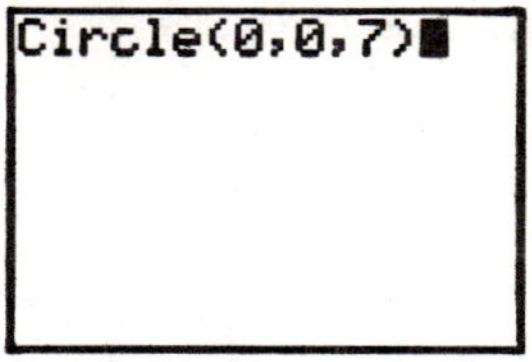

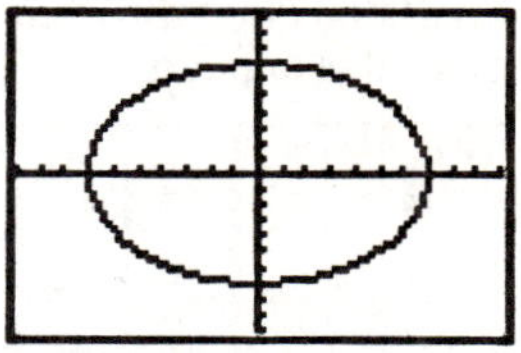

Tip: When you use **Circle(** on the home screen or from a program, the current window values may distort the drawn circle. Use **ZSquare** (Chapter 3) before drawing the circle to adjust the window variables and make the circle circular.

Placing Text Directly on a Graph

To place text on a graph when the graph is displayed, follow these steps.

1. Select **0:Text(** from the DRAW menu.
2. Place the cursor where you want the text to begin.
3. Enter the characters. Press [ALPHA] or [2nd] [A-LOCK] to enter letters and θ. You may enter TI-83 Plus functions, variables, and instructions. The font is proportional, so the exact number of characters you can place on the graph varies. As you type, the characters are placed on top of the graph.

To cancel **Text(**, press [CLEAR].

Placing Text on a Graph from the Home Screen or a Program

Text(places on the current graph the characters comprising *value*, which can include TI-83 Plus functions and instructions. The top-left corner of the first character is at pixel (*row*,*column*), where *row* is an integer between 0 and 57 and *column* is an integer between 0 and 94. Both *row* and *column* can be expressions.

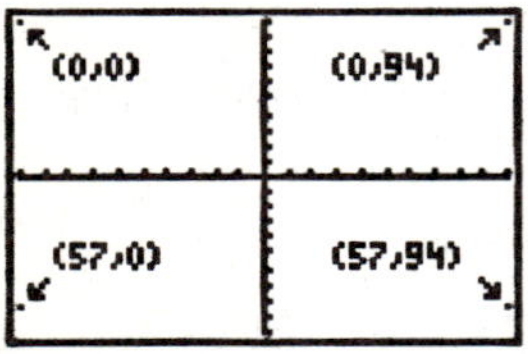

Text(*row*,*column*,*value*,*value* . . .**)**

value can be text enclosed in quotation marks ("), or it can be an expression. The TI-83 Plus will evaluate an expression and display the result with up to 10 characters.

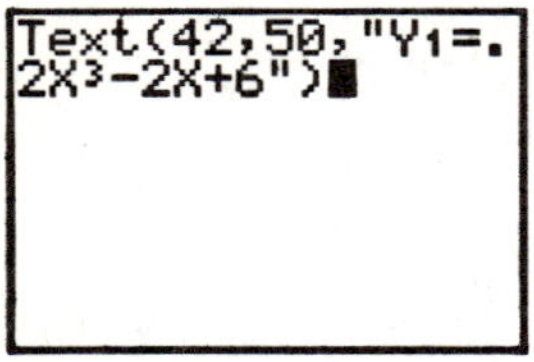

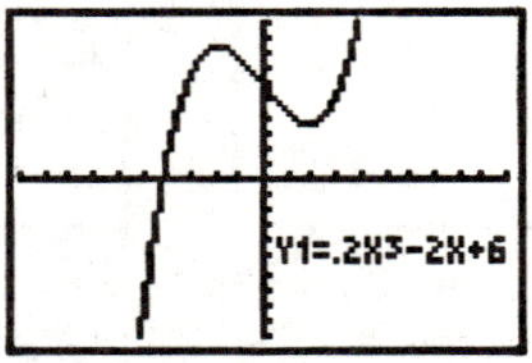

Split Screen

On a **Horiz** split screen, the maximum value for *row* is 25. On a **G-T** split screen, the maximum value for *row* is 45, and the maximum value for *column* is 46.

Using Pen to Draw on a Graph

Pen draws directly on a graph only. You cannot execute **Pen** from the home screen or a program.

To draw on a displayed graph, follow these steps.

1. Select **A:Pen** from the DRAW menu.
2. Place the cursor on the point where you want to begin drawing. Press [ENTER] to turn on the pen.
3. Move the cursor. As you move the cursor, you draw on the graph, shading one pixel at a time.
4. Press [ENTER] to turn off the pen.

For example, **Pen** was used to create the arrow pointing to the local minimum of the selected function.

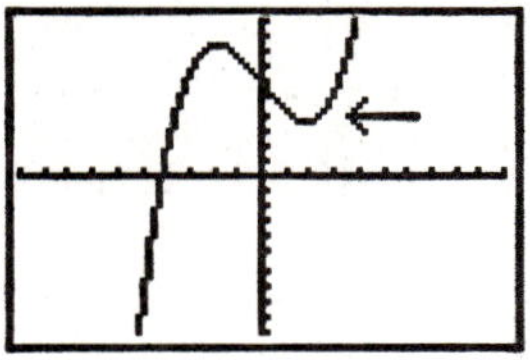

To continue drawing on the graph, move the cursor to a new position where you want to begin drawing again, and then repeat steps 2, 3, and 4. To cancel **Pen**, press [CLEAR].

DRAW POINTS Menu

To display the DRAW POINTS menu, press [2nd] [DRAW] [▶]. The TI-83 Plus's interpretation of these instructions depends on whether you accessed this menu from the home screen or the program editor or directly from a graph.

DRAW POINTS STO	
1: Pt-On(	Turns on a point.
2: Pt-Off(	Turns off a point.
3: Pt-Change(	Toggles a point on or off.
4: Pxl-On(	Turns on a pixel.
5: Pxl-Off(	Turns off a pixel.
6: Pxl-Change(	Toggles a pixel on or off.
7: pxl-Test(	Returns 1 if pixel on, 0 if pixel off.

Drawing Points Directly on a Graph with Pt-On(

To draw a point on a graph, follow these steps.

1. Select **1:Pt-On(** from the DRAW POINTS menu.

2. Move the cursor to the position where you want to draw the point.

3. Press [ENTER] to draw the point.

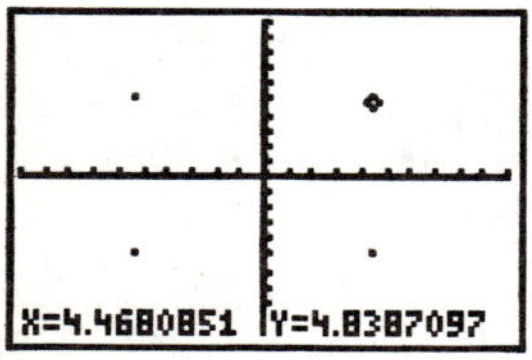

To continue drawing points, repeat steps 2 and 3. To cancel **Pt-On(**, press [CLEAR].

Erasing Points with Pt-Off(

To erase (turn off) a drawn point on a graph, follow these steps.

1. Select **2:Pt-Off(** (point off) from the DRAW POINTS menu.
2. Move the cursor to the point you want to erase.
3. Press [ENTER] to erase the point.

To continue erasing points, repeat steps 2 and 3. To cancel **Pt-Off(**, press [CLEAR].

Changing Points with Pt-Change(

To change (toggle on or off) a point on a graph, follow these steps.

1. Select **3:Pt-Change(** (point change) from the DRAW POINTS menu.
2. Move the cursor to the point you want to change.
3. Press [ENTER] to change the point's on/off status.

To continue changing points, repeat steps 2 and 3. To cancel **Pt-Change(**, press [CLEAR].

Drawing Points from the Home Screen or a Program

Pt-On((point on) turns on the point at (**X**=x,**Y**=y). **Pt-Off(** turns the point off. **Pt-Change(** toggles the point on or off. *mark* is optional; it determines the point's appearance; specify **1**, **2**, or **3**, where:

1 = • (dot; default) **2** = □ (box) **3** = + (cross)

Pt-On(x,y[,*mark*]**)**
Pt-Off(x,y[,*mark*]**)**
Pt-Change(x,y**)**

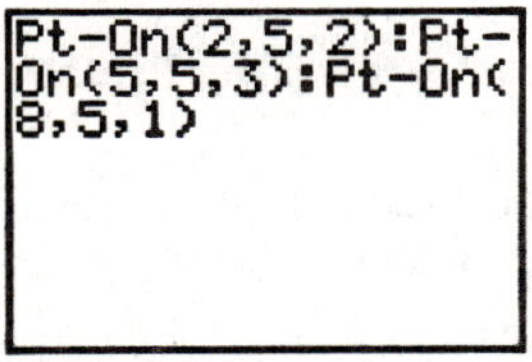

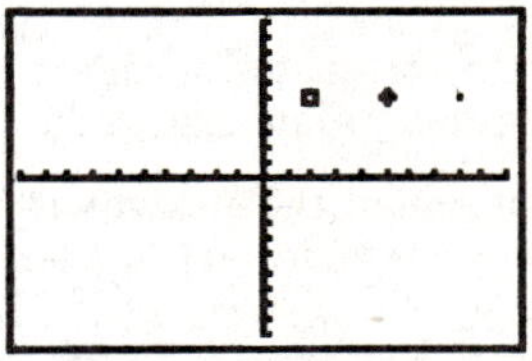

Note: If you specified *mark* to turn on a point with **Pt-On(**, you must specify *mark* when you turn off the point with **Pt-Off(**. **Pt-Change(** does not have the *mark* option.

TI-83 Plus Pixels

A pixel is a square dot on the TI-83 Plus display. The **Pxl-** (pixel) instructions let you turn on, turn off, or reverse a pixel (dot) on the graph using the cursor. When you select a pixel instruction from the DRAW POINTS menu, the TI-83 Plus returns to the home screen or the program editor. The pixel instructions are not interactive.

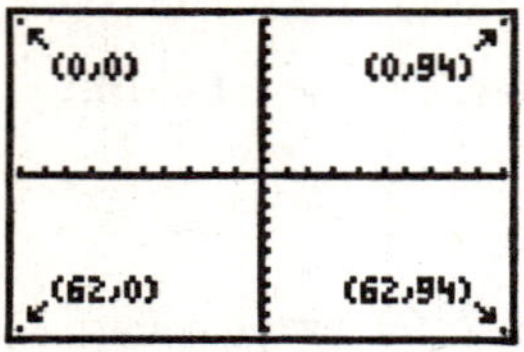

Turning On and Off Pixels with Pxl-On(and Pxl-Off(

Pxl-On((pixel on) turns on the pixel at (*row*,*column*), where *row* is an integer between 0 and 62 and *column* is an integer between 0 and 94.

Pxl-Off(turns the pixel off. **Pxl-Change(** toggles the pixel on and off.

Pxl-On(*row*,*column***)**
Pxl-Off(*row*,*column***)**
Pxl-Change(*row*,*column***)**

Using pxl-Test(

pxl-Test((pixel test) returns 1 if the pixel at (*row*,*column*) is turned on or 0 if the pixel is turned off on the current graph. *row* must be an integer between 0 and 62. *column* must be an integer between 0 and 94.

pxl-Test(*row*,*column***)**

Split Screen

On a **Horiz** split screen, the maximum value for *row* is 30 for **Pxl-On(**, **Pxl-Off(**, **Pxl-Change(**, and **pxl-Test(**.

On a **G-T** split screen, the maximum value for *row* is 50 and the maximum value for *column* is 46 for **Pxl-On(**, **Pxl-Off(**, **Pxl-Change(**, and **pxl-Test(**.

DRAW STO Menu

To display the DRAW STO menu, press [2nd] [DRAW] [◄]. When you select an instruction from the DRAW STO menu, the TI-83 Plus returns to the home screen or the program editor. The picture and graph database instructions are not interactive.

DRAW POINTS STO	
1:StorePic	Stores the current picture.
2:RecallPic	Recalls a saved picture.
3:StoreGDB	Stores the current graph database.
4:RecallGDB	Recalls a saved graph database.

Storing a Graph Picture

You can store up to 10 graph pictures, each of which is an image of the current graph display, in picture variables **Pic1** through **Pic9**, or **Pic0**. Later, you can superimpose the stored picture onto a displayed graph from the home screen or a program.

A picture includes drawn elements, plotted functions, axes, and tick marks. The picture does not include axes labels, lower and upper bound indicators, prompts, or cursor coordinates. Any parts of the display hidden by these items are stored with the picture.

To store a graph picture, follow these steps.

1. Select **1:StorePic** from the DRAW STO menu. **StorePic** is pasted to the current cursor location.

2. Enter the number (from **1** to **9**, or **0**) of the picture variable to which you want to store the picture. For example, if you enter **3**, the TI-83 Plus will store the picture to **Pic3**.

```
StorePic 3
```

Note: You also can select a variable from the PICTURE secondary menu ([VARS] **4**). The variable is pasted next to **StorePic.**

3. Press [ENTER] to display the current graph and store the picture.

Recalling a Graph Picture

To recall a graph picture, follow these steps.

1. Select **2:RecallPic** from the DRAW STO menu. **RecallPic** is pasted to the current cursor location.
2. Enter the number (from **1** to **9**, or **0**) of the picture variable from which you want to recall a picture. For example, if you enter **3**, the TI-83 Plus will recall the picture stored to **Pic3**.

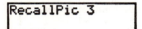

   ```
   RecallPic 3
   ```

   Note: You also can select a variable from the PICTURE secondary menu ([VARS] **4**). The variable is pasted next to **RecallPic**.
3. Press [ENTER] to display the current graph with the picture superimposed on it.

Note: Pictures are drawings. You cannot trace a curve that is part of a picture.

Deleting a Graph Picture

To delete graph pictures from memory, use the MEMORY MANAGEMENT /DELETE secondary menu (Chapter 18).

What Is a Graph Database?

A graph database (GDB) contains the set of elements that defines a particular graph. You can recreate the graph from these elements. You can store up to 10 GDBs in variables **GDB1** through **GDB9**, or **GDB0** and recall them to recreate graphs.

A GDB stores five elements of a graph.

- Graphing mode
- Window variables
- Format settings
- All functions in the Y= editor and the selection status of each
- Graph style for each Y= function

GDBs do not contain drawn items or stat plot definitions.

Storing a Graph Database

To store a graph database, follow these steps.

1. Select **3:StoreGDB** from the DRAW STO menu. **StoreGDB** is pasted to the current cursor location.

2. Enter the number (from **1** to **9**, or **0**) of the GDB variable to which you want to store the graph database. For example, if you enter **7**, the TI-83 Plus will store the GDB to **GDB7**.

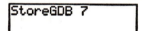

```
StoreGDB 7
```

Note: You also can select a variable from the GDB secondary menu ([VARS] **3**). The variable is pasted next to **StoreGDB.**

3. Press [ENTER] to store the current database to the specified GDB variable.

Recalling a Graph Database

CAUTION: When you recall a GDB, it replaces all existing Y= functions. Consider storing the current Y= functions to another database before recalling a stored GDB.

To recall a graph database, follow these steps.

1. Select **4:RecallGDB** from the DRAW STO menu. **RecallGDB** is pasted to the current cursor location.

2. Enter the number (from **1** to **9**, or **0**) of the GDB variable from which you want to recall a GDB. For example, if you enter **7**, the TI-83 Plus will recall the GDB stored to **GDB7**.

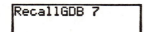

   ```
   RecallGDB 7
   ```

   Note: You also can select a variable from the GDB secondary menu ([VARS] **3**). The variable is pasted next to **RecallGDB.**

3. Press [ENTER] to replace the current GDB with the recalled GDB. The new graph is not plotted. The TI-83 Plus changes the graphing mode automatically, if necessary.

Deleting a Graph Database

To delete a GDB from memory, use the MEMORY MANAGEMENT/DELETE secondary menu (Chapter 18).

9 Split Screen

Contents

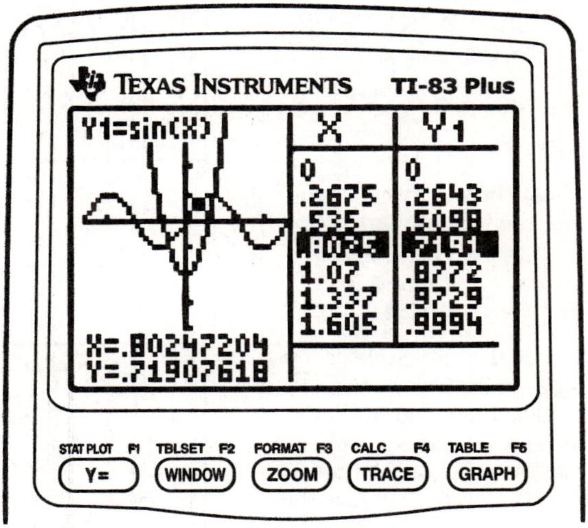

Getting Started is a fast-paced introduction. Read the chapter for details.

Use G-T (graph-table) split-screen mode to explore the unit circle and its relationship to the numeric values for the commonly used trigonometric angles of 0°, 30°, 45°, 60°, 90°, and so on.

1. Press [MODE] to display the mode screen. Press [▼] [▼] [▶] [ENTER] to select **Degree** mode. Press [▼] [▶] [ENTER] to select **Par** (parametric) graphing mode.

   Press [▼] [▼] [▼] [▼] [▶] [▶] [ENTER] to select **G-T** (graph-table) split-screen mode.

2. Press [2nd] [FORMAT] to display the format screen. Press [▼] [▼] [▼] [▼] [▼] [▶] [ENTER] to select **ExprOff.**

3. Press [Y=] to display the Y= editor for **Par** graphing mode. Press [COS] [X,T,Θ,n] [)] [ENTER] to store **cos(T)** to **X1T**. Press [SIN] [X,T,Θ,n] [)] [ENTER] to store **sin(T)** to **Y1T**.

4. Press [WINDOW] to display the window editor. Enter these values for the window variables.

Tmin=0	**Xmin=⁻2.3**	**Ymin=⁻2.5**
Tmax=360	**Xmax=2.3**	**Ymax=2.5**
Tstep=15	**Xscl=1**	**Yscl=1**

5. Press [TRACE]. On the left, the unit circle is graphed parametrically in **Degree** mode and the trace cursor is activated. When **T=0** (from the graph trace coordinates), you can see from the table on the right that the value of **X1T** (**cos(T)**) is **1** and **Y1T** (**sin(T)**) is **0**. Press [▶] to move the cursor to the next 15° angle increment. As you trace around the circle in steps of 15°, an approximation of the standard value for each angle is highlighted in the table.

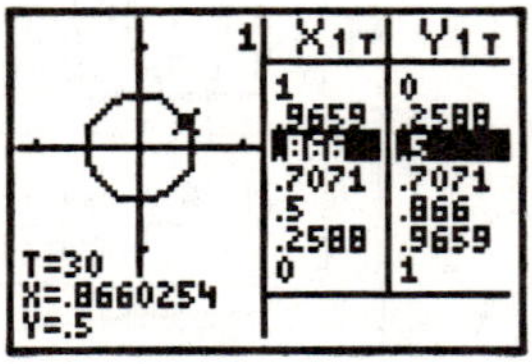

Setting a Split-Screen Mode

To set a split-screen mode, press [MODE], and then move the cursor to the bottom line of the mode screen.

- Select **Horiz** (horizontal) to display the graph screen and another screen split horizontally.
- Select **G-T** (graph-table) to display the graph screen and table screen split vertically.

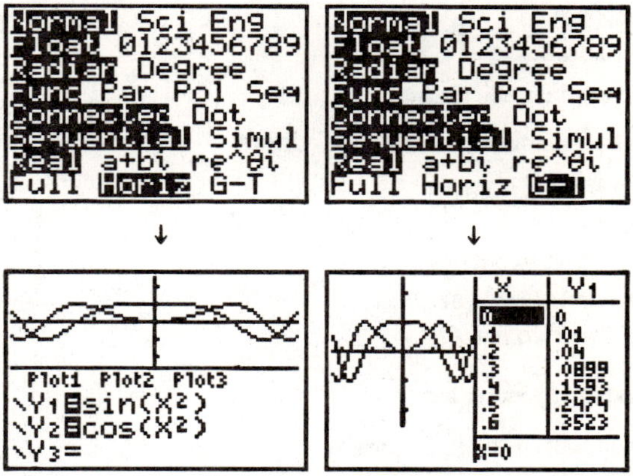

The split screen is activated when you press any key that applies to either half of the split screen.

Some screens are never displayed as split screens. For example, if you press [MODE] in **Horiz** or **G-T** mode, the mode screen is displayed as a full screen. If you then press a key that displays either half of a split screen, such as [TRACE], the split screen returns.

When you press a key or key combination in either **Horiz** or **G-T** mode, the cursor is placed in the half of the display for which that key applies. For example, if you press [TRACE], the cursor is placed in the half in which the graph is displayed. If you press [2nd] [TABLE], the cursor is placed in the half in which the table is displayed.

The TI-83 Plus will remain in split-screen mode until you change back to **Full** screen mode.

Horiz Mode

In **Horiz** (horizontal) split-screen mode, a horizontal line splits the screen into top and bottom halves.

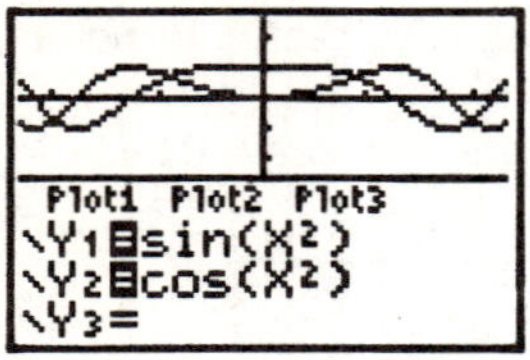

The top half displays the graph.

The bottom half displays any of these editors.

- Home screen (four lines)
- Y= editor (four lines)
- Stat list editor (two rows)
- Window editor (three settings)
- Table editor (two rows)

Moving from Half to Half in Horiz Mode

To use the top half of the split screen:

- Press [GRAPH] or [TRACE].
- Select a ZOOM or CALC operation.

To use the bottom half of the split screen:

- Press any key or key combination that displays the home screen.
- Press [Y=] (Y= editor).
- Press [STAT] [ENTER] (stat list editor).
- Press [WINDOW] (window editor).
- Press [2nd] [TABLE] (table editor).

Full Screens in Horiz Mode

All other screens are displayed as full screens in **Horiz** split-screen mode.

To return to the **Horiz** split screen from a full screen when in **Horiz** mode, press any key or key combination that displays the graph, home screen, Y= editor, stat list editor, window editor, or table editor.

G-T Mode

In **G-T** (graph-table) split-screen mode, a vertical line splits the screen into left and right halves.

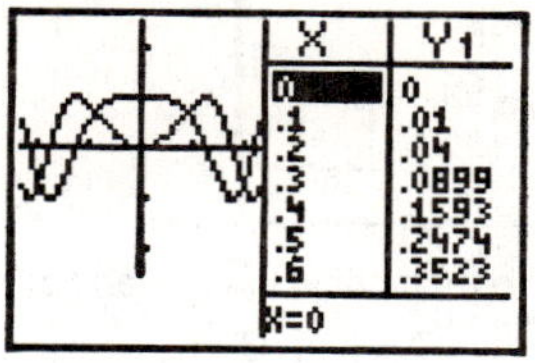

The left half displays the graph.

The right half displays the table.

Moving from Half to Half in G-T Mode

To use the left half of the split screen:

- Press [GRAPH] or [TRACE].
- Select a ZOOM or CALC operation.

To use the right half of the split screen, press [2nd] [TABLE].

Using [TRACE] in G-T Mode

As you move the trace cursor along a graph in the split screen's left half in **G-T** mode, the table on the right half automatically scrolls to match the current cursor values.

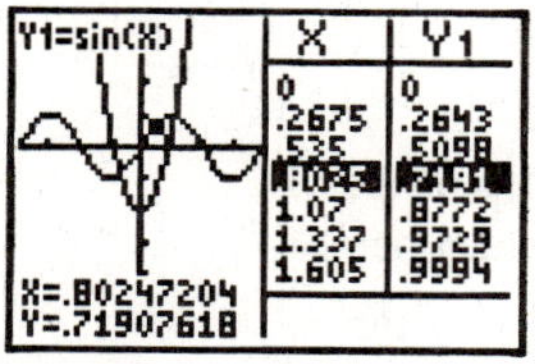

Note: When you trace in **Par** graphing mode, both components of an equation (**X*n*T** and **Y*n*T**) are displayed in the two columns of the table. As you trace, the current value of the independent variable **T** is displayed on the graph.

Full Screens in G-T Mode

All screens other than the graph and the table are displayed as full screens in **G-T** split-screen mode.

To return to the **G-T** split screen from a full screen when in **G-T** mode, press any key or key combination that displays the graph or the table.

TI-83 Plus Pixels in Horiz and G-T Modes

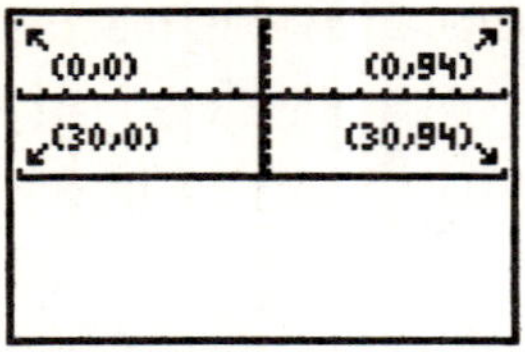

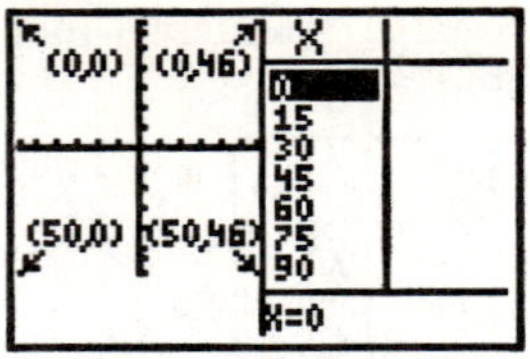

Note: Each set of numbers in parentheses above represents the row and column of a corner pixel, which is turned on.

DRAW POINTS Menu Pixel Instructions

For **Pxl-On(**, **Pxl-Off(**, **Pxl-Change(**, and **pxl-Test(**:

- In **Horiz** mode, *row* must be ≤30; *column* must be ≤94.
- In **G-T** mode, *row* must be ≤50; *column* must be ≤46.

Pxl-On(*row*,*column***)**

DRAW Menu Text(Instruction

For the **Text(** instruction:

- In **Horiz** mode, *row* must be ≤25; *column* must be ≤94.
- In **G-T** mode, *row* must be ≤45; *column* must be ≤46.

Text(*row*,*column*,**"***text***")**

PRGM I/O Menu Output(Instruction

For the **Output(** instruction:

- In **Horiz** mode, *row* must be ≤4; *column* must be ≤16.
- In **G-T** mode, *row* must be ≤8; *column* must be ≤16.

Output(*row*,*column*,**"***text***")**

Setting a Split-Screen Mode from the Home Screen or a Program

To set **Horiz** or **G-T** from a program, follow these steps.

1. Press [MODE] while the cursor is on a blank line in the program editor.
2. Select **Horiz** or **G-T**.

The instruction is pasted to the cursor location. The mode is set when the instruction is encountered during program execution. It remains in effect after execution.

Note: You also can paste **Horiz** or **G-T** to the home screen or program editor from the CATALOG (Chapter 15).

10 Matrices

Contents

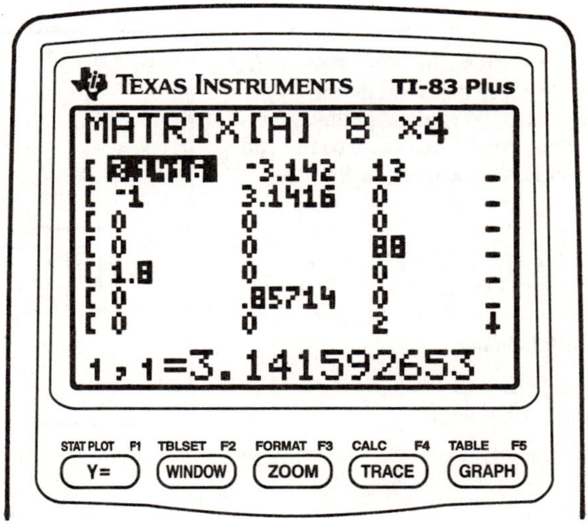

Getting Started is a fast-paced introduction. Read the chapter for details.

Find the solution of $X + 2Y + 3Z = 3$ and $2X + 3Y + 4Z = 3$. On the TI-83 Plus, you can solve a system of linear equations by entering the coefficients as elements in a matrix, and then using rref(to obtain the reduced row-echelon form.

1. Press [2nd] [MATRX]. Press [▶] [▶] to display the MATRX EDIT menu. Press **1** to select **1: [A]**.

2. Press **2** [ENTER] **4** [ENTER] to define a 2×4 matrix. The rectangular cursor indicates the current element. Ellipses (...) indicate additional columns beyond the screen.

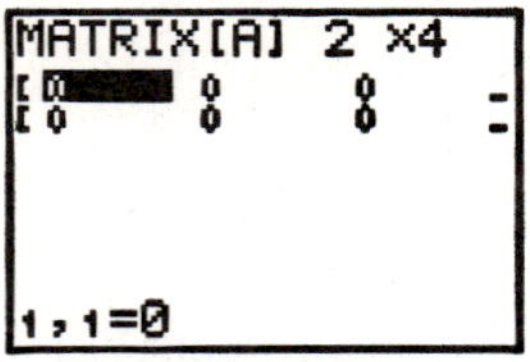

3. Press **1** [ENTER] to enter the first element. The rectangular cursor moves to the second column of the first row.

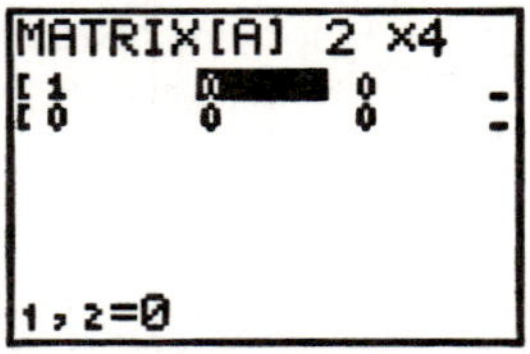

4. Press **2** [ENTER] **3** [ENTER] **3** [ENTER] to complete the first row for $X + 2Y + 3Z = 3$.

5. Press **2** [ENTER] **3** [ENTER] **4** [ENTER] **3** [ENTER] to enter the second row for $2X + 3Y + 4Z = 3$.

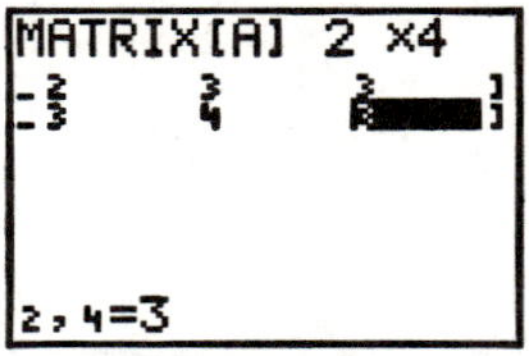

6. Press [2nd] [QUIT] to return to the home screen. If necessary, press [CLEAR] to clear the home screen. Press [2nd] [MATRX] [▶] to display the MATRX MATH menu. Press [▲] to wrap to the end of the menu. Select **B:rref(** to copy **rref(** to the home screen.

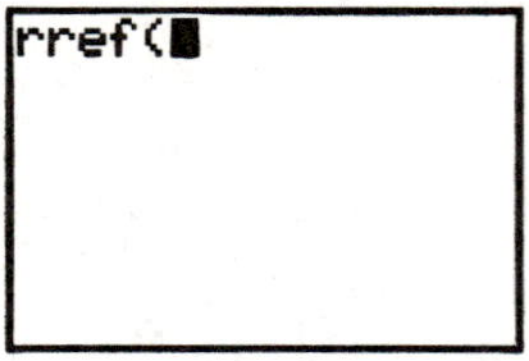

7. Press [2nd] [MATRX] **1** to select **1: [A]** from the MATRX NAMES menu. Press [)] [ENTER]. The reduced row-echelon form of the matrix is displayed and stored in **Ans**.

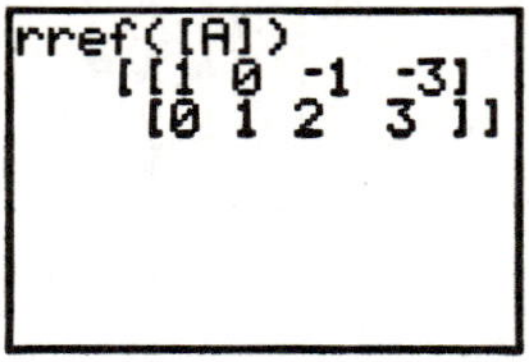

$1X - 1Z = -3$ so $X = -3 + Z$
$1Y + 2Z = 3$ so $Y = 3 - 2Z$

What Is a Matrix?

A matrix is a two-dimensional array. You can display, define, or edit a matrix in the matrix editor. The TI-83 Plus has 10 matrix variables, **[A]** through **[J]**. You can define a matrix directly in an expression. A matrix, depending on available memory, may have up to 99 rows or columns. You can store only real numbers in TI-83 Plus matrices.

Selecting a Matrix

Before you can define or display a matrix in the editor, you first must select the matrix name. To do so, follow these steps.

1. Press [2nd] [MATRX] [◄] to display the MATRX EDIT menu. The dimensions of any previously defined matrices are displayed.

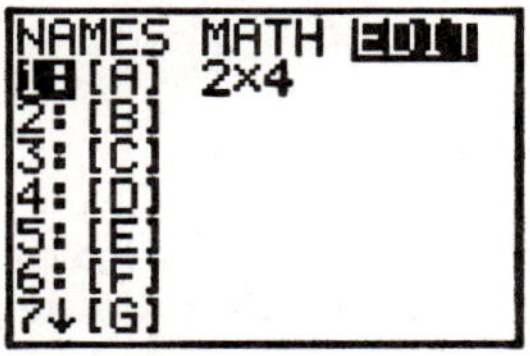

2. Select the matrix you want to define. The MATRX EDIT screen is displayed.

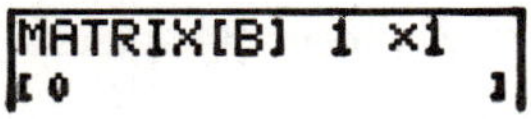

Accepting or Changing Matrix Dimensions

The dimensions of the matrix (*row* × *column*) are displayed on the top line. The dimensions of a new matrix are **1 ×1**. You must accept or change the dimensions each time you edit a matrix. When you select a matrix to define, the cursor highlights the row dimension.

- To accept the row dimension, press [ENTER].
- To change the row dimension, enter the number of rows (up to **99**), and then press [ENTER].

The cursor moves to the column dimension, which you must accept or change the same way you accepted or changed the row dimension. When you press [ENTER], the rectangular cursor moves to the first matrix element.

Displaying Matrix Elements

After you have set the dimensions of the matrix, you can view the matrix and enter values for the matrix elements. In a new matrix, all values are zero.

Select the matrix from the MATRX EDIT menu and enter or accept the dimensions. The center portion of the matrix editor displays up to seven rows and three columns of a matrix, showing the values of the elements in abbreviated form if necessary. The full value of the current element, which is indicated by the rectangular cursor, is displayed on the bottom line.

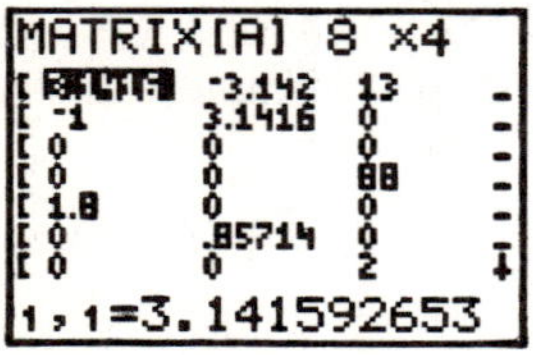

This is an 8 × 4 matrix. Ellipses in the left or right column indicate additional columns. ↑ or ↓ in the right column indicate additional rows.

Deleting a Matrix

To delete matrices from memory, use the MEMORY MANAGEMENT/DELETE secondary menu (Chapter 18).

Viewing a Matrix

The matrix editor has two contexts, viewing and editing. In viewing context, you can use the cursor keys to move quickly from one matrix element to the next. The full value of the highlighted element is displayed on the bottom line.

Select the matrix from the **MATRX EDIT** menu, and then enter or accept the dimensions.

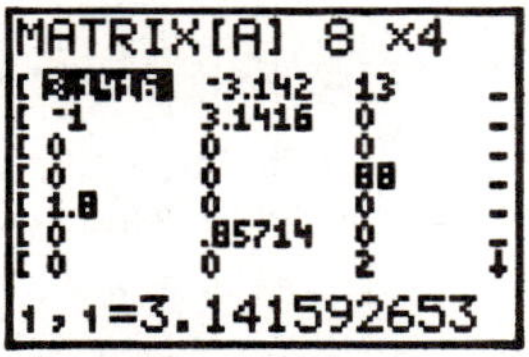

Viewing-Context Keys

Key	Function
[◄] or [►]	Moves the rectangular cursor within the current row
[▼] or [▲]	Moves the rectangular cursor within the current column; on the top row, [▲] moves the cursor to the column dimension; on the column dimension, [▲] moves the cursor to the row dimension
[ENTER]	Switches to editing context; activates the edit cursor on the bottom line
[CLEAR]	Switches to editing context; clears the value on the bottom line
Any entry character	Switches to editing context; clears the value on the bottom line; copies the character to the bottom line
[2nd] [INS]	Nothing
[DEL]	Nothing

Editing a Matrix Element

In editing context, an edit cursor is active on the bottom line. To edit a matrix element value, follow these steps.

1. Select the matrix from the MATRX EDIT menu, and then enter or accept the dimensions.
2. Press [◄], [▲], [►], and [▼] to move the cursor to the matrix element you want to change.
3. Switch to editing context by pressing [ENTER], [CLEAR], or an entry key.
4. Change the value of the matrix element using the editing-context keys described below. You may enter an expression, which is evaluated when you leave editing context.

   Note: You can press [CLEAR] [ENTER] to restore the value at the rectangular cursor if you make a mistake.

5. Press [ENTER], [▲], or [▼] to move to another element.

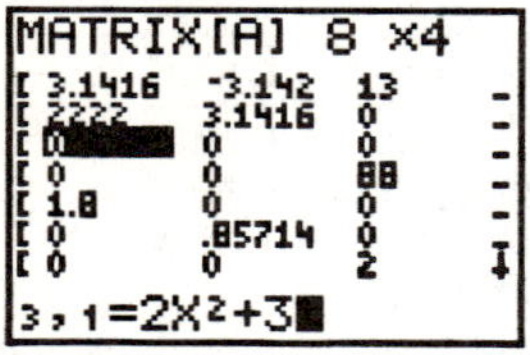

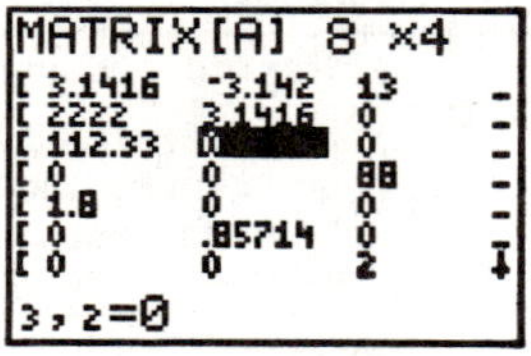

Editing-Context Keys

Key	Function
[◄] or [►]	Moves the edit cursor within the value
[▼] or [▲]	Stores the value displayed on the bottom line to the matrix element; switches to viewing context and moves the rectangular cursor within the column
[ENTER]	Stores the value displayed on the bottom line to the matrix element; switches to viewing context and moves the rectangular cursor to the next row element
[CLEAR]	Clears the value on the bottom line
Any entry character	Copies the character to the location of the edit cursor on the bottom line
[2nd] [INS]	Activates the insert cursor
[DEL]	Deletes the character under the edit cursor on the bottom line

Using a Matrix in an Expression

To use a matrix in an expression, you can do any of the following.

- Copy the name from the MATRX NAMES menu.
- Recall the contents of the matrix into the expression with [2nd] [RCL] (Chapter 1).
- Enter the matrix directly (see below).

Entering a Matrix in an Expression

You can enter, edit, and store a matrix in the matrix editor. You also can enter a matrix directly in an expression.

To enter a matrix in an expression, follow these steps.

1. Press [2nd] [[ ] to indicate the beginning of the matrix.
2. Press [2nd] [[ ] to indicate the beginning of a row.
3. Enter a value, which can be an expression, for each element in the row. Separate the values with commas.
4. Press [2nd] [ ]] to indicate the end of a row.
5. Repeat steps 2 through 4 to enter all of the rows.
6. Press [2nd] [ ]] to indicate the end of the matrix.

   Note: The closing]] are not necessary at the end of an expression or preceding ➔.

   The resulting matrix is displayed in the form:

   [[$element_{1,1}$,...,$element_{1,n}$**]**,...,**[**$element_{m,1}$,...,$element_{m,n}$**]]**

   Any expressions are evaluated when the entry is executed.

```
2*[[1,2,3][4,5,6
]]
        [[2 4  6 ]
         [8 10 12]]
```

Note: The commas that you must enter to separate elements are not displayed on output.

Displaying a Matrix

To display the contents of a matrix on the home screen, select the matrix from the MATRX NAMES menu, and then press [ENTER].

```
[A]
        [[7 8 9]
         [3 2 1]]
```

Ellipses in the left or right column indicate additional columns. ↑ or ↓ in the right column indicate additional rows. Press [▶], [◀], [▼], and [▲] to scroll the matrix.

```
…46.0000   161.0↑
…116.0000  -188.…
…49.0000   -62.0…
…235.0000  -96.0…
…2.0000    65.00…
…47.0000   136.0…
…3.0000    -69.0↓
```

Copying One Matrix to Another

To copy a matrix, follow these steps.

1. Press [2nd] [MATRX] to display the MATRX NAMES menu.
2. Select the name of the matrix you want to copy.
3. Press [STO▶].
4. Press [2nd] [MATRX] again and select the name of the new matrix to which you want to copy the existing matrix.
5. Press [ENTER] to copy the matrix to the new matrix name.

```
[A]→[B]
        [[7 8 9]
         [3 2 1]]
```

Accessing a Matrix Element

On the home screen or from within a program, you can store a value to, or recall a value from, a matrix element. The element must be within the currently defined matrix dimensions. Select *matrix* from the MATRX NAMES menu.

[*matrix*](*row*,*column*)

```
0→[B](2,3):[B]
        [[7 8 9]
         [3 2 0]]
[B](2,3)
               0
```

Using Math Functions with Matrices

You can use many of the math functions on the TI-83 Plus keyboard, the MATH menu, the MATH NUM menu, and the MATH TEST menu with matrices. However, the dimensions must be appropriate. Each of the functions below creates a new matrix; the original matrix remains the same.

+ (Add), – (Subtract), * (Multiply)

To add ([+]) or subtract ([–]) matrices, the dimensions must be the same. The answer is a matrix in which the elements are the sum or difference of the individual corresponding elements.

matrixA+matrixB
matrixA–matrixB

To multiply ([×]) two matrices together, the column dimension of *matrixA* must match the row dimension of *matrixB*.

*matrixA***matrixB*

```
[A]
              [[2 2]
               [3 4]]
[B]
              [[0 5]
               [4 3]]
```

```
[A]+[B]
              [[2 7]
               [7 7]]
[A]*[B]
            [[8  16]
             [16 27]]
```

Multiplying a *matrix* by a *value* or a *value* by a *matrix* returns a matrix in which each element of *matrix* is multiplied by *value*.

*matrix***value*
*value***matrix*

```
[A]*3
             [[6 6 ]
              [9 12]]
```

- (Negation)

Negating a matrix ([(-)]) returns a matrix in which the sign of every element is changed (reversed).

-matrix

```
[A]
             [[2 -2]
              [3 4 ]]
-[A]
            [[-2 2 ]
             [-3 -4]]
```

abs(

abs((absolute value, MATH NUM menu) returns a matrix containing the absolute value of each element of *matrix*.

abs(*matrix***)**

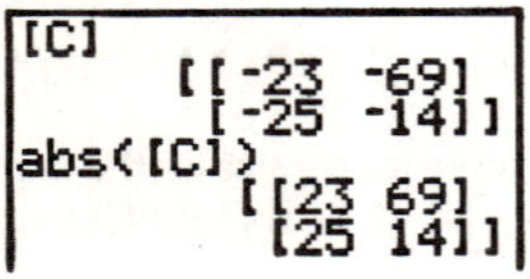

round(

round((MATH NUM menu) returns a matrix. It rounds every element in *matrix* to *#decimals* (≤ 9). If *#decimals* is omitted, the elements are rounded to 10 digits.

round(*matrix*[,*#decimals*]**)**

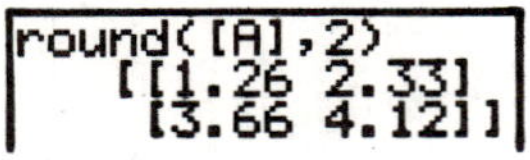

$^{-1}$ (Inverse)

Use the $^{-1}$ function ([x-1]) to invert a matrix (**^-1** is not valid). *matrix* must be square. The determinant cannot equal zero.

matrix^{-1}

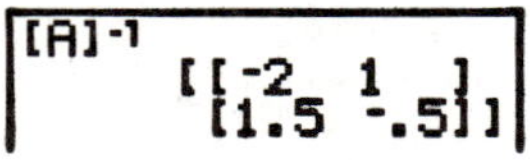

Powers

To raise a matrix to a power, *matrix* must be square. You can use **²** ([x²]), **³** (MATH menu), or **^***power* ([^]) for integer *power* between **0** and **255**.

matrix^2
matrix^3
matrix^*power*

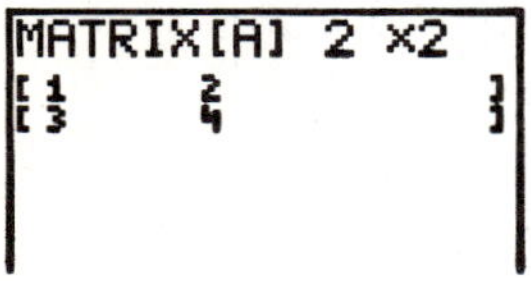

Relational Operations

To compare two matrices using the relational operations = and ≠ (TEST menu), they must have the same dimensions. = and ≠ compare *matrixA* and *matrixB* on an element-by-element basis. The other relational operations are not valid with matrices.

matrixA=*matrixB* returns **1** if every comparison is true; it returns **0** if any comparison is false.

matrixA≠*matrixB* returns **1** if at least one comparison is false; it returns **0** if no comparison is false.

```
[A]
          [[1 2 3]
           [3 2 1]]
[B]
          [[3 2 1]
           [1 2 3]]
```

```
[A]=[B]
                0
[A]≠[B]
                1
```

iPart(, fPart(, int(

iPart((integer part), **fPart(** (fractional part), and **int(** (greatest integer) are on the MATH NUM menu.

iPart(returns a matrix containing the integer part of each element of *matrix*.

fPart(returns a matrix containing the fractional part of each element of *matrix*.

int(returns a matrix containing the greatest integer of each element of *matrix*.

iPart(*matrix***)**
fPart(*matrix***)**
int(*matrix***)**

```
[D]
 [[1.25  3.333]
  [100.5 47.15]]
```

```
iPart([D])
       [[1    3 ]
        [100 47]]
fPart([D])
     [[.25 .333]
      [.5  .15 ]]
```

MATRX MATH Menu

To display the MATRX MATH menu, press [2nd] [MATRX] [▶].

NAMES MATH EDIT	
1:det(	Calculates the determinant.
2:^{T}	Transposes the matrix.
3:dim(	Returns the matrix dimensions.
4:Fill(	Fills all elements with a constant.
5:identity(	Returns the identity matrix.
6:randM(	Returns a random matrix.
7:augment(	Appends two matrices.
8:Matr▶list(	Stores a matrix to a list.
9:List▶matr(	Stores a list to a matrix.
0:cumSum(	Returns the cumulative sums of a matrix.
A:ref(	Returns the row-echelon form of a matrix.
B:rref(	Returns the reduced row-echelon form.
C:rowSwap(	Swaps two rows of a matrix.
D:row+(	Adds two rows; stores in the second row.
E:*row(	Multiplies the row by a number.
F:*row+(	Multiplies the row, adds to the second row.

det(

det((determinant) returns the determinant (a real number) of a square *matrix*.

det(*matrix***)**

^{T} (Transpose)

^{T} (transpose) returns a matrix in which each element (row, column) is swapped with the corresponding element (column, row) of *matrix*.

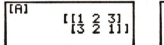

matrix^T

```
[A]
             [[1 2 3]
              [3 2 1]]
```

```
[A]T
             [[1 3]
              [2 2]
              [3 1]]
```

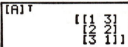

Accessing Matrix Dimensions with dim(

dim((dimension) returns a list containing the dimensions (**{***rows columns***}**) of *matrix*.

dim(*matrix***)**

Note: dim(*matrix***)→L***n***:L***n***(1)** returns the number of rows. **dim(***matrix***)→L***n***:L***n***(2)** returns the number of columns.

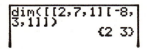

```
dim([[2,7,1][-8,
3,1]])
               {2 3}
```

```
dim([[2,7,1][-8,
3,11]])→L1:L1(1)
                   2
```

Creating a Matrix with dim(

Use **dim(** with [STO▶] to create a new *matrixname* of dimensions *rows* × *columns* with **0** as each element.

{*rows*,*columns***}→dim(***matrixname***)**

```
{2,2}→dim([E])
            {2 2}
[E]
          [[0 0]
           [0 0]]
```

Redimensioning a Matrix with dim(

Use **dim(** with [STO▶] to redimension an existing *matrixname* to dimensions *rows* × *columns*. The elements in the old *matrixname* that are within the new dimensions are not changed. Additional created elements are zeros. Matrix elements that are outside the new dimensions are deleted.

{*rows*,*columns***}→dim(***matrixname***)**

Fill(

Fill(stores *value* to every element in *matrixname*.

Fill(*value*,*matrixname***)**

```
Fill(5,[E])
             Done
[E]
          [[5 5]
           [5 5]]
```

Identity(

Identity(returns the identity matrix of *dimension* rows × *dimension* columns.

Identity(*dimension***)**

randM(

randM((create random matrix) returns a *rows* × *columns* random matrix of integers ≥ ⁻9 and ≤ 9. The seed value stored to the **rand** function controls the values (Chapter 2).

randM(*rows*,*columns***)**

```
0→rand:randM(2,2
)
         [[0 -7]
          [8 8 ]]
```

augment(

augment(appends *matrixA* to *matrixB* as new columns. *matrixA* and *matrixB* both must have the same number of rows.

augment(*matrixA*,*matrixB***)**

```
[[1,2][3,4]]→[A]
:[[5,6][7,8]]→[B
]:augment([A],[B
])
       [[1 2 5 6]
        [3 4 7 8]]
```

Matr▸list(

Matr▸list((matrix stored to list) fills each *listname* with elements from each column in *matrix*. **Matr▸list(** ignores extra *listname* arguments. Likewise, **Matr▸list(** ignores extra *matrix* columns.

Matr▸list(*matrix*,*listnameA*,...,*listname n***)**

```
[A]
          [[1 2 3]
           [4 5 6]]
Matr▸list([A],L1
,L2,L3)
             Done
```

→

```
L1
            {1 4}
L2
            {2 5}
L3
            {3 6}
```

Matr▸list(also fills a *listname* with elements from a specified *column#* in *matrix*. To fill a list with a specific column from *matrix*, you must enter *column#* after *matrix*.

Matr▸list(*matrix*,*column#*,*listname***)**

```
[A]
          [[1 2 3]
           [4 5 6]]
Matr▸list([A],3,
L1)
             Done
```

→

```
L1
            {3 6}
```

List▸matr(

List▸matr((lists stored to matrix) fills *matrixname* column by column with the elements from each *list*. If dimensions of all *lists* are not equal, **List▸matr(** fills each extra *matrixname* row with **0**. Complex lists are not valid.

List▸matr(*listA*,...,*list n*,*matrixname***)**

```
{1,2,3}→LX
          {1 2 3}
{4,5,6}→LY
          {4 5 6}
{7,8,9}→LB
          {7 8 9}
```

→

```
List▸matr(LX,LY,
LB,[C])
             Done
[C]
        [[1 4 7]
         [2 5 8]
         [3 6 9]]
```

cumSum(

cumSum(returns cumulative sums of the elements in *matrix*, starting with the first element. Each element is the cumulative sum of the column from top to bottom.

cumSum(*matrix***)**

```
[D]
            [[1 2]
             [3 4]
             [5 6]]
```

```
cumSum([D])
            [[1 2 ]
             [4 6 ]
             [9 12]]
```

Row Operations

MATRX MATH menu items **A** through **F** are row operations. You can use a row operation in an expression. Row operations do not change *matrix* in memory. You can enter all row numbers and values as expressions. You can select the matrix from the MATRX NAMES menu.

ref(, rref(

ref((row-echelon form) returns the row-echelon form of a real *matrix*. The number of columns must be greater than or equal to the number of rows.

ref(*matrix***)**

rref((reduced row-echelon form) returns the reduced row-echelon form of a real *matrix*. The number of columns must be greater than or equal to the number of rows.

rref(*matrix***)**

```
[B]
          [[4 5 6]
           [7 8 9]]
```

```
ref([B])
[[1 1.142857143…
 [0 1          …
rref([B])
         [[1 0 -1]
          [0 1 2 ]]
```

rowSwap(

rowSwap(returns a matrix. It swaps *rowA* and *rowB* of *matrix*.

rowSwap(*matrix,rowA,rowB***)**

```
[F]
        [[2 3 6 9]
         [5 8 4 7]
         [2 5 1 0]
         [6 3 8 5]]
```

```
rowSwap([F],2,4)
        [[2 3 6 9]
         [6 3 8 5]
         [2 5 1 0]
         [5 8 4 7]]
```

row+(

row+((row addition) returns a matrix. It adds *rowA* and *rowB* of *matrix* and stores the results in *rowB*.

row+(*matrix,rowA,rowB***)**

```
[[2,5,7][8,9,4]]
→[D]
          [[2 5 7]
           [8 9 4]]
```

```
row+([D],1,2)
        [[2  5  7 ]
         [10 14 11]]
```

*row(

***row(** (row multiplication) returns a matrix. It multiplies *row* of *matrix* by *value* and stores the results in *row*.

row(value,matrix,row***)**

*row+(

***row+(** (row multiplication and addition) returns a matrix. It multiplies *rowA* of *matrix* by *value*, adds it to *rowB*, and stores the results in *rowB*.

row+(value,matrix,rowA,rowB***)**

```
[[1,2,3][4,5,6]]
→[E]
          [[1 2 3]
           [4 5 6]]
```

```
*row+(3,[E],1,2)
        [[1 2  3 ]
         [7 11 15]]
```

11 Lists

Contents

Getting Started is a fast-paced introduction. Read the chapter for details.

Calculate the first eight terms of the sequence 1/A². Store the results to a user-created list. Then display the results in fraction form. Begin this example on a blank line on the home screen.

1. Press [2nd] [LIST] [▶] to display the LIST OPS menu.

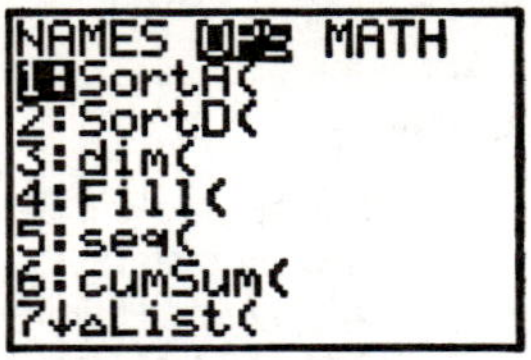

2. Press **5** to select **5:seq(**, which pastes **seq(** to the current cursor location.

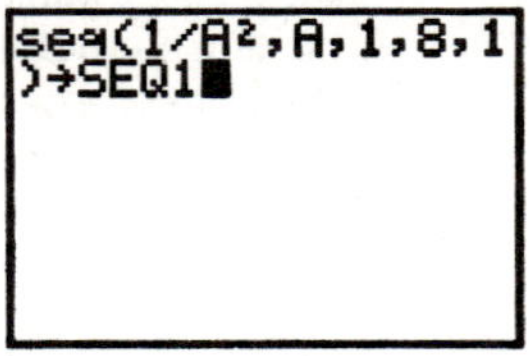

3. Press **1** [÷] [ALPHA] [A] [x²] [,] [ALPHA] [A] [,] **1** [,] **8** [,] **1** [)] to enter the sequence.
4. Press [STO▶], and then press [2nd] [ALPHA] to turn on alpha-lock. Press [S] [E] [Q], and then press [ALPHA] to turn off alpha-lock. Press **1** to complete the list name.
5. Press [ENTER] to generate the list and store it in **SEQ1**. The list is displayed on the home screen. An ellipsis (...) indicates that the list continues beyond the viewing window. Press [▶] repeatedly (or press and hold [▶]) to scroll the list and view all the list elements.

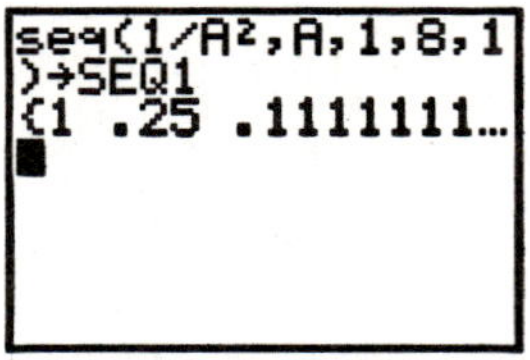

6. Press [2nd] [LIST] to display the LIST NAMES menu. Press **7** to select **7:seq(** to paste ʟ**SEQ1** to the current cursor location. (If **SEQ1** is not item **7** on your LIST NAMES menu, move the cursor to **SEQ1** before you press [ENTER].)

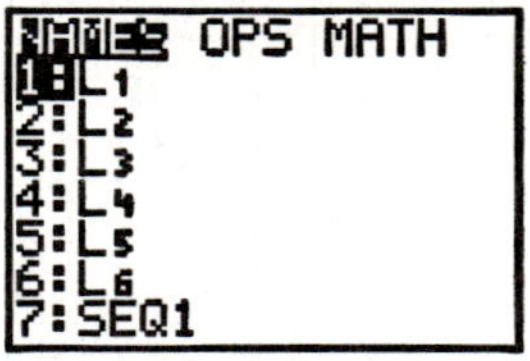

7. Press [MATH] to display the MATH menu. Press **1** to select **1:▶Frac**, which pastes **▶Frac** to the current cursor location.
8. Press [ENTER] to show the sequence in fraction form. Press [▶] repeatedly (or press and hold [▶]) to scroll the list and view all the list elements.

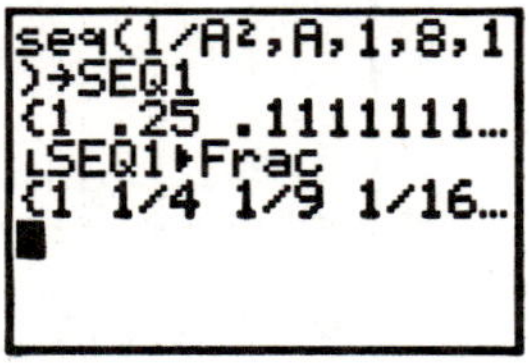

Using TI-83 Plus List Names L1 through L6

The TI-83 Plus has six list names in memory: **L1**, **L2**, **L3**, **L4**, **L5**, and **L6**. The list names **L1** through **L6** are on the keyboard above the numeric keys [1] through [6]. To paste one of these names to a valid screen, press [2nd], and then press the appropriate key. **L1** through **L6** are stored in stat list editor columns **1** through **6** when you reset memory.

Creating a List Name on the Home Screen

To create a list name on the home screen, follow these steps.

1. Press [2nd] [{], enter one or more list elements, and then press [2nd] [}]. Separate list elements with commas. List elements can be real numbers, complex numbers, or expressions.

```
{1,2,3,4}
```

2. Press [STO▸].
3. Press [ALPHA] [*letter from A to Z or θ*] to enter the first letter of the name.
4. Enter zero to four letters, θ, or numbers to complete the name.

```
{1,2,3,4}→TEST
```

5. Press [ENTER]. The list is displayed on the next line. The list name and its elements are stored in memory. The list name becomes an item on the LIST NAMES menu.

```
{1,2,3,4}→TEST
        {1 2 3 4}
```

```
NAMES OPS MATH
1:SEQ1
2:T123
3:TEST
```

Note: If you want to view a user-created list in the stat list editor, you must store it in the stat list editor (Chapter 12).

You also can create a list name in these four places.

- At the **Name=** prompt in the stat list editor
- At an **Xlist:**, **Ylist:**, or **Data List:** prompt in the stat plot editor
- At a **List:**, **List1:**, **List2:**, **Freq:**, **Freq1:**, **Freq2:**, **XList:**, or **YList:** prompt in the inferential stat editors
- On the home screen using **SetUpEditor**

You can create as many list names as your TI-83 Plus memory has space to store.

Storing Elements to a List

You can store list elements in either of two ways.

- Use braces and [STO▸] on the home screen.

```
{4+2i,5-3i}→L6
      {4+2i 5-3i}
```

- Use the stat list editor (Chapter 12).

The maximum dimension of a list is 999 elements.

Tip: When you store a complex number to a list, the entire list is converted to a list of complex numbers. To convert the list to a list of real numbers, display the home screen, and then enter **real(***listname***)→***listname*.

Displaying a List on the Home Screen

To display the elements of a list on the home screen, enter the name of the list (preceded by ʟ if necessary; see page 11-16), and then press [ENTER]. An ellipsis indicates that the list continues beyond the viewing window. Press [▸] repeatedly (or press and hold [▸]) to scroll the list and view all the list elements.

```
L1
          {2 5 10}
ʟDATA
{2.154 50.47 9.…
```

Copying One List to Another

To copy a list, store it to another list.

```
ʟTEST
        {1 2 3 4}
ʟTEST→TEST2
        {1 2 3 4}
```

Accessing a List Element

You can store a value to or recall a value from a specific list *element*. You can store to any element within the current list dimension or one element beyond.

listname(*element*)

```
{1,2,3}→L3
          {1 2 3}
4→L3(4):L3
        {1 2 3 4}
L3(2)
                2
```

Deleting a List from Memory

To delete lists from memory, including **L1** through **L6**, use the MEMORY MANAGEMENT/DELETE secondary menu (Chapter 18). Resetting memory restores **L1** through **L6**. Removing a list from the stat list editor does not delete it from memory.

Using Lists in Graphing

You can use lists to graph a family of curves (Chapter 3).

Entering List Names

Using the LIST NAMES Menu

To display the LIST NAMES menu, press [2nd] [LIST]. Each item is a user-created list name. LIST NAMES menu items are sorted automatically in alphanumerical order. Only the first 10 items are labeled, using **1** through **9**, then **0**. To jump to the first list name that begins with a particular alpha character or θ, press [ALPHA] [*letter from A to Z or θ*].

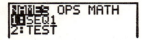

```
NAMES OPS MATH
1:SEQ1
2:TEST
```

Tip: From the top of a menu, press [▲] to move to the bottom. From the bottom, press [▼] to move to the top.

Note: The LIST NAMES menu omits list names **L1** through **L6**. Enter **L1** through **L6** directly from the keyboard (page 11-3).

When you select a list name from the LIST NAMES menu, the list name is pasted to the current cursor location.

- The list name symbol **ʟ** precedes a list name when the name is pasted where non-list name data also is valid, such as the home screen.

```
ʟTEST
        {1 2 3 4}
```

- The **ʟ** symbol does not precede a list name when the name is pasted where a list name is the only valid input, such as the stat list editor's **Name=** prompt or the stat plot editor's **XList:** and **YList:** prompts.

Entering a User-Created List Name Directly

To enter an existing list name directly, follow these steps.

1. Press [2nd] [LIST] [▶] to display the LIST OPS menu.
2. Select **B:ʟ**, which pastes **ʟ** to the current cursor location. **ʟ** is not always necessary (page 11-16).

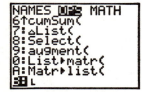

```
NAMES OPS MATH
6↑cumSum(
7:ΔList(
8:Select(
9:augment(
0:List▶matr(
A:Matr▶list(
B:ʟ
```

Note: You also can paste **ʟ** to the current cursor location from the CATALOG (Chapter 15).

3. Enter the characters that comprise the list name.

```
ʟT123█
```

Attaching a Formula to a List Name

You can attach a formula to a list name so that each list element is a result of the formula. When executed, the attached formula must resolve to a list.

When anything in the attached formula changes, the list to which the formula is attached is updated automatically.

- When you edit an element of a list that is referenced in the formula, the corresponding element in the list to which the formula is attached is updated.
- When you edit the formula itself, all elements in the list to which the formula is attached are updated.

For example, the first screen below shows that elements are stored to **L3**, and the formula **L3+10** is attached to the list name **ʟADD10**. The quotation marks designate the formula to be attached to **ʟADD10**. Each element of **ʟADD10** is the sum of an element in **L3** and 10.

```
{1,2,3}→L3
        {1 2 3}
"L3+10"→ʟADD10
L3+10
ʟADD10
      {11 12 13}
```

The next screen shows another list, **L4**. The elements of **L4** are the sum of the same formula that is attached to **L3**. However, quotation marks are not entered, so the formula is not attached to **L4**.

On the next line, **⁻6→L3(1):L3** changes the first element in **L3** to **⁻6**, and then redisplays **L3**.

```
L3+10→L4
      {11 12 13}
-6→L3(1):L3
        {-6 2 3}
```

The last screen shows that editing **L3** updated **ʟADD10**, but did not change **L4**. This is because the formula **L3+10** is attached to **ʟADD10**, but it is not attached to **L4**.

```
ʟADD10
       {4 12 13}
L4
      {11 12 13}
```

Note: To view a formula that is attached to a list name, use the stat list editor (Chapter 12).

Attaching a Formula to a List on the Home Screen or in a Program

To attach a formula to a list name from a blank line on the home screen or from a program, follow these steps.

1. Press [ALPHA] ["], enter the formula (which must resolve to a list), and press [ALPHA] ["] again.

   Note: When you include more than one list name in a formula, each list must have the same dimension.

2. Press [STO▸].
3. Enter the name of the list to which you want to attach the formula.
   - Press [2nd], and then enter a TI-83 Plus list name **L1** through **L6**.
   - Press [2nd] [LIST] and select a user-created list name from the LIST NAMES menu.
   - Enter a user-created list name directly using **ʟ** (page 11-16).
4. Press [ENTER].

```
{4,8,9}→L1
            {4 8 9}
"5*L1"→ʟLIST
5*L1
ʟLIST
         {20 40 45}
```

Note: The stat list editor displays a formula-lock symbol next to each list name that has an attached formula. Chapter 12 describes how to use the stat list editor to attach formulas to lists, edit attached formulas, and detach formulas from lists.

Detaching a Formula from a List

You can detach (clear) an attached formula from a list in several ways.

For example:

- Enter ""→*listname* on the home screen.
- Edit any element of a list to which a formula is attached.
- Use the stat list editor (Chapter 12).
- Use **ClrList** or **ClrAllList** to detach a formula from a list (Chapter 18).

Using a List in an Expression

You can use lists in an expression in any of three ways. When you press [ENTER], any expression is evaluated for each list element, and a list is displayed.

- Use **L1–L6** or any user-created list name in an expression.

```
{2,5,10}→L1
         {2 5 10}
20/L1
         {10 4 2}
```

- Enter the list elements directly (step 1 on page 11-3).

```
20/{2,5,10}
         {10 4 2}
```

- Use [2nd] [RCL] to recall the contents of the list into an expression at the cursor location (Chapter 1).

```
Rcl L1
```

→

```
{2,5,10}²
       {4 25 100}
```

Note: You must paste user-created list names to the **Rcl** prompt by selecting them from the LIST NAMES menu. You cannot enter them directly using **L**.

Using Lists with Math Functions

You can use a list to input several values for some math functions. Other chapters and Appendix A specify whether a list is valid. The function is evaluated for each list element, and a list is displayed.

- When you use a list with a function, the function must be valid for every element in the list. In graphing, an invalid element, such as **⁻1** in **√({1,0,⁻1})**, is ignored.

```
√({1,0,⁻1})
```

This returns an error.

```
 Plot1 Plot2 Plot3
\Y1■X√({1,0,⁻1})
```

This graphs **X*√(1)** and **X*√(0)**, but skips **X*√(⁻1)**.

- When you use two lists with a two-argument function, the dimension of each list must be the same. The function is evaluated for corresponding elements.

```
{1,2,3}+{4,5,6}
          {5 7 9}
```

- When you use a list and a value with a two-argument function, the value is used with each element in the list.

```
{1,2,3}+4
          {5 6 7}
```

LIST OPS Menu

To display the LIST OPS menu, press [2nd] [LIST] [▶].

NAMES OPS MATH	
1:SortA(	Sorts lists in ascending order.
2:SortD(	Sorts lists in descending order.
3:dim(	Sets the list dimension.
4:Fill(	Fills all elements with a constant.
5:seq(	Creates a sequence.
6:cumSum(	Returns a list of cumulative sums.
7:ΔList(	Returns difference of successive elements.
8:Select(	Selects specific data points.
9:augment(	Concatenates two lists.
0:List▶matr(	Stores a list to a matrix.
A:Matr▶list(	Stores a matrix to a list.
B:ʟ	Designates the list-name data type.

SortA(, SortD(

SortA((sort ascending) sorts list elements from low to high values. **SortD(** (sort descending) sorts list elements from high to low values. Complex lists are sorted based on magnitude (modulus).

With one list, **SortA(** and **SortD(** sort the elements of *listname* and update the list in memory.

SortA(*listname***)**

```
{5,6,4}→L3
          {5 6 4}
SortA(L3)
             Done
L3
          {4 5 6}
```

SortD(*listname***)**

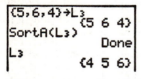

```
SortD(L3)
             Done
L3
          {6 5 4}
```

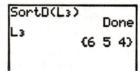

With two or more lists, **SortA(** and **SortD(** sort *keylistname*, and then sort each *dependlist* by placing its elements in the same order as the corresponding elements in *keylistname*. All lists must have the same dimension.

SortA(*keylistname,dependlist1*[*,dependlist2,...,dependlist n*]**)**
SortD(*keylistname,dependlist1*[*,dependlist2,...,dependlist n*]**)**

```
{5,6,4}→L4
          {5 6 4}
{1,2,3}→L5
          {1 2 3}
```

```
SortA(L4,L5)
             Done
L4
          {4 5 6}
L5
          {3 1 2}
```

Note: In the example, **5** is the first element in **L4**, and **1** is the first element in **L5**. After **SortA(L4,L5)**, **5** becomes the second element of **L4**, and likewise, **1** becomes the second element of **L5**.

Note: SortA(and **SortD(** are the same as **SortA(** and **SortD(** on the STAT EDIT menu (Chapter 12).

Using dim(to Find List Dimensions

dim((dimension) returns the length (number of elements) of *list*.

dim(*list***)**

```
dim({1,3,5,7})
                   4
```

Using dim(to Create a List

You can use **dim(** with [STO▸] to create a new *listname* with dimension *length* from 1 to 999. The elements are zeros.

length→**dim(***listname***)**

```
3→dim(L2)
                   3
L2
             {0 0 0}
```

Using dim(to Redimension a List

You can use **dim** with [STO▸] to redimension an existing *listname* to dimension *length* from 1 to 999.

- The elements in the old *listname* that are within the new dimension are not changed.
- Extra list elements are filled by **0**.
- Elements in the old list that are outside the new dimension are deleted.

length→**dim(***listname***)**

```
{4,8,6}→L1
             {4 8 6}
4→dim(L1)
                   4
L1
           {4 8 6 0}
```

```
3→dim(L1)
                   3
L1
             {4 8 6}
```

Fill(

Fill(replaces each element in *listname* with *value*.

Fill(*value*,*listname***)**

```
{3,4,5}→L3
             {3 4 5}
Fill(8,L3)
                Done
L3
             {8 8 8}
```

```
Fill(4+3i,L3)
                Done
L3
{4+3i 4+3i 4+3i}
```

Note: dim(and **Fill(** are the same as **dim(** and **Fill(** on the MATRX MATH menu (Chapter 10).

seq(

seq((sequence) returns a list in which each element is the result of the evaluation of *expression* with regard to *variable* for the values ranging from *begin* to *end* at steps of *increment*. *variable* need not be defined in memory. *increment* can be negative; the default value for *increment* is 1. **seq(** is not valid within *expression*. Complex lists are not valid.

seq(*expression*,*variable*,*begin*,*end*[,*increment*]**)**

```
seq(A²,A,1,11,3)
   {1 16 49 100}
```

cumSum(

cumSum((cumulative sum) returns the cumulative sums of the elements in *list*, starting with the first element. *list* elements can be real or complex numbers.

cumSum(*list***)**

```
cumSum({1,2,3,4,
5})
   {1 3 6 10 15}
```

ΔList(

ΔList(returns a list containing the differences between consecutive elements in *list*. **ΔList** subtracts the first element in *list* from the second element, subtracts the second element from the third, and so on. The list of differences is always one element shorter than the original *list*. *list* elements can be a real or complex numbers.

ΔList(*list***)**

```
{20,30,45,70}→LD
IST
   {20 30 45 70}
ΔList(LDIST)
      {10 15 25}
```

Select(

Select(selects one or more specific data points from a scatter plot or xyLine plot (only), and then stores the selected data points to two new lists, *xlistname* and *ylistname*. For example, you can use **Select(** to select and then analyze a portion of plotted CBL or CBR data.

Select(*xlistname*,*ylistname***)**

Note: Before you use **Select(**, you must have selected (turned on) a scatter plot or xyLine plot. Also, the plot must be displayed in the current viewing window (page 11-13).

Before Using Select(

Before using **Select(**, follow these steps.

1. Create two list names and enter the data.
2. Turn on a stat plot, select (scatter plot) or (xyLine), and enter the two list names for **Xlist:** and **Ylist:** (Chapter 12).
3. Use **ZoomStat** to plot the data (Chapter 3).

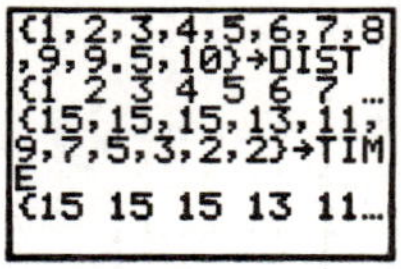

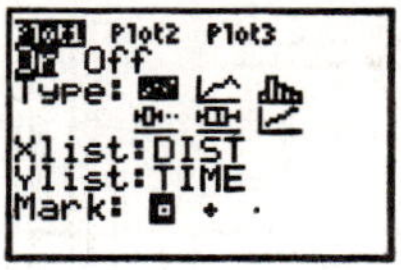

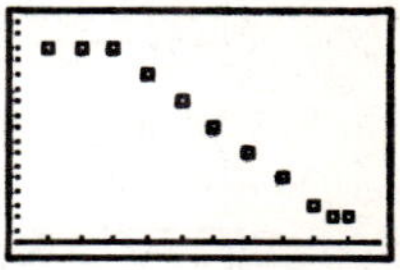

Using Select(to Select Data Points from a Plot

To select data points from a scatter plot or xyLine plot, follow these steps.

1. Press [2nd] [LIST] [▶] **8** to select **8:Select(** from the LIST OPS menu. **Select(** is pasted to the home screen.
2. Enter *xlistname*, press [,], enter *ylistname*, and then press [)] to designate list names into which you want the selected data to be stored.

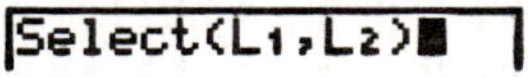

3. Press [ENTER]. The graph screen is displayed with **Left Bound?** in the bottom-left corner.

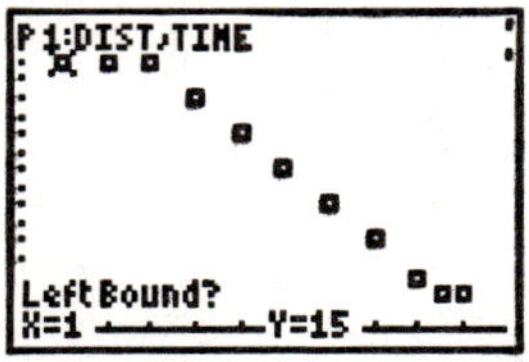

4. Press [▲] or [▼] (if more than one stat plot is selected) to move the cursor onto the stat plot from which you want to select data points.
5. Press [◀] and [▶] to move the cursor to the stat plot data point that you want as the left bound.

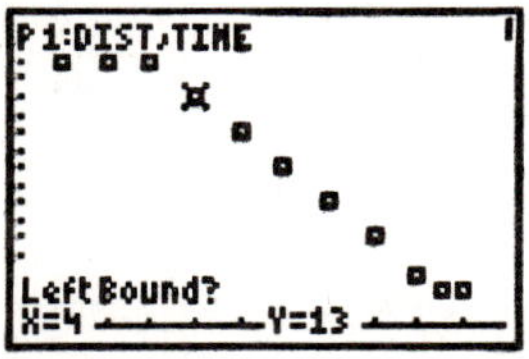

Using Select(to Select Data Points from a Plot (continued)

6. Press [ENTER]. A ▸ indicator on the graph screen shows the left bound. **Right Bound?** is displayed in the bottom-left corner.

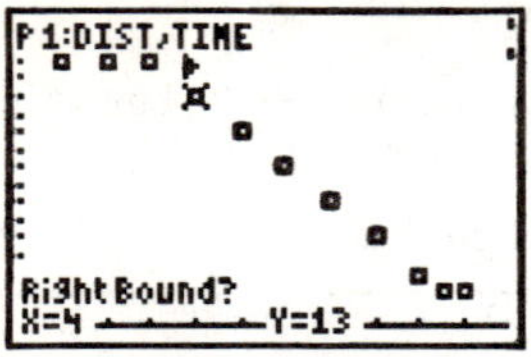

7. Press [◂] or [▸] to move the cursor to the stat plot point that you want for the right bound, and then press [ENTER].

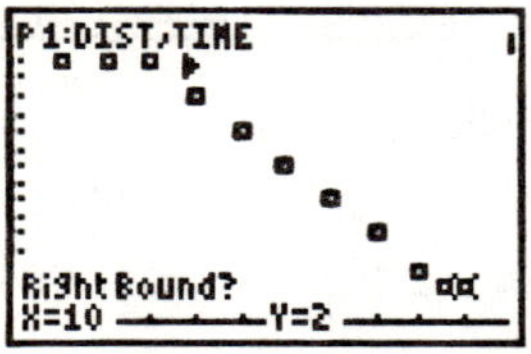

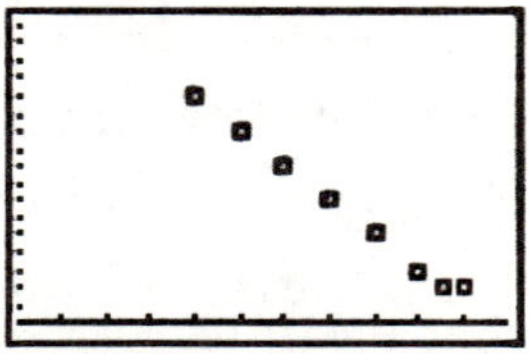

The x-values and y-values of the selected points are stored in *xlistname* and *ylistname*. A new stat plot of *xlistname* and *ylistname* replaces the stat plot from which you selected data points. The list names are updated in the stat plot editor.

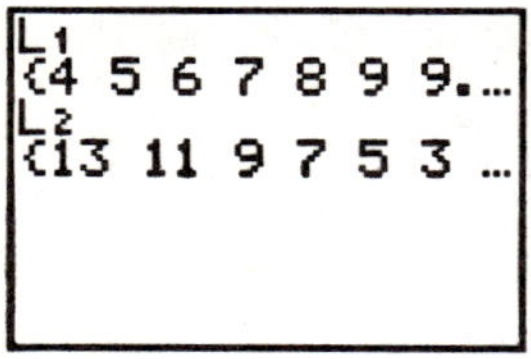

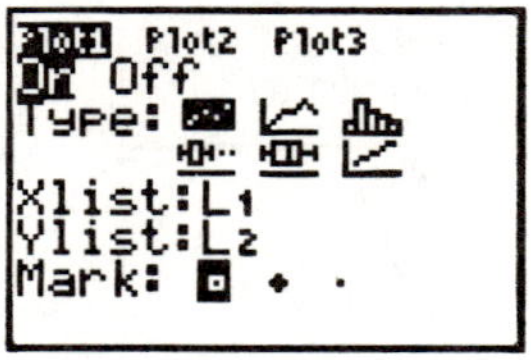

Note: The two new lists (*xlistname* and *ylistname*) will include the points you select as left bound and right bound. Also, *left-bound x-value* ≤ *right-bound x-value* must be true.

augment(

augment(concatenates the elements of *listA* and *listB*. The list elements can be real or complex numbers.

augment(*listA*,*listB***)**

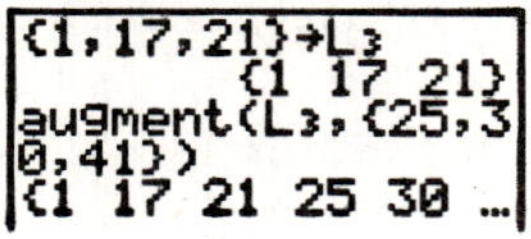

List▸matr(

List▸matr((lists stored to matrix) fills *matrixname* column by column with the elements from each list. If the dimensions of all lists are not equal, then **List▸matr(** fills each extra *matrixname* row with **0**. Complex lists are not valid.

List▸matr(*list1*,*list2*, . . . ,*list n*,*matrixname***)**

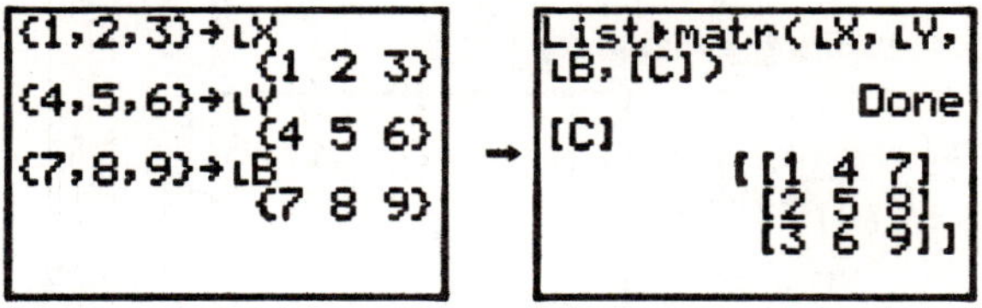

Matr▸list(

Matr▸list((matrix stored to lists) fills each *listname* with elements from each column in *matrix*. If the number of *listname* arguments exceeds the number of columns in *matrix*, then **Matr▸list(** ignores extra *listname* arguments. Likewise, if the number of columns in *matrix* exceeds the number of *listname* arguments, then **Matr▸list(** ignores extra *matrix* columns.

Matr▸list(*matrix*,*listname1*,*listname2*, . . . ,*listname n***)**

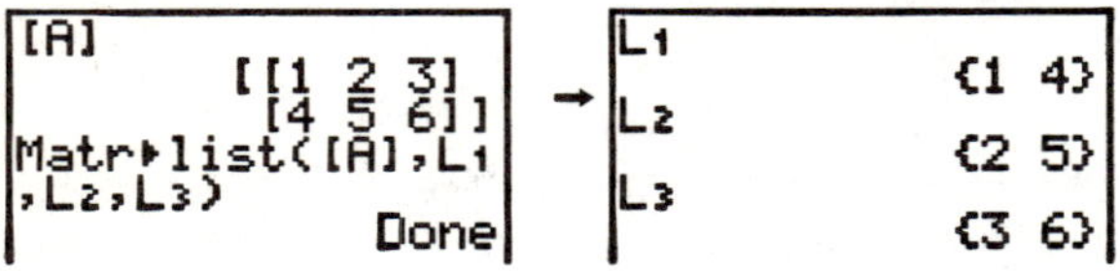

Matr▸list(also fills a *listname* with elements from a specified *column#* in *matrix*. To fill a list with a specific column from *matrix*, you must enter a *column#* after *matrix*.

Matr▸list(*matrix*,*column#*,*listname***)**

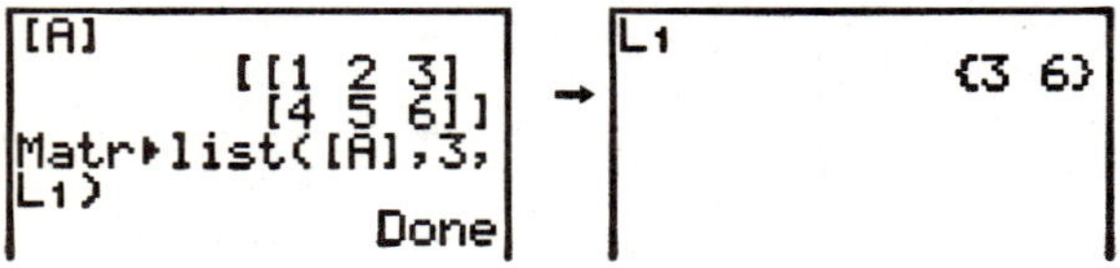

ʟ preceding one to five characters identifies those characters as a user-created *listname*. *listname* may comprise letters, θ, and numbers, but it must begin with a letter from A to Z or θ.

ʟ*listname*

Generally, **ʟ** must precede a user-created list name when you enter a user-created list name where other input is valid, for example, on the home screen. Without the **ʟ**, the TI-83 Plus may misinterpret a user-created list name as implied multiplication of two or more characters.

ʟ need not precede a user-created list name where a list name is the only valid input, for example, at the **Name=** prompt in the stat list editor or the **Xlist:** and **Ylist:** prompts in the stat plot editor. If you enter **ʟ** where it is not necessary, the TI-83 Plus will ignore the entry.

LIST MATH Menu

To display the LIST MATH menu, press [2nd] [LIST] [◄].

NAMES OPS MATH	
1: min(	Returns minimum element of a list.
2: max(	Returns maximum element of a list.
3: mean(	Returns mean of a list.
4: median(	Returns median of a list.
5: sum(	Returns sum of elements in a list.
6: prod(	Returns product of elements in list.
7: stdDev(	Returns standard deviation of a list.
8: variance(	Returns the variance of a list.

min(, max(

min((minimum) and **max(** (maximum) return the smallest or largest element of *listA*. If two lists are compared, it returns a list of the smaller or larger of each pair of elements in *listA* and *listB*. For a complex list, the element with smallest or largest magnitude (modulus) is returned.

min(*listA*[,*listB*]**)**
max(*listA*[,*listB*]**)**

```
min({1,2,3},{3,2
,1})
         {1 2 1}
max({1,2,3},{3,2
,1})
         {3 2 3}
```

Note: min(and **max(** are the same as **min(** and **max(** on the MATH NUM menu.

mean(, median(

mean(returns the mean value of *list*. **median(** returns the median value of *list*. The default value for *freqlist* is 1. Each *freqlist* element counts the number of consecutive occurrences of the corresponding element in *list*. Complex lists are not valid.

mean(*list*[,*freqlist*]**)**
median(*list*[,*freqlist*]**)**

```
mean({1,2,3},{3,
2,1})
     1.666666667
median({1,2,3})
               2
```

sum(, prod(

sum((summation) returns the sum of the elements in *list*. *start* and *end* are optional; they specify a range of elements. *list* elements can be real or complex numbers.

prod(returns the product of all elements of *list*. *start* and *end* elements are optional; they specify a range of list elements. *list* elements can be real or complex numbers.

sum(*list*[,*start*,*end*]) **prod(***list*[,*start*,*end*])

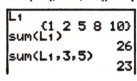

```
L1
      {1 2 5 8 10}
sum(L1)
                26
sum(L1,3,5)
                23
```

```
L1
      {1 2 5 8 10}
prod(L1)
               800
prod(L1,3,5)
               400
```

Sums and Products of Numeric Sequences

You can combine **sum(** or **prod(** with **seq(** to obtain:

$$\sum_{x=lower}^{upper} expression(x) \qquad \prod_{x=lower}^{upper} expression(x)$$

To evaluate $\Sigma\, 2^{(N-1)}$ from N=1 to 4:

```
sum(seq(2^(N-1),
N,1,4,1))
                15
```

stdDev(, variance(

stdDev(returns the standard deviation of the elements in *list*. The default value for *freqlist* is 1. Each *freqlist* element counts the number of consecutive occurrences of the corresponding element in *list*. Complex lists are not valid.

variance(returns the variance of the elements in *list*. The default value for *freqlist* is 1. Each *freqlist* element counts the number of consecutive occurrences of the corresponding element in *list*. Complex lists are not valid.

stdDev(*list*[,*freqlist*]) **variance(***list*[,*freqlist*])

```
stdDev({1,2,5,-6
,3,-2})
       3.937003937
```

```
variance({1,2,5,
-6,3,-2})
              15.5
```

12 Statistics

Contents

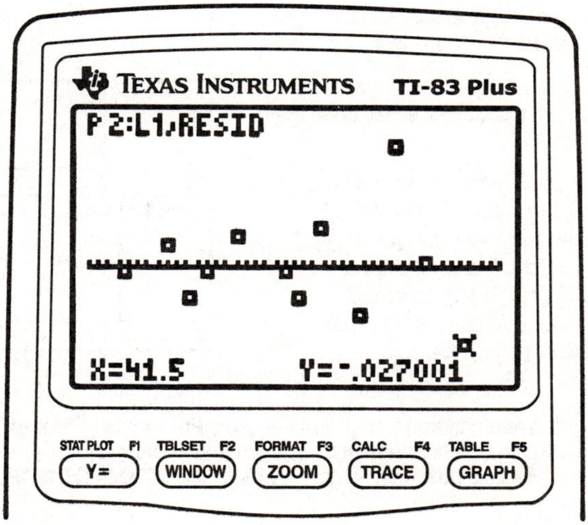

Getting Started: Pendulum Lengths and Periods

Getting Started is a fast-paced introduction. Read the chapter for details.

A group of students is attempting to determine the mathematical relationship between the length of a pendulum and its period (one complete swing of a pendulum). The group makes a simple pendulum from string and washers and then suspends it from the ceiling. They record the pendulum's period for each of 12 string lengths.*

Length (cm)	Time (sec)
6.5	0.51
11.0	0.68
13.2	0.73
15.0	0.79
18.0	0.88
23.1	0.99
24.4	1.01
26.6	1.08
30.5	1.13
34.3	1.26
37.6	1.28
41.5	1.32

1. Press [MODE] [▼] [▼] [▼] [ENTER] to set **Func** graphing mode.

2. Press [STAT] **5** to select **5:SetUpEditor**. **SetUpEditor** is pasted to the home screen.

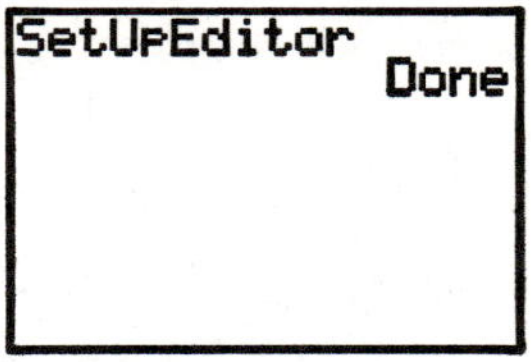

   Press [ENTER]. This removes lists from stat list editor columns **1** through **20**, and then stores lists **L1** through **L6** in columns **1** through **6**.

   Note: Removing lists from the stat list editor does not delete them from memory.

3. Press [STAT] **1** to select **1:Edit** from the STAT EDIT menu. The stat list editor is displayed. If elements are stored in **L1** and **L2**, press [▲] to move the cursor onto **L1**, and then press [CLEAR] [ENTER] [▶] [▲] [CLEAR] [ENTER] to clear both lists. Press [◀] to move the rectangular cursor back to the first row in **L1**.

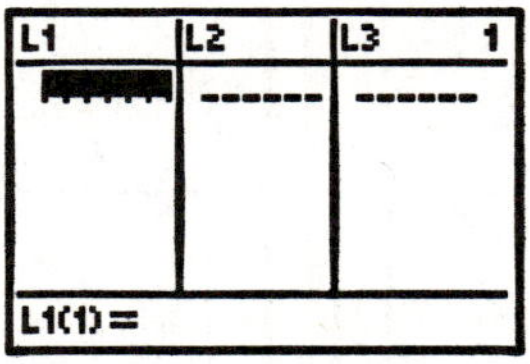

* This example is quoted and adapted from *Contemporary Precalculus Through Applications*, by the North Carolina School of Science and Mathematics, by permission of Janson Publications, Inc., Dedham, MA. 1-800-322-MATH. © 1992. All rights reserved.

4. Press **6** [.] **5** [ENTER] to store the first pendulum string length (6.5 cm) in **L1**. The rectangular cursor moves to the next row. Repeat this step to enter each of the 12 string length values in the table on page 12-2.

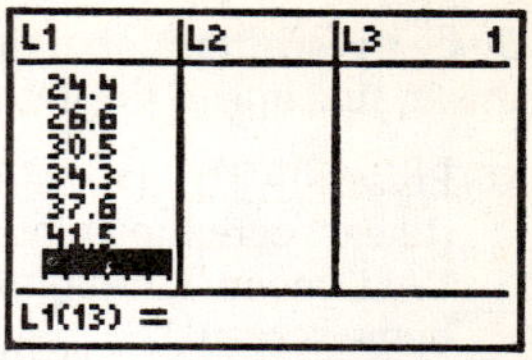

5. Press [▸] to move the rectangular cursor to the first row in **L2**.

   Press [.] **51** [ENTER] to store the first time measurement (.51 sec) in **L2**. The rectangular cursor moves to the next row. Repeat this step to enter each of the 12 time values in the table on page 12-2.

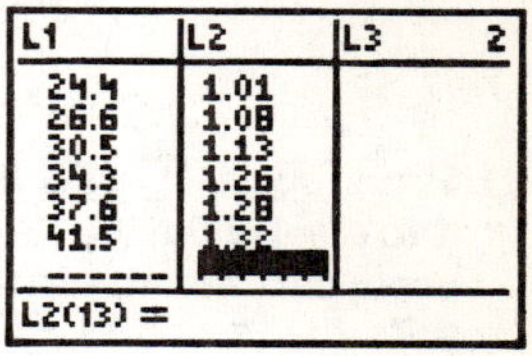

6. Press [Y=] to display the Y= editor.

   If necessary, press [CLEAR] to clear the function **Y1**. As necessary, press [▴], [ENTER], and [▸] to turn off **Plot1**, **Plot2**, and **Plot3** from the top line of the Y= editor (Chapter 3). As necessary, press [▾], [◂], and [ENTER] to deselect functions.

7. Press [2nd] [STAT PLOT] **1** to select **1:Plot1** from the STAT PLOTS menu. The stat plot editor is displayed for plot 1.

8. Press [ENTER] to select **On**, which turns on plot 1. Press [▾] [ENTER] to select ⋰ (scatter plot). Press [▾] [2nd] [L1] to specify **Xlist:L1** for plot 1. Press [▾] [2nd] [L2] to specify **Ylist:L2** for plot 1. Press [▾] [▸] [ENTER] to select **+** as the **Mark** for each data point on the scatter plot.

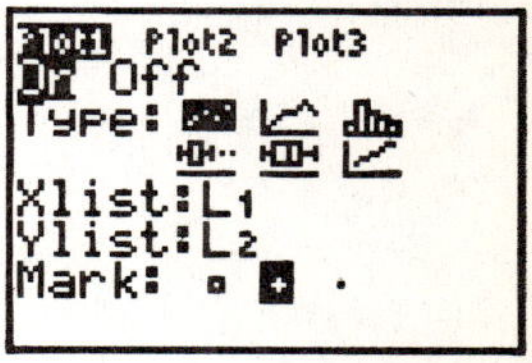

9. Press [ZOOM] **9** to select **9:ZoomStat** from the ZOOM menu. The window variables are adjusted automatically, and plot 1 is displayed. This is a scatter plot of the time-versus-length data.

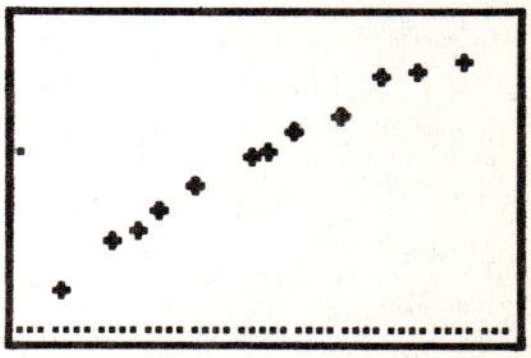

Since the scatter plot of time-versus-length data appears to be approximately linear, fit a line to the data.

10. Press [STAT] [▶] **4** to select **4:LinReg(ax+b)** (linear regression model) from the STAT CALC menu. **LinReg(ax+b)** is pasted to the home screen.

11. Press [2nd] [L1] [,] [2nd] [L2] [,]. Press [VARS] [▶] **1** to display the VARS Y-VARS FUNCTION secondary menu, and then press **1** to select **1:Y1**. **L1**, **L2**, and **Y1** are pasted to the home screen as arguments to **LinReg(ax+b)**.

12. Press [ENTER] to execute **LinReg(ax+b)**. The linear regression for the data in **L1** and **L2** is calculated. Values for **a** and **b** are displayed on the home screen. The linear regression equation is stored in **Y1**. Residuals are calculated and stored automatically in the list name **RESID**, which becomes an item on the LIST NAMES menu.

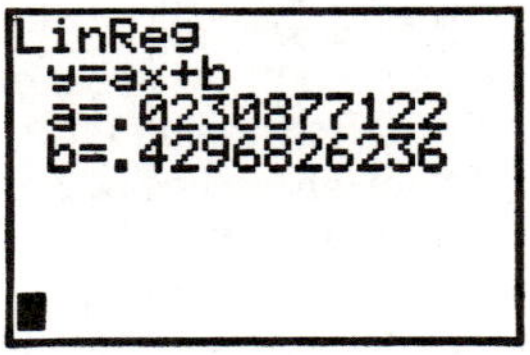

13. Press [GRAPH]. The regression line and the scatter plot are displayed.

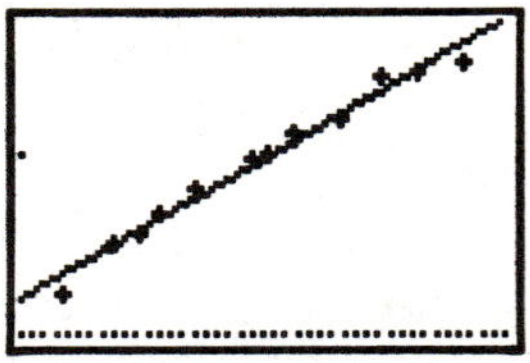

The regression line appears to fit the central portion of the scatter plot well. However, a residual plot may provide more information about this fit.

14. Press [STAT] **1** to select **1:Edit**. The stat list editor is displayed.

    Press [▶] and [▲] to move the cursor onto **L3**.

    Press [2nd] [INS]. An unnamed column is displayed in column **3**; **L3**, **L4**, **L5**, and **L6** shift right one column. The **Name=** prompt is displayed in the entry line, and alpha-lock is on.

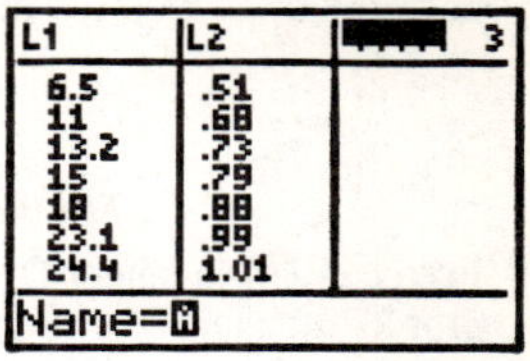

15. Press [2nd] [LIST] to display the **LIST NAMES** menu.

    If necessary, press [▼] to move the cursor onto the list name **RESID**.

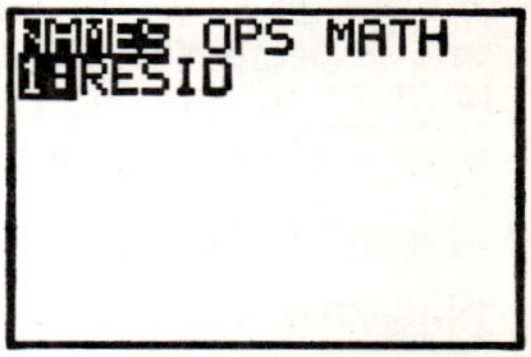

16. Press [ENTER] to select **RESID** and paste it to the stat list editor's **Name=** prompt.

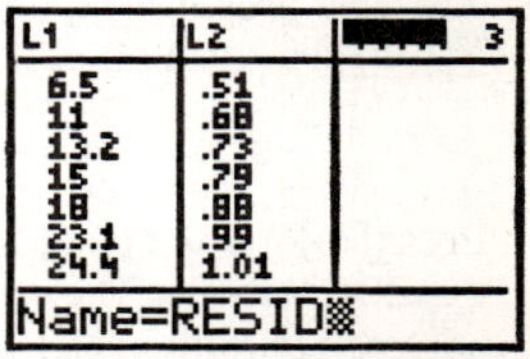

17. Press [ENTER]. **RESID** is stored in column **3** of the stat list editor.

    Press [▼] repeatedly to examine the residuals.

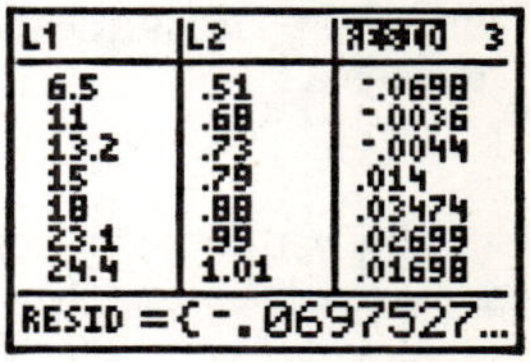

Notice that the first three residuals are negative. They correspond to the shortest pendulum string lengths in **L1**. The next five residuals are positive, and three of the last four are negative. The latter correspond to the longer string lengths in **L1**. Plotting the residuals will show this pattern more clearly.

18. Press [2nd] [STAT PLOT] **2** to select **2:Plot2** from the STAT PLOTS menu. The stat plot editor is displayed for plot 2.

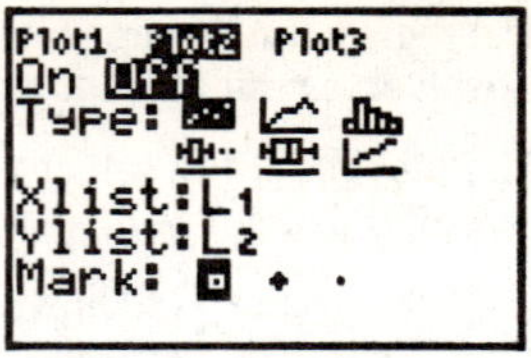

19. Press [ENTER] to select **On**, which turns on plot 2.

    Press [▼] [ENTER] to select (scatter plot). Press [▼] [2nd] [L1] to specify **Xlist:L1** for plot 2. Press [▼] [R] [E] [S] [I] [D] (alpha-lock is on) to specify **Ylist:RESID** for plot 2. Press [▼] [ENTER] to select □ as the mark for each data point on the scatter plot.

20. Press [Y=] to display the Y= editor.

    Press [◄] to move the cursor onto the = sign, and then press [ENTER] to deselect **Y1**. Press [▲] [ENTER] to turn off plot 1.

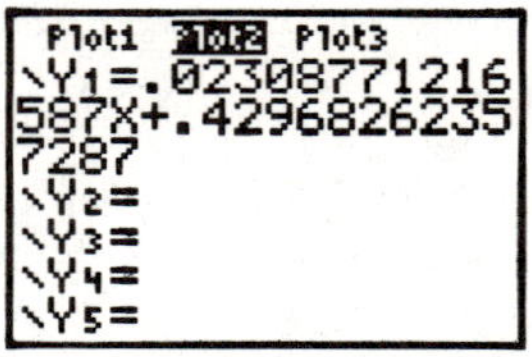

21. Press [ZOOM] **9** to select **9:ZoomStat** from the ZOOM menu. The window variables are adjusted automatically, and plot 2 is displayed. This is a scatter plot of the residuals.

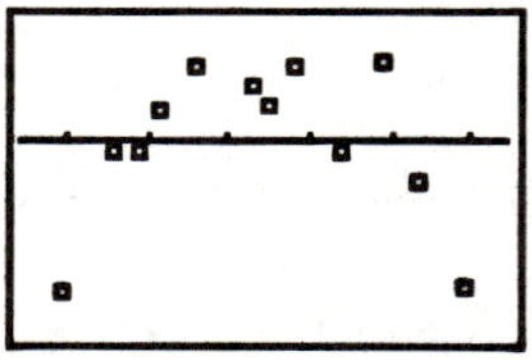

Notice the pattern of the residuals: a group of negative residuals, then a group of positive residuals, and then another group of negative residuals.

The residual pattern indicates a curvature associated with this data set for which the linear model did not account. The residual plot emphasizes a downward curvature, so a model that curves down with the data would be more accurate. Perhaps a function such as square root would fit. Try a power regression to fit a function of the form $y = a * x^b$.

22. Press [Y=] to display the Y= editor.

    Press [CLEAR] to clear the linear regression equation from **Y1**. Press [▲] [ENTER] to turn on plot 1. Press [▶] [ENTER] to turn off plot 2.

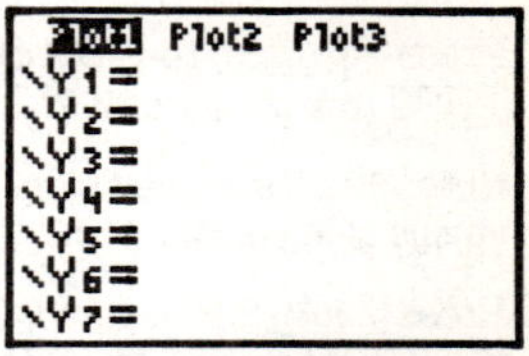

23. Press [ZOOM] **9** to select **9:ZoomStat** from the ZOOM menu. The window variables are adjusted automatically, and the original scatter plot of time-versus-length data (plot 1) is displayed.

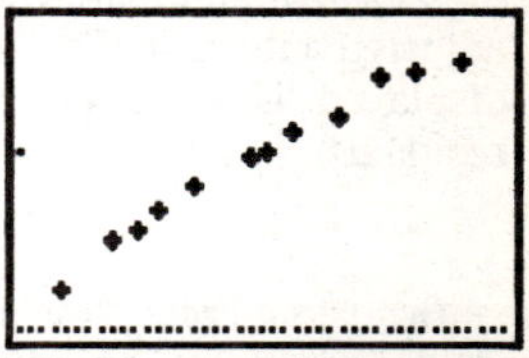

24. Press [STAT] [▶] [ALPHA] [A] to select **A:PwrReg** from the STAT CALC menu. **PwrReg** is pasted to the home screen.

    Press [2nd] [L1] [,] [2nd] [L2] [,]. Press [VARS] [▶] **1** to display the VARS Y-VARS FUNCTION secondary menu, and then press **1** to select **1:Y1**. **L1**, **L2**, and **Y1** are pasted to the home screen as arguments to **PwrReg**.

25. Press [ENTER] to calculate the power regression. Values for **a** and **b** are displayed on the home screen. The power regression equation is stored in **Y1**. Residuals are calculated and stored automatically in the list name **RESID**.

26. Press [GRAPH]. The regression line and the scatter plot are displayed.

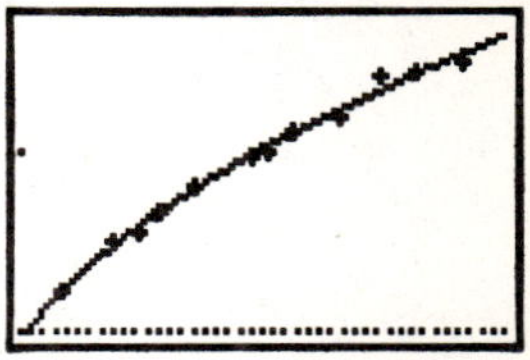

The new function $y=.192x^{.522}$ appears to fit the data well. To get more information, examine a residual plot.

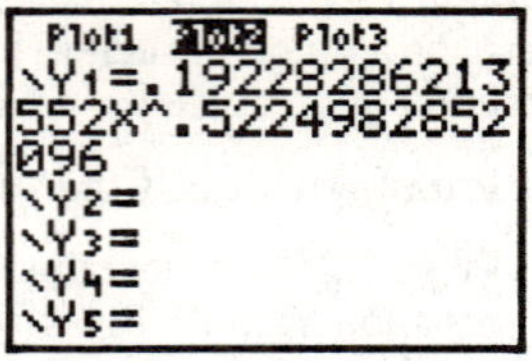

27. Press [Y=] to display the Y= editor.

    Press [◄] [ENTER] to deselect **Y1**.

    Press [▲] [ENTER] to turn off plot 1. Press [►] [ENTER] to turn on plot 2.

    Note: Step 19 defined plot 2 to plot residuals (**RESID**) versus string length (**L1**).

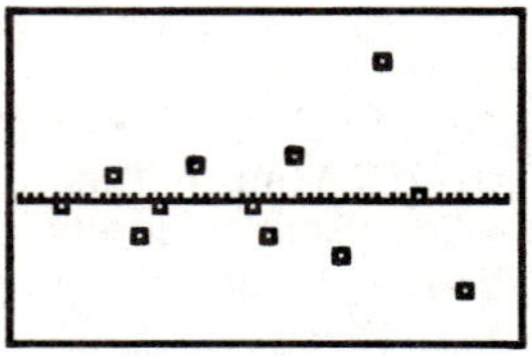

28. Press [ZOOM] **9** to select **9:ZoomStat** from the ZOOM menu. The window variables are adjusted automatically, and plot 2 is displayed. This is a scatter plot of the residuals.

The new residual plot shows that the residuals are random in sign, with the residuals increasing in magnitude as the string length increases.

To see the magnitudes of the residuals, continue with these steps.

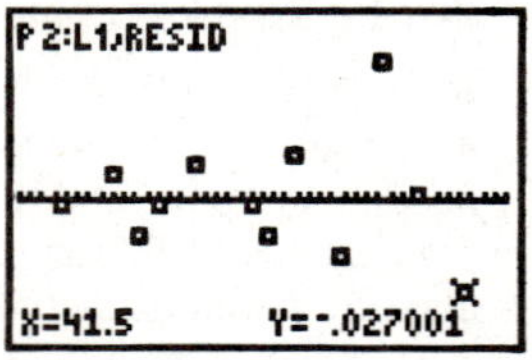

29. Press [TRACE].

    Press [►] and [◄] to trace the data. Observe the values for **Y** at each point.

    With this model, the largest positive residual is about 0.041 and the smallest negative residual is about ⁻0.027. All other residuals are less than 0.02 in magnitude.

Now that you have a good model for the relationship between length and period, you can use the model to predict the period for a given string length. To predict the periods for a pendulum with string lengths of 20 cm and 50 cm, continue with these steps.

30. Press [VARS] [▶] **1** to display the VARS Y-VARS FUNCTION secondary menu, and then press **1** to select **1:Y1**. **Y1** is pasted to the home screen.

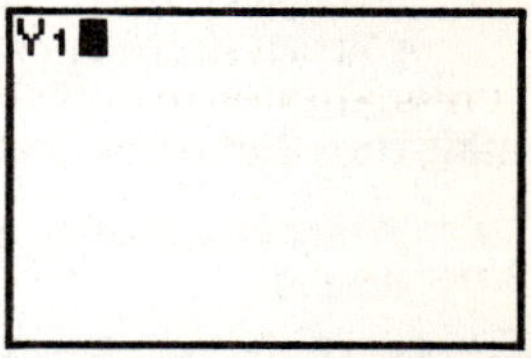

31. Press [(] **20** [)] to enter a string length of 20 cm.

    Press [ENTER] to calculate the predicted time of about 0.92 seconds.

    Based on the residual analysis, we would expect the prediction of about 0.92 seconds to be within about 0.02 seconds of the actual value.

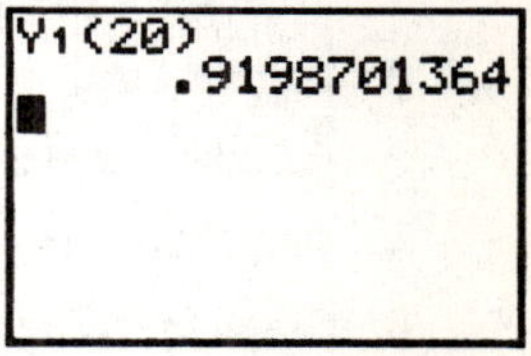

32. Press [2nd] [ENTRY] to recall the Last Entry.

    Press [◀] [◀] [◀] **5** to change the string length to 50 cm.

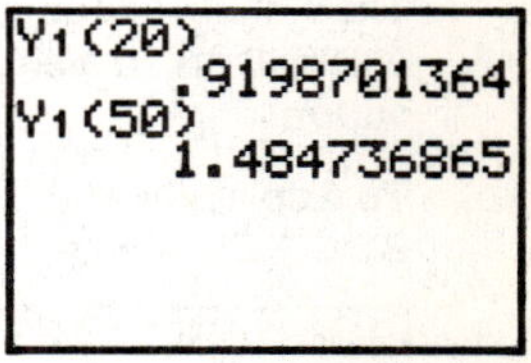

33. Press [ENTER] to calculate the predicted time of about 1.48 seconds.

    Since a string length of 50 cm exceeds the lengths in the data set, and since residuals appear to be increasing as string length increases, we would expect more error with this estimate.

    Note: You also can make predictions using the table with the TABLE SETUP settings **Indpnt:Ask** and **Depend:Auto** (Chapter 7).

Using Lists to Store Data

Data for statistical analyses is stored in lists, which you can create and edit using the stat list editor. The TI-83 Plus has six list variables in memory, **L1** through **L6**, to which you can store data for statistical calculations. Also, you can store data to list names that you create (Chapter 11).

Setting Up a Statistical Analysis

To set up a statistical analysis, follow these steps. Read the chapter for details.

1. Enter the statistical data into one or more lists.
2. Plot the data.
3. Calculate the statistical variables or fit a model to the data.
4. Graph the regression equation for the plotted data.
5. Graph the residuals list for the given regression model.

Displaying the Stat List Editor

The stat list editor is a table where you can store, edit, and view up to 20 lists that are in memory. Also, you can create list names from the stat list editor.

To display the stat list editor, press [STAT], and then select **1:Edit** from the STAT EDIT menu.

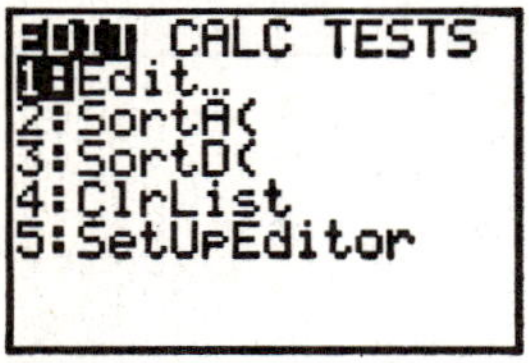

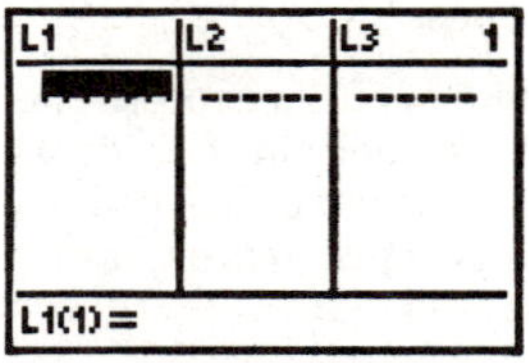

The top line displays list names. **L1** through **L6** are stored in columns **1** through **6** after a memory reset. The number of the current column is displayed in the top-right corner.

The bottom line is the entry line. All data entry occurs on this line. The characteristics of this line change according to the current context (page 12-17).

The center area displays up to seven elements of up to three lists; it abbreviates values when necessary. The entry line displays the full value of the current element.

Entering a List Name in the Stat List Editor

To enter a list name in the stat list editor, follow these steps.

1. Display the **Name=** prompt in the entry line in either of two ways.
   - Move the cursor onto the list name in the column where you want to insert a list, and then press [2nd] [INS]. An unnamed column is displayed and the remaining lists shift right one column.
   - Press [▲] until the cursor is on the top line, and then press [▶] until you reach the unnamed column.

   Note: If list names are stored to all 20 columns, you must remove a list name to make room for an unnamed column.

   The **Name=** prompt is displayed and alpha-lock is on.

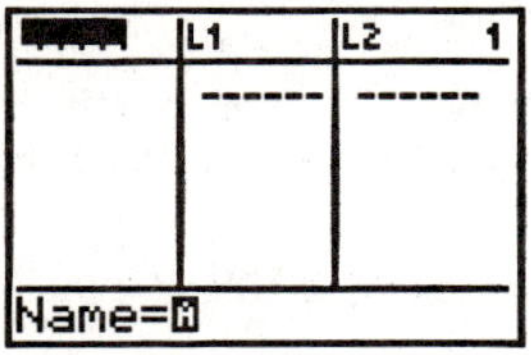

2. Enter a valid list name in any of four ways.
   - Select a name from the LIST NAMES menu (Chapter 11).
   - Enter **L1**, **L2**, **L3**, **L4**, **L5**, or **L6** from the keyboard.
   - Enter an existing user-created list name directly from the keyboard.
   - Enter a new user-created list name (page 12-12).

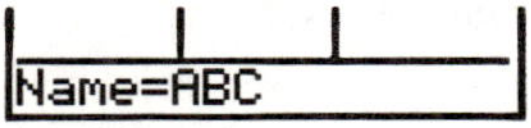

3. Press [ENTER] or [▼] to store the list name and its elements, if any, in the current column of the stat list editor.

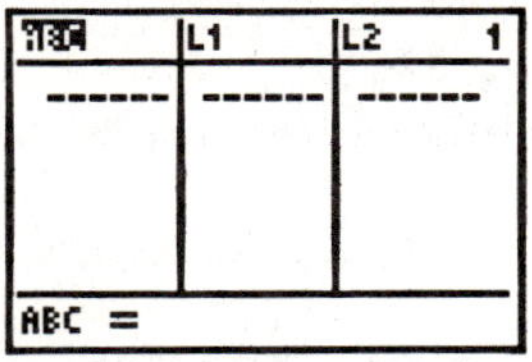

To begin entering, scrolling, or editing list elements, press [▼]. The rectangular cursor is displayed.

Note: If the list name you entered in step 2 already was stored in another stat list editor column, then the list and its elements, if any, move to the current column from the previous column. Remaining list names shift accordingly.

Creating a Name in the Stat List Editor

To create a name in the stat list editor, follow these steps.

1. Follow step 1 on page 12-11 to display the **Name=** prompt.
2. Press [*letter from A to Z or θ*] to enter the first letter of the name. The first character cannot be a number.
3. Enter zero to four letters, θ, or numbers to complete the new user-created list name. List names can be one to five characters long.
4. Press [ENTER] or [▼] to store the list name in the current column of the stat list editor. The list name becomes an item on the **LIST NAMES** menu (Chapter 11).

Removing a List from the Stat List Editor

To remove a list from the stat list editor, move the cursor onto the list name and then press [DEL]. The list is not deleted from memory; it is only removed from the stat list editor.

Note1: To delete a list name from memory, use the MEMORY MANAGEMENT/ DELETE secondary menu (Chapter 18).

Note2: If you archive a list, it will be removed from the stat list editor.

Removing All Lists and Restoring L1 through L6

You can remove all user-created lists from the stat list editor and restore list names **L1** through **L6** to columns **1** through **6** in either of two ways.

- Use **SetUpEditor** with no arguments (page 12-21).
- Reset all memory (Chapter 18).

Clearing All Elements from a List

You can clear all elements from a list in any of five ways.

- Use **ClrList** to clear specified lists (page 12-20).
- In the stat list editor, press [▲] to move the cursor onto a list name, and then press [CLEAR] [ENTER].
- In the stat list editor, move the cursor onto each element, and then press [DEL] one by one.
- On the home screen or in the program editor, enter **0→dim(***listname***)** to set the dimension of *listname* to 0 (Chapter 11).
- Use **ClrAllLists** to clear all lists in memory (Chapter 18).

Editing a List Element

To edit a list element, follow these steps.

1. Move the rectangular cursor onto the element you want to edit.
2. Press [ENTER] to move the cursor to the entry line.

   Note: If you want to replace the current value, you can enter a new value without first pressing [ENTER]. When you enter the first character, the current value is cleared automatically.

3. Edit the element in the entry line.

   - Press one or more keys to enter the new value. When you enter the first character, the current value is cleared automatically.
   - Press [▶] to move the cursor to the character before which you want to insert, press [2nd] [INS], and then enter one or more characters.
   - Press [▶] to move the cursor to a character you want to delete, and then press [DEL] to delete the character.

   To cancel any editing and restore the original element at the rectangular cursor, press [CLEAR] [ENTER].

   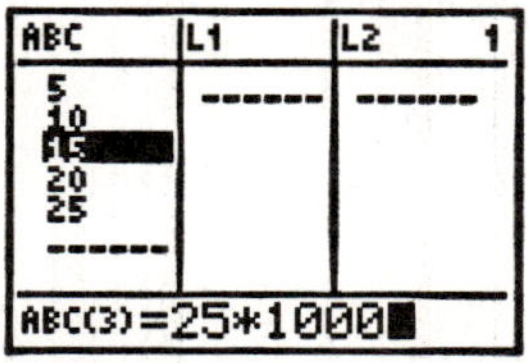

   Note: You can enter expressions and variables for elements.

4. Press [ENTER], [▲], or [▼] to update the list. If you entered an expression, it is evaluated. If you entered only a variable, the stored value is displayed as a list element.

   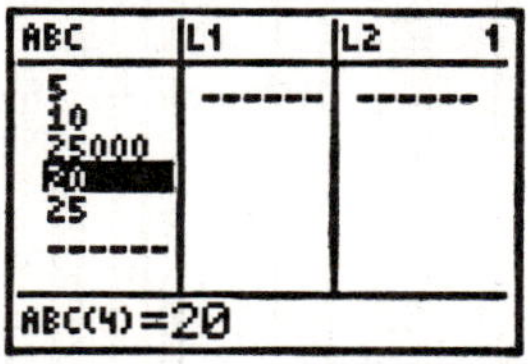

When you edit a list element in the stat list editor, the list is updated in memory immediately.

Attaching a Formula to a List Name in Stat List Editor

You can attach a formula to a list name in the stat list editor, and then display and edit the calculated list elements. When executed, the attached formula must resolve to a list. Chapter 11 describes in detail the concept of attaching formulas to list names.

To attach a formula to a list name that is stored in the stat list editor, follow these steps.

1. Press [STAT] [ENTER] to display the stat list editor.
2. Press [▲] to move the cursor to the top line.
3. Press [◄] or [►], if necessary, to move the cursor onto the list name to which you want to attach the formula.

   Note: If a formula in quotation marks is displayed on the entry line, then a formula is already attached to the list name. To edit the formula, press [ENTER], and then edit the formula.

4. Press [ALPHA] ["], enter the formula, and press [ALPHA] ["].

   Note: If you do not use quotation marks, the TI-83 Plus calculates and displays the same initial list of answers, but does not attach the formula for future calculations.

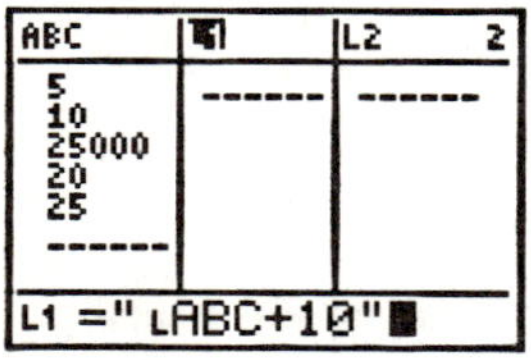

   Note: Any user-created list name referenced in a formula must be preceded by an ʟ symbol (Chapter 11).

5. Press [ENTER]. The TI-83 Plus calculates each list element and stores it to the list name to which the formula is attached. A lock symbol is displayed in the stat list editor, next to the list name to which the formula is attached.

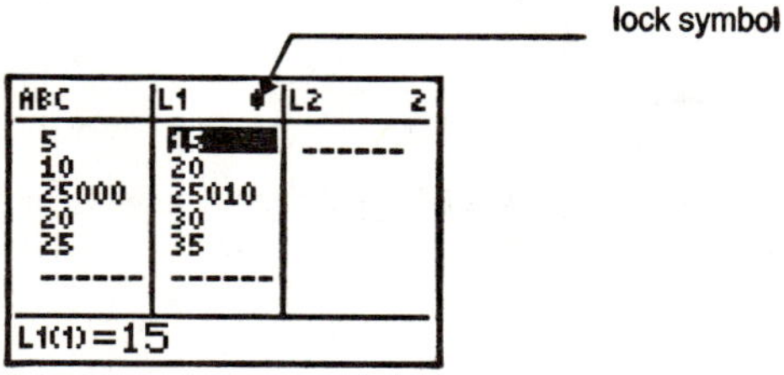

Using the Stat List Editor When Formula-Generated Lists Are Displayed

When you edit an element of a list referenced in an attached formula, the TI-83 Plus updates the corresponding element in the list to which the formula is attached (Chapter 11).

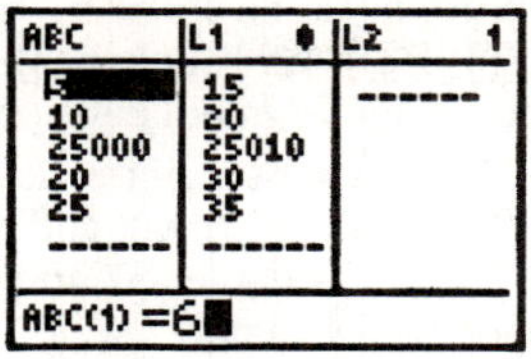

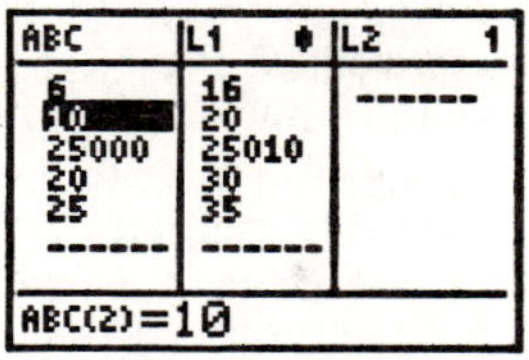

When a list with a formula attached is displayed in the stat list editor and you edit or enter elements of another displayed list, then the TI-83 Plus takes slightly longer to accept each edit or entry than when no lists with formulas attached are in view.

Tip: To speed editing time, scroll horizontally until no lists with formulas are displayed, or rearrange the stat list editor so that no lists with formulas are displayed.

Handling Errors Resulting from Attached Formulas

On the home screen, you can attach to a list a formula that references another list with dimension 0 (Chapter 11). However, you cannot display the formula-generated list in the stat list editor or on the home screen until you enter at least one element to the list that the formula references.

All elements of a list referenced by an attached formula must be valid for the attached formula. For example, if **Real** number mode is set and the attached formula is **log(L1)**, then each element of **L1** must be greater than **0**, since the logarithm of a negative number returns a complex result.

Tip: If an error menu is returned when you attempt to display a formula-generated list in the stat list editor, you can select **2:Goto**, write down the formula that is attached to the list, and then press [CLEAR] [ENTER] to detach (clear) the formula. You then can use the stat list editor to find the source of the error. After making the appropriate changes, you can reattach the formula to a list.

If you do not want to clear the formula, you can select **1:Quit**, display the referenced list on the home screen, and find and edit the source of the error. To edit an element of a list on the home screen, store the new value to *listname*(*element#*) (Chapter 11).

Detaching a Formula from a List Name

You can detach (clear) a formula from a list name in several ways.

For example:

- In the stat list editor, move the cursor onto the name of the list to which a formula is attached. Press [ENTER] [CLEAR] [ENTER]. All list elements remain, but the formula is detached and the lock symbol disappears.
- In the stat list editor, move the cursor onto an element of the list to which a formula is attached. Press [ENTER], edit the element, and then press [ENTER]. The element changes, the formula is detached, and the lock symbol disappears. All other list elements remain.
- Use **ClrList** (page 12-20). All elements of one or more specified lists are cleared, each formula is detached, and each lock symbol disappears. All list names remain.
- Use **ClrAllLists** (Chapter 18). All elements of all lists in memory are cleared, all formulas are detached from all list names, and all lock symbols disappear. All list names remain.

Editing an Element of a Formula-Generated List

As described above, one way to detach a formula from a list name is to edit an element of the list to which the formula is attached. The TI-83 Plus protects against inadvertently detaching the formula from the list name by editing an element of the formula-generated list.

Because of the protection feature, you must press [ENTER] before you can edit an element of a formula-generated list.

The protection feature does not allow you to delete an element of a list to which a formula is attached. To delete an element of a list to which a formula is attached, you must first detach the formula in any of the ways described above.

Stat List Editor Contexts

The stat list editor has four contexts.

- View-elements context
- View-names context
- Edit-elements context
- Enter-name context

The stat list editor is first displayed in view-elements context. To switch through the four contexts, select **1:Edit** from the STAT EDIT menu and follow these steps.

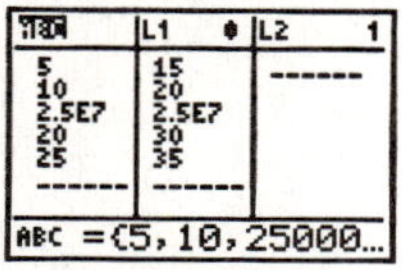

1. Press [▲] to move the cursor onto a list name. You are now in view-names context. Press [▶] and [◀] to view list names stored in other stat list editor columns.

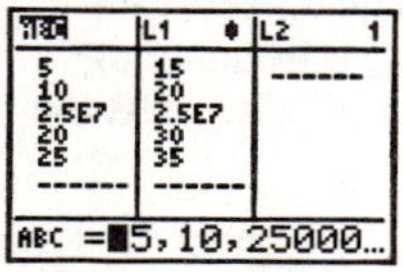

2. Press [ENTER]. You are now in edit-elements context. You may edit any element in a list. All elements of the current list are displayed in braces (**{ }**)in the entry line. Press [▶] and [◀] to view more list elements.

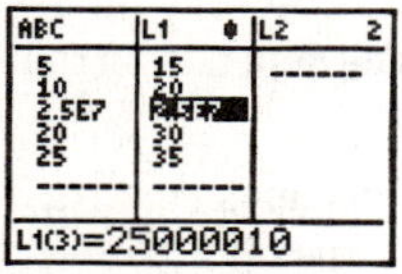

3. Press [ENTER] again. You are now in view-elements context. Press [▶], [◀], [▼], and [▲] to view other list elements. The current element's full value is displayed in the entry line.

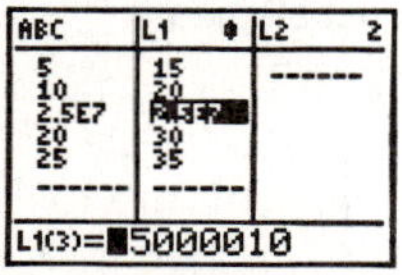

4. Press [ENTER] again. You are now in edit-elements context. You may edit the current element in the entry line.

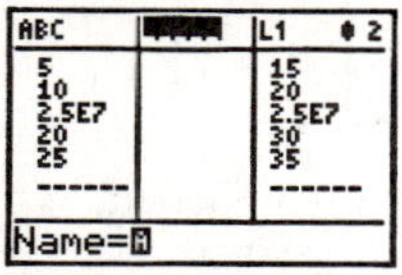

5. Press [▲] until the cursor is on a list name, then press [2nd] [INS]. You are now in enter-name context.

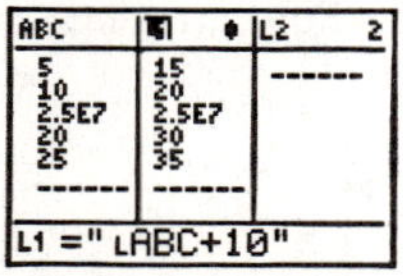

6. Press [CLEAR]. You are now in view-names context.

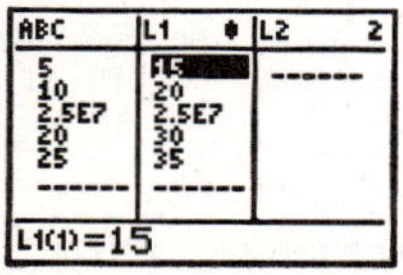

7. Press [▼]. You are now back in view-elements context.

View-Elements Context

In view-elements context, the entry line displays the list name, the current element's place in that list, and the full value of the current element, up to 12 characters at a time. An ellipsis (...) indicates that the element continues beyond 12 characters.

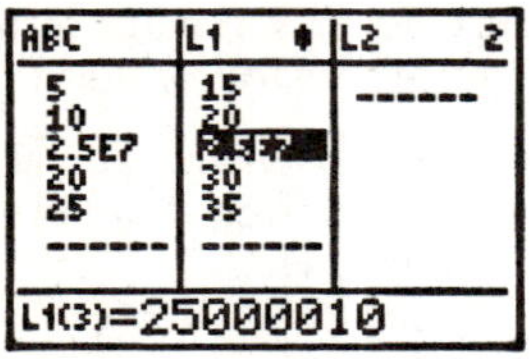

To page down the list six elements, press [ALPHA] [▼]. To page up six elements, press [ALPHA] [▲]. To delete a list element, press [DEL]. Remaining elements shift up one row. To insert a new element, press [2nd] [INS]. **0** is the default value for a new element.

Edit-Elements Context

In edit-elements context, the data displayed in the entry line depends on the previous context.

- When you switch to edit-elements context from view-elements context, the full value of the current element is displayed. You can edit the value of this element, and then press [▼] and [▲] to edit other list elements.

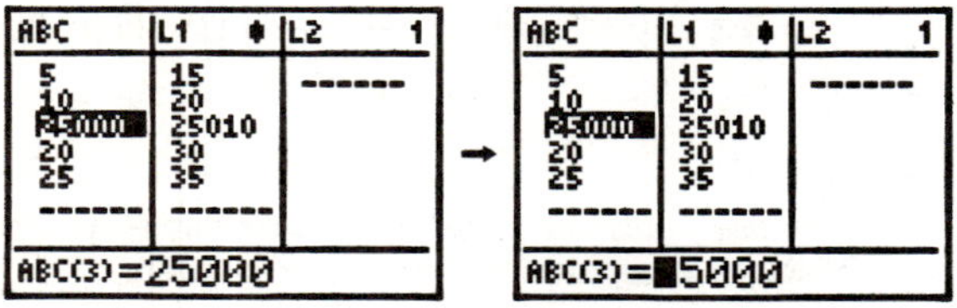

- When you switch to edit-elements context from view-names context, the full values of all elements in the list are displayed. An ellipsis indicates that list elements continue beyond the screen. You can press [▶] and [◀] to edit any element in the list.

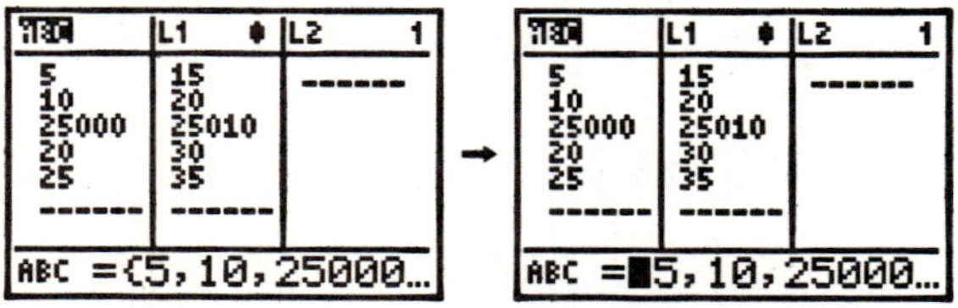

Note: In edit-elements context, you can attach a formula to a list name only if you switched to it from view-names context.

View-Names Context

In view-names context, the entry line displays the list name and the list elements.

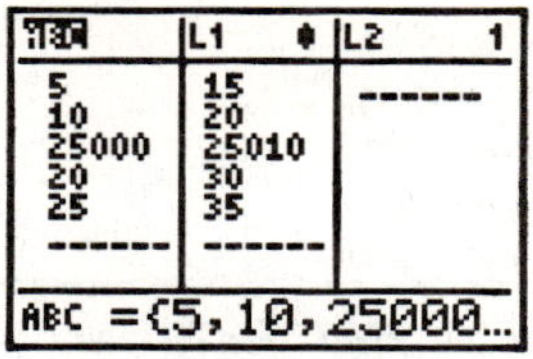

To remove a list from the stat list editor, press [DEL]. Remaining lists shift to the left one column. The list is not deleted from memory.

To insert a name in the current column, press [2nd] [INS]. Remaining columns shift to the right one column.

Enter-Name Context

In enter-name context, the **Name=** prompt is displayed in the entry line, and alpha-lock is on.

At the **Name=** prompt, you can create a new list name, paste a list name from **L1** to **L6** from the keyboard, or paste an existing list name from the LIST NAMES menu (Chapter 11). The **ʟ** symbol is not required at the **Name=** prompt.

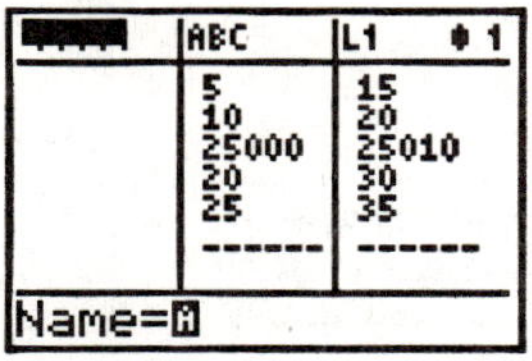

To leave enter-name context without entering a list name, press [CLEAR]. The stat list editor switches to view-names context.

STAT EDIT Menu

To display the STAT EDIT menu, press [STAT].

EDIT CALC TESTS	
1: Edit...	Displays the stat list editor.
2: SortA(	Sorts a list in ascending order.
3: SortD(	Sorts a list in descending order.
4: ClrList	Deletes all elements of a list.
5: SetUpEditor	Stores lists in the stat list editor.

Note: Chapter 13: Inferential Statistics describes the STAT TESTS menu items.

SortA(, SortD(

SortA((sort ascending) sorts list elements from low to high values. **SortD(** (sort descending) sorts list elements from high to low values. Complex lists are sorted based on magnitude (modulus). **SortA(** and **SortD(** each can sort in either of two ways.

- With one *listname*, **SortA(** and **SortD(** sort the elements in *listname* and update the list in memory.
- With two or more lists, **SortA(** and **SortD(** sort *keylistname*, and then sort each *dependlist* by placing its elements in the same order as the corresponding elements in *keylistname*. This lets you sort two-variable data on **X** and keep the data pairs together. All lists must have the same dimension.

The sorted lists are updated in memory.

SortA(*listname***)**
SortD(*listname***)**
SortA(*keylistname,dependlist1*[*,dependlist2,...,dependlist n*]**)**
SortD(*keylistname,dependlist1*[*,dependlist2,...,dependlist n*]**)**

```
{5,4,3}→L3
           {5 4 3}
{1,2,3}→L4
           {1 2 3}
SortA(L3,L4)
              Done
```

```
L3
           {3 4 5}
L4
           {3 2 1}
■
```

Note: SortA(and **SortD(** are the same as **SortA(** and **SortD(** on the LIST OPS menu.

ClrList

ClrList clears (deletes) from memory the elements of one or more *listnames*. **ClrList** also detaches any formula attached to a *listname*.

ClrList *listname1,listname2,...,listname n*

Note: To clear from memory all elements of all list names, use **ClrAllLists** (Chapter 18).

SetUpEditor

With **SetUpEditor** you can set up the stat list editor to display one or more *listnames* in the order that you specify. You can specify zero to 20 *listnames.*

Additionally, if you want to use *listnames* which happen to be archived, the SetUp Editor will automatically unarchive the *listnames* and place them in the stat list editor at the same time.

SetUpEditor [*listname1*,*listname2*,...,*listname n*]

SetUpEditor with one to 20 *listnames* removes all list names from the stat list editor and then stores *listnames* in the stat list editor columns in the specified order, beginning in column **1**.

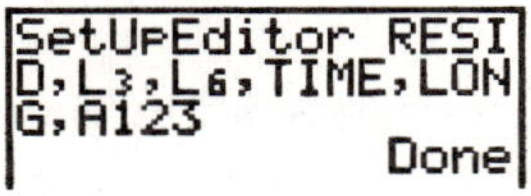

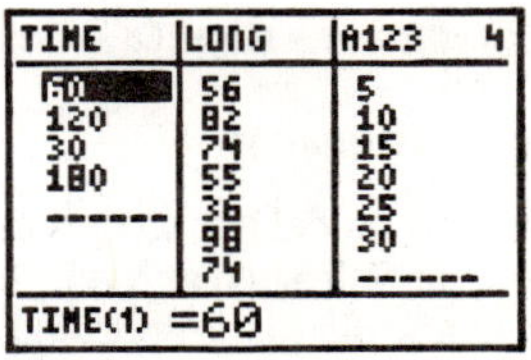

If you enter a *listname* that is not stored in memory already, then *listname* is created and stored in memory; it becomes an item on the **LIST NAMES** menu.

Restoring L1 through L6 to the Stat List Editor

SetUpEditor with no *listnames* removes all list names from the stat list editor and restores list names **L1** through **L6** in the stat list editor columns **1** through **6**.

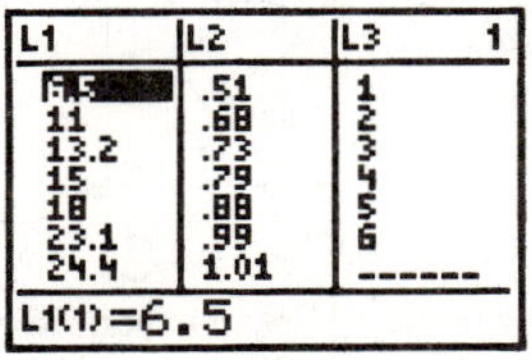

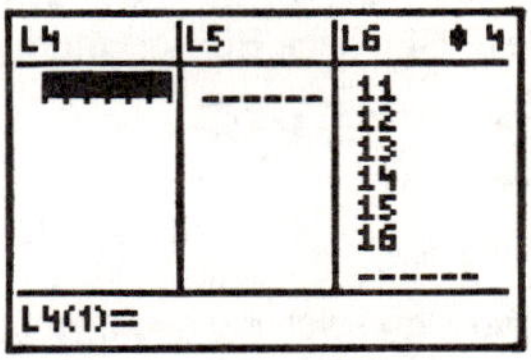

Regression Model Features

STAT CALC menu items **3** through **C** are regression models (page 12-24). The automatic residual list and automatic regression equation features apply to all regression models. Diagnostics display mode applies to some regression models.

Automatic Residual List

When you execute a regression model, the automatic residual list feature computes and stores the residuals to the list name **RESID**. **RESID** becomes an item on the LIST NAMES menu (Chapter 11).

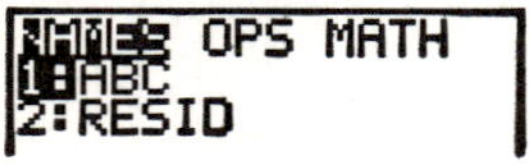

The TI-83 Plus uses the formula below to compute **RESID** list elements. The next section describes the variable **RegEQ**.

RESID = *Ylistname* − **RegEQ**(*Xlistname*)

Automatic Regression Equation

Each regression model has an optional argument, *regequ*, for which you can specify a Y= variable such as **Y1**. Upon execution, the regression equation is stored automatically to the specified Y= variable and the Y= function is selected.

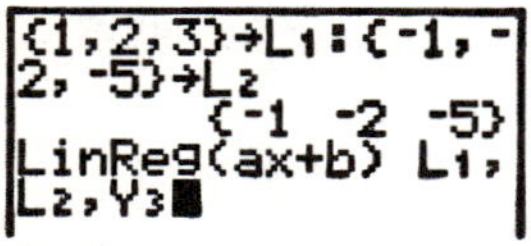

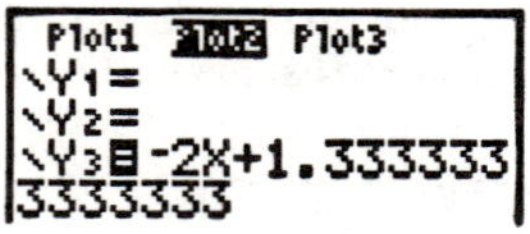

Regardless of whether you specify a Y= variable for *regequ*, the regression equation always is stored to the TI-83 Plus variable **RegEQ**, which is item **1** on the VARS Statistics EQ secondary menu.

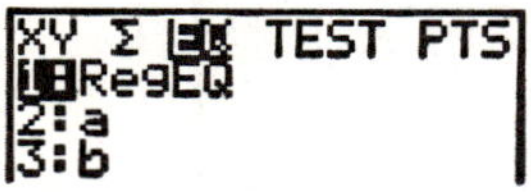

Note: For the regression equation, you can use the fixed-decimal mode setting to control the number of digits stored after the decimal point (Chapter 1). However, limiting the number of digits to a small number could affect the accuracy of the fit.

Diagnostics Display Mode

When you execute some regression models, the TI-83 Plus computes and stores diagnostics values for **r** (correlation coefficient) and r^2 (coefficient of determination) or for R^2 (coefficient of determination).

r and r^2 are computed and stored for these regression models.

LinReg(ax+b)	**LnReg**	**PwrReg**
LinReg(a+bx)	**ExpReg**	

R^2 is computed and stored for these regression models.

QuadReg	**CubicReg**	**QuartReg**

The **r** and r^2 that are computed for **LnReg**, **ExpReg**, and **PwrReg** are based on the linearly transformed data. For example, for **ExpReg** (y=ab^x), **r** and r^2 are computed on ln y=ln a+x(ln b).

By default, these values are not displayed with the results of a regression model when you execute it. However, you can set the diagnostics display mode by executing the **DiagnosticOn** or **DiagnosticOff** instruction. Each instruction is in the CATALOG (Chapter 15).

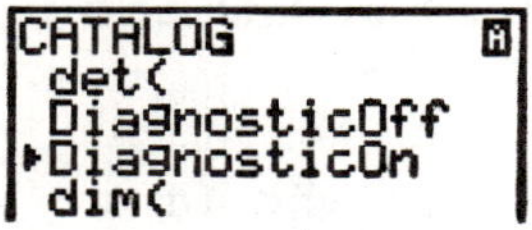

Note: To set **DiagnosticOn** or **DiagnosticOff** from the home screen, press [2nd] [CATALOG], and then select the instruction for the mode you want. The instruction is pasted to the home screen. Press [ENTER] to set the mode.

When **DiagnosticOn** is set, diagnostics are displayed with the results when you execute a regression model.

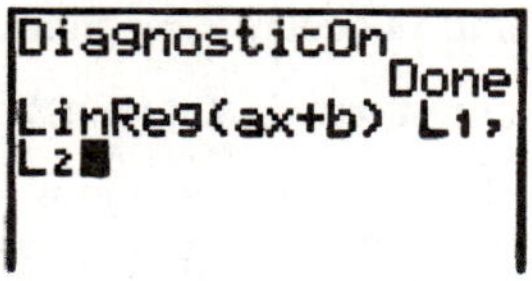

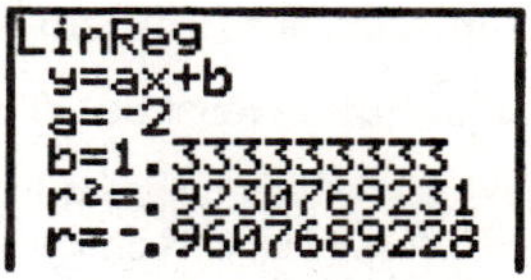

When **DiagnosticOff** is set, diagnostics are not displayed with the results when you execute a regression model.

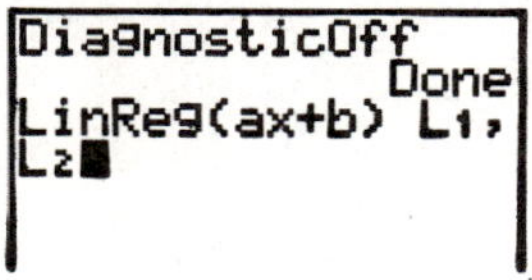

STAT CALC Menu

To display the STAT CALC menu, press [STAT] [▸].

EDIT CALC TESTS	
1: 1-Var Stats	Calculates 1-variable statistics.
2: 2-Var Stats	Calculates 2-variable statistics.
3: Med-Med	Calculates a median-median line.
4: LinReg(ax+b)	Fits a linear model to data.
5: QuadReg	Fits a quadratic model to data.
6: CubicReg	Fits a cubic model to data.
7: QuartReg	Fits a quartic model to data.
8: LinReg(a+bx)	Fits a linear model to data.
9: LnReg	Fits a logarithmic model to data.
0: ExpReg	Fits an exponential model to data.
A: PwrReg	Fits a power model to data.
B: Logistic	Fits a logistic model to data.
C: SinReg	Fits a sinusoidal model to data.

For each STAT CALC menu item, if neither *Xlistname* nor *Ylistname* is specified, then the default list names are **L1** and **L2**. If you do not specify *freqlist*, then the default is **1** occurrence of each list element.

Frequency of Occurrence for Data Points

For most STAT CALC menu items, you can specify a list of data occurrences, or frequencies (*freqlist*).

Each element in *freqlist* indicates how many times the corresponding data point or data pair occurs in the data set you are analyzing.

For example, if **L1={15,12,9,14}** and **ʟFREQ={1,4,1,3}**, then the TI-83 Plus interprets the instruction **1-Var Stats L1**, **ʟFREQ** to mean that **15** occurs once, **12** occurs four times, **9** occurs once, and **14** occurs three times.

Each element in *freqlist* must be ≥ 0, and at least one element must be > **0**.

Noninteger *freqlist* elements are valid. This is useful when entering frequencies expressed as percentages or parts that add up to 1. However, if *freqlist* contains noninteger frequencies, **Sx** and **Sy** are undefined; values are not displayed for **Sx** and **Sy** in the statistical results.

1-Var Stats

1-Var Stats (one-variable statistics) analyzes data with one measured variable. Each element in *freqlist* is the frequency of occurrence for each corresponding data point in *Xlistname*. *freqlist* elements must be real numbers > 0.

1-Var Stats [*Xlistname*,*freqlist*]

```
1-Var Stats L1,L
2█
```

2-Var Stats

2-Var Stats (two-variable statistics) analyzes paired data. *Xlistname* is the independent variable. *Ylistname* is the dependent variable. Each element in *freqlist* is the frequency of occurrence for each data pair (*Xlistname*,*Ylistname*).

2-Var Stats [*Xlistname*,*Ylistname*,*freqlist*]

Med-Med (ax+b)

Med-Med (median-median) fits the model equation y=ax+b to the data using the median-median line (resistant line) technique, calculating the summary points **x1**, **y1**, **x2**, **y2**, **x3**, and **y3**. **Med-Med** displays values for **a** (slope) and **b** (y-intercept).

Med-Med [*Xlistname*,*Ylistname*,*freqlist*,*regequ*]

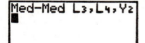

```
Med-Med L3,L4,Y2
█
```

```
Med-Med
 y=ax+b
 a=.875
 b=1.541666667
```

LinReg (ax+b)

LinReg(ax+b) (linear regression) fits the model equation y=ax+b to the data using a least-squares fit. It displays values for **a** (slope) and **b** (y-intercept); when **DiagnosticOn** is set, it also displays values for r^2 and **r**.

LinReg(ax+b) [*Xlistname*,*Ylistname*,*freqlist*,*regequ*]

QuadReg (ax2+bx+c)

QuadReg (quadratic regression) fits the second-degree polynomial $y=ax^2+bx+c$ to the data. It displays values for **a**, **b**, and **c**; when **DiagnosticOn** is set, it also displays a value for R^2. For three data points, the equation is a polynomial fit; for four or more, it is a polynomial regression. At least three data points are required.

QuadReg [*Xlistname*,*Ylistname*,*freqlist*,*regequ*]

CubicReg—(ax^3+bx^2+cx+d)

CubicReg (cubic regression) fits the third-degree polynomial $y=ax^3+bx^2+cx+d$ to the data. It displays values for **a**, **b**, **c**, and **d**; when **DiagnosticOn** is set, it also displays a value for $\mathbf{R^2}$. For four points, the equation is a polynomial fit; for five or more, it is a polynomial regression. At least four points are required.

CubicReg [*Xlistname,Ylistname,freqlist,regequ*]

QuartReg—($ax^4+bx^3+cx^2+ dx+e$)

QuartReg (quartic regression) fits the fourth-degree polynomial $y=ax^4+bx^3+cx^2+dx+e$ to the data. It displays values for **a**, **b**, **c**, **d**, and **e**; when **DiagnosticOn** is set, it also displays a value for $\mathbf{R^2}$. For five points, the equation is a polynomial fit; for six or more, it is a polynomial regression. At least five points are required.

QuartReg [*Xlistname,Ylistname,freqlist,regequ*]

LinReg—(a+bx)

LinReg(a+bx) (linear regression) fits the model equation y=a+bx to the data using a least-squares fit. It displays values for **a** (y-intercept) and **b** (slope); when **DiagnosticOn** is set, it also displays values for $\mathbf{r^2}$ and **r**.

LinReg(a+bx) [*Xlistname,Ylistname,freqlist,regequ*]

LnReg—(a+b ln(x))

LnReg (logarithmic regression) fits the model equation y=a+b ln(x) to the data using a least-squares fit and transformed values ln(x) and y. It displays values for **a** and **b**; when **DiagnosticOn** is set, it also displays values for $\mathbf{r^2}$ and **r**.

LnReg [*Xlistname,Ylistname,freqlist,regequ*]

ExpReg—(ab^x)

ExpReg (exponential regression) fits the model equation $y=ab^x$ to the data using a least-squares fit and transformed values x and ln(y). It displays values for **a** and **b**; when **DiagnosticOn** is set, it also displays values for $\mathbf{r^2}$ and **r**.

ExpReg [*Xlistname,Ylistname,freqlist,regequ*]

PwrReg—(ax^b)

PwrReg (power regression) fits the model equation y=ax^b to the data using a least-squares fit and transformed values ln(x) and ln(y). It displays values for **a** and **b**; when **DiagnosticOn** is set, it also displays values for **r^2** and **r.**

PwrReg [*Xlistname*,*Ylistname*,*freqlist*,*regequ*]

Logistic—c/(1+a∗e^{-bx})

Logistic fits the model equation y=c/(1+a∗e^{-bx}) to the data using an iterative least-squares fit. It displays values for **a**, **b**, and **c**.

Logistic [*Xlistname*,*Ylistname*,*freqlist*,*regequ*]

SinReg—a sin(bx+c)+d

SinReg (sinusoidal regression) fits the model equation y=a sin(bx+c)+d to the data using an iterative least-squares fit. It displays values for **a**, **b**, **c**, and **d**. At least four data points are required. At least two data points per cycle are required in order to avoid aliased frequency estimates.

SinReg [*iterations*,*Xlistname*,*Ylistname*,*period*,*regequ*]

iterations is the maximum number of times the algorithm will iterate to find a solution. The value for *iterations* can be an integer ≥ 1 and ≤ 16; if not specified, the default is 3. The algorithm may find a solution before *iterations* is reached. Typically, larger values for *iterations* result in longer execution times and better accuracy for **SinReg**, and vice versa.

A *period* guess is optional. If you do not specify *period*, the difference between time values in *Xlistname* must be equal and the time values must be ordered in ascending sequential order. If you specify *period*, the algorithm may find a solution more quickly, or it may find a solution when it would not have found one if you had omitted a value for *period*. If you specify *period*, the differences between time values in *Xlistname* can be unequal.

Note: The output of **SinReg** is always in radians, regardless of the **Radian/Degree** mode setting.

A **SinReg** example is shown on the next page.

SinReg Example: Daylight Hours in Alaska for One Year

Compute the regression model for the number of hours of daylight in Alaska during one year.

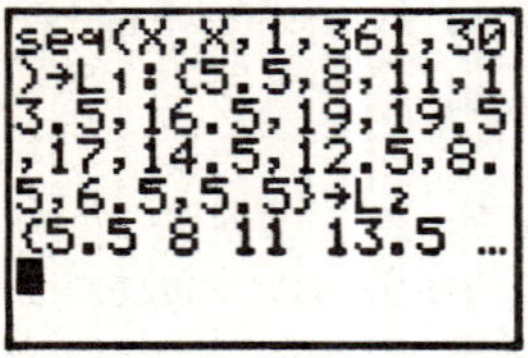

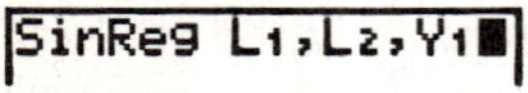

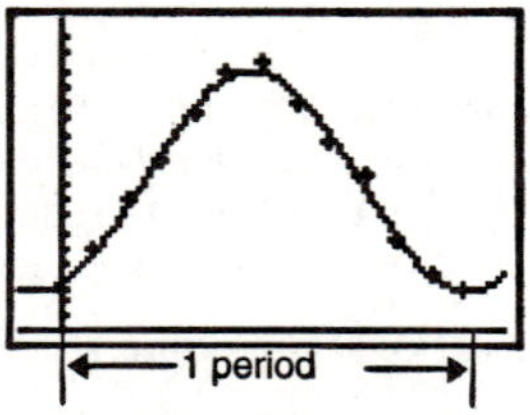

With noisy data, you will achieve better convergence results when you specify an accurate estimate for *period*. You can obtain a *period* guess in either of two ways.

- Plot the data and trace to determine the x-distance between the beginning and end of one complete period, or cycle. The illustration above and to the right graphically depicts a complete period, or cycle.
- Plot the data and trace to determine the x-distance between the beginning and end of N complete periods, or cycles. Then divide the total distance by N.

After your first attempt to use **SinReg** and the default value for *iterations* to fit the data, you may find the fit to be approximately correct, but not optimal. For an optimal fit, execute **SinReg 16,***Xlistname***,***Ylistname***,2π / *b*** where *b* is the value obtained from the previous **SinReg** execution.

Statistical Variables

The statistical variables are calculated and stored as indicated below. To access these variables for use in expressions, press [VARS], and select **5:Statistics**. Then select the VARS menu shown in the column below under VARS menu. If you edit a list or change the type of analysis, all statistical variables are cleared.

Variables	1-Var Stats	2-Var Stats	Other	VARS menu
mean of x values	$\bar{x}$	$\bar{x}$		XY
sum of x values	Σx	Σx		Σ
sum of x^2 values	Σx^2	Σx^2		Σ
sample standard deviation of x	Sx	Sx		XY
population standard deviation of **x**	σx	σx		XY
number of data points	n	n		XY
mean of **y** values		$\bar{y}$		XY
sum of **y** values		Σy		Σ
sum of y^2 values		Σy^2		Σ
sample standard deviation of **y**		Sy		XY
population standard deviation of **y**		σy		XY
sum of **x * y**		Σxy		Σ
minimum of **x** values	minX	minX		XY
maximum of **x** values	maxX	maxX		XY
minimum of **y** values		minY		XY
maximum of **y** values		maxY		XY
1st quartile	Q_1			PTS
median	Med			PTS
3rd quartile	Q_3			PTS
regression/fit coefficients			a, b	EQ
polynomial, **Logistic**, and **SinReg** coefficients			a, b, c, d, e	EQ
correlation coefficient			r	EQ
coefficient of determination			r^2, R^2	EQ
regression equation			RegEQ	EQ
summary points (**Med-Med** only)			x1, y1, x2, y2, x3, y3	PTS

Q_1 and Q_3

The first quartile (**Q_1**) is the median of points between **minX** and **Med** (median). The third quartile (**Q_3**) is the median of points between **Med** and **maxX**.

Entering Stat Data

You can enter statistical data, calculate statistical results, and fit models to data from a program. You can enter statistical data into lists directly within the program (Chapter 11).

```
PROGRAM:STATS
:{1,2,3}→L1
:{-1,-2,-5}→L2
```

Statistical Calculations

To perform a statistical calculation from a program, follow these steps.

1. On a blank line in the program editor, select the type of calculation from the STAT CALC menu.
2. Enter the names of the lists to use in the calculation. Separate the list names with a comma.
3. Enter a comma and then the name of a Y= variable, if you want to store the regression equation to a Y= variable.

```
PROGRAM:STATS
:{1,2,3}→L1
:{-1,-2,-5}→L2
:LinReg(ax+b) L1
,L2,Y2
:■
```

Steps for Plotting Statistical Data in Lists

You can plot statistical data that is stored in lists. The six types of plots available are scatter plot, xyLine, histogram, modified box plot, regular box plot, and normal probability plot. You can define up to three plots.

To plot statistical data in lists, follow these steps.

1. Store the stat data in one or more lists.
2. Select or deselect Y= functions as appropriate.
3. Define the stat plot.
4. Turn on the plots you want to display.
5. Define the viewing window.
6. Display and explore the graph.

(Scatter)

Scatter plots plot the data points from **Xlist** and **Ylist** as coordinate pairs, showing each point as a box (□), cross (+), or dot (•). **Xlist** and **Ylist** must be the same length. You can use the same list for **Xlist** and **Ylist**.

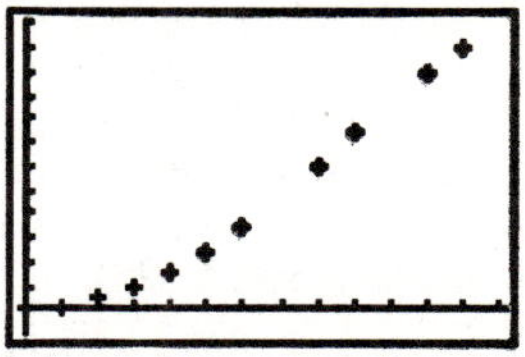

(xyLine)

xyLine is a scatter plot in which the data points are plotted and connected in order of appearance in **Xlist** and **Ylist**. You may want to use **SortA(** or **SortD(** to sort the lists before you plot them (page 12-20).

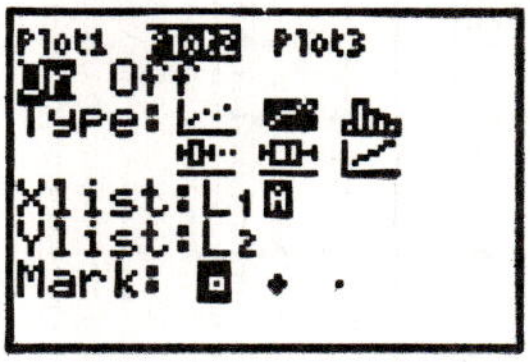

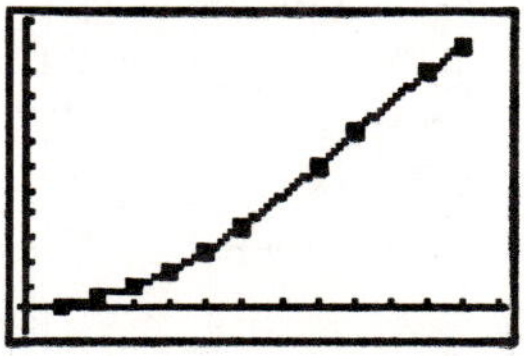

(Histogram)

Histogram plots one-variable data. The **Xscl** window variable value determines the width of each bar, beginning at **Xmin**. **ZoomStat** adjusts **Xmin**, **Xmax**, **Ymin**, and **Ymax** to include all values, and also adjusts **Xscl**. The inequality (**Xmax** − **Xmin**) / **Xscl** ≤ 47 must be true. A value that occurs on the edge of a bar is counted in the bar to the right.

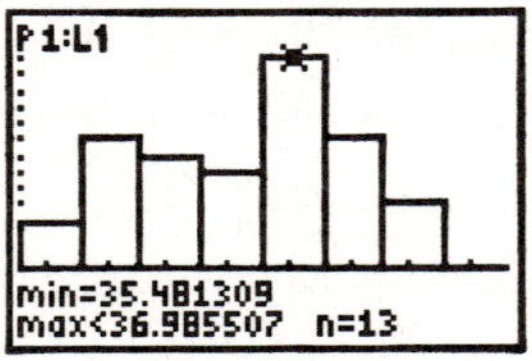

—(ModBoxplot)

ModBoxplot (modified box plot) plots one-variable data, like the regular box plot, except points that are 1.5 ∗ Interquartile Range beyond the quartiles. (The Interquartile Range is defined as the difference between the third quartile **Q3** and the first quartile **Q1**.) These points are plotted individually beyond the whisker, using the **Mark** (□ or + or •) you select. You can trace these points, which are called outliers.

The prompt for outlier points is **x=**, except when the outlier is the maximum point (**maxX**) or the minimum point (**minX**). When outliers exist, the end of each whisker will display **x=**. When no outliers exist, **minX** and **maxX** are the prompts for the end of each whisker. **Q1**, **Med** (median), and **Q3** define the box (page 12-29).

Box plots are plotted with respect to **Xmin** and **Xmax**, but ignore **Ymin** and **Ymax**. When two box plots are plotted, the first one plots at the top of the screen and the second plots in the middle. When three are plotted, the first one plots at the top, the second in the middle, and the third at the bottom.

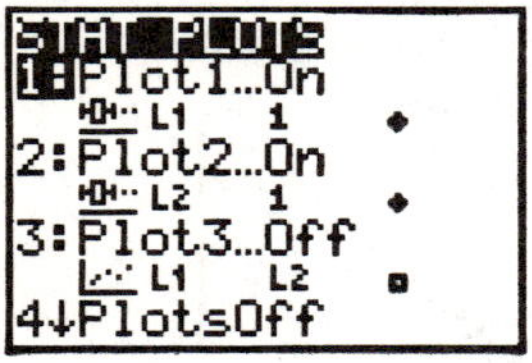

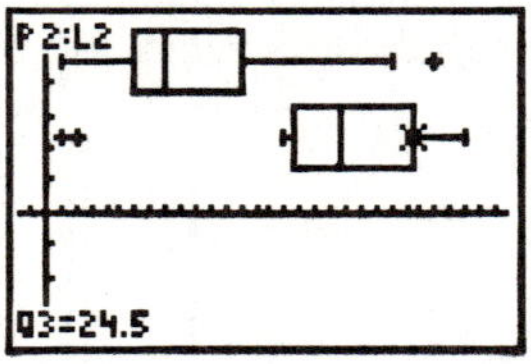

(Boxplot)

Boxplot (regular box plot) plots one-variable data. The whiskers on the plot extend from the minimum data point in the set (**minX**) to the first quartile (**Q1**) and from the third quartile (**Q3**) to the maximum point (**maxX**). The box is defined by **Q1**, **Med** (median), and **Q3** (page 12-29).

Box plots are plotted with respect to **Xmin** and **Xmax**, but ignore **Ymin** and **Ymax**. When two box plots are plotted, the first one plots at the top of the screen and the second plots in the middle. When three are plotted, the first one plots at the top, the second in the middle, and the third at the bottom.

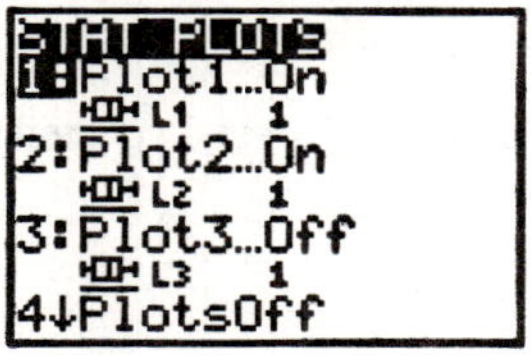

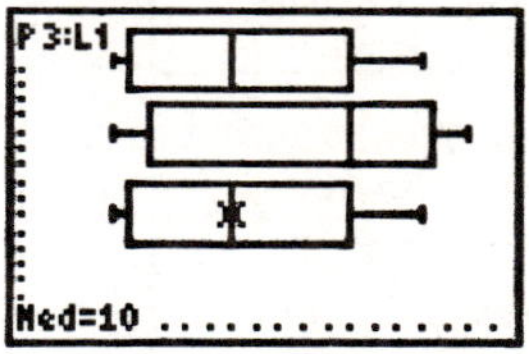

(NormProbPlot)

NormProbPlot (normal probability plot) plots each observation **X** in **Data List** versus the corresponding quantile **z** of the standard normal distribution. If the plotted points lie close to a straight line, then the plot indicates that the data are normal.

Enter a valid list name in the **Data List** field. Select **X** or **Y** for the **Data Axis** setting.

- If you select **X**, the TI-83 Plus plots the data on the x-axis and the z-values on the y-axis.
- If you select **Y**, the TI-83 Plus plots the data on the y-axis and the z-values on the x-axis.

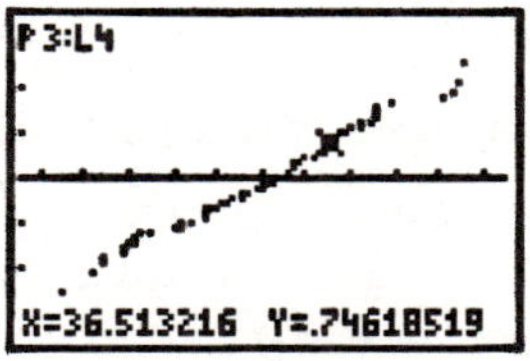

Defining the Plots

To define a plot, follow these steps.

1. Press [2nd] [STAT PLOT]. The **STAT PLOTS** menu is displayed with the current plot definitions.

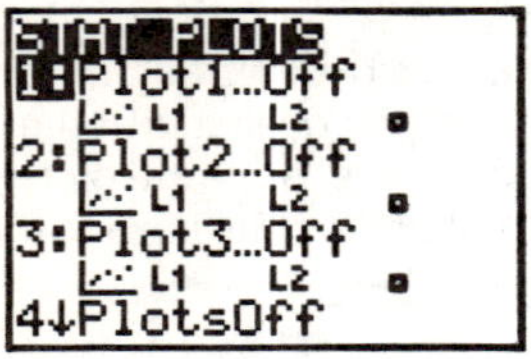

2. Select the plot you want to use. The stat plot editor is displayed for the plot you selected.

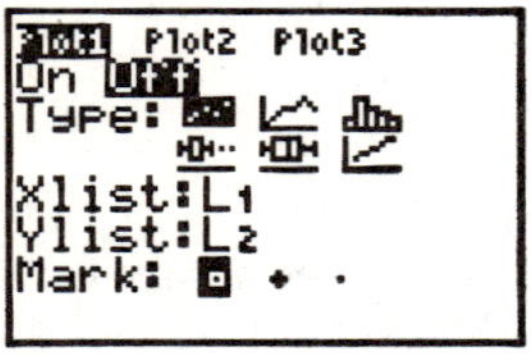

3. Press [ENTER] to select **On** if you want to plot the statistical data immediately. The definition is stored whether you select **On** or **Off**.

4. Select the type of plot. Each type prompts for the options checked in this table.

Plot Type	XList	YList	Mark	Freq	Data List	Data Axis
Scatter	☑	☑	☑	☐	☐	☐
xyLine	☑	☑	☑	☐	☐	☐
Histogram	☑	☐	☐	☑	☐	☐
ModBoxplot	☑	☐	☑	☑	☐	☐
Boxplot	☑	☐	☐	☑	☐	☐
NormProbPlot	☐	☐	☑	☐	☑	☑

5. Enter list names or select options for the plot type.

- **Xlist** (list name containing independent data)
- **Ylist** (list name containing dependent data)
- **Mark** (□ or **+** or •)
- **Freq** (frequency list for **Xlist** elements; default is **1**)
- **Data List** (list name for **NormProbPlot**)
- **Data Axis** (axis on which to plot **Data List**)

Displaying Other Stat Plot Editors

Each stat plot has a unique stat plot editor. The name of the current stat plot (**Plot1**, **Plot2**, or **Plot3**) is highlighted in the top line of the stat plot editor. To display the stat plot editor for a different plot, press [▲], [▶], and [◀] to move the cursor onto the name in the top line, and then press [ENTER]. The stat plot editor for the selected plot is displayed, and the selected name remains highlighted.

Turning On and Turning Off Stat Plots

PlotsOn and **PlotsOff** allow you to turn on or turn off stat plots from the home screen or a program. With no plot number, **PlotsOn** turns on all plots and **PlotsOff** turns off all plots. With one or more plot numbers (**1**, **2**, and **3**), **PlotsOn** turns on specified plots, and **PlotsOff** turns off specified plots.

PlotsOff [1,2,3]
PlotsOn [1,2,3]

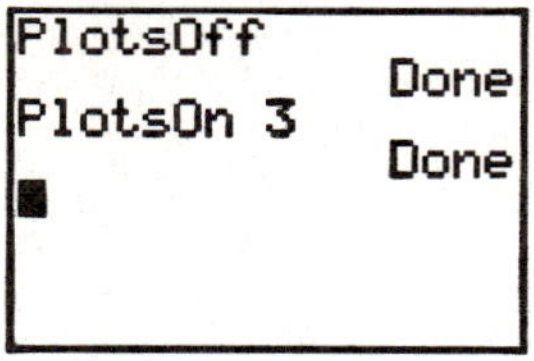

Note: You also can turn on and turn off stat plots in the top line of the Y= editor (Chapter 3).

Defining the Viewing Window

Stat plots are displayed on the current graph. To define the viewing window, press [WINDOW] and enter values for the window variables. **ZoomStat** redefines the viewing window to display all statistical data points.

Tracing a Stat Plot

When you trace a scatter plot or xyLine, tracing begins at the first element in the lists.

When you trace a histogram, the cursor moves from the top center of one column to the top center of the next, starting at the first column.

When you trace a box plot, tracing begins at **Med** (the median). Press [◄] to trace to **Q1** and **minX**. Press [►] to trace to **Q3** and **maxX**.

When you press [▲] or [▼] to move to another plot or to another **Y=** function, tracing moves to the current or beginning point on that plot (not the nearest pixel).

The **ExprOn/ExprOff** format setting applies to stat plots (Chapter 3).When **ExprOn** is selected, the plot number and plotted data lists are displayed in the top-left corner.

Defining a Stat Plot in a Program

To display a stat plot from a program, define the plot, and then display the graph.

To define a stat plot from a program, begin on a blank line in the program editor and enter data into one or more lists; then, follow these steps.

1. Press [2nd] [STAT PLOT] to display the STAT PLOTS menu.

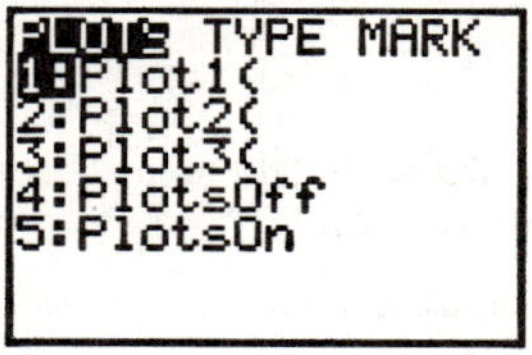

2. Select the plot to define, which pastes **Plot1(**, **Plot2(**, or **Plot3(** to the cursor location.

3. Press [2nd] [STAT PLOT] [▸] to display the STAT TYPE menu.

4. Select the type of plot, which pastes the name of the plot type to the cursor location.

Defining a Stat Plot in a Program (cont.)

5. Press [,]. Enter the list names, separated by commas.
6. Press [,] [2nd] [STAT PLOT] [◄] to display the STAT PLOT MARK menu. (This step is not necessary if you selected **3:Histogram** or **5:Boxplot** in step 4.)

```
PLOTS TYPE MARK
1:□
2:+
3:·
```

Select the type of mark (□ or + or ·) for each data point. The selected mark symbol is pasted to the cursor location.

7. Press [)] [ENTER] to complete the command line.

```
PROGRAM:PLOT
:{1,2,3,4}→L1
:{5,6,7,8}→L2
:Plot2(Scatter,L
1,L2,□)
:█
```

Displaying a Stat Plot from a Program

To display a plot from a program, use the **DispGraph** instruction (Chapter 16) or any of the ZOOM instructions (Chapter 3).

```
PROGRAM:PLOT
:{1,2,3,4}→L1
:{5,6,7,8}→L2
:Plot2(Scatter,L
1,L2,□)
:DispGraph
:█
```

```
PROGRAM:PLOT
:{1,2,3,4}→L1
:{5,6,7,8}→L2
:Plot2(Scatter,L
1,L2,□)
:ZoomStat
:█
```

13 Inferential Statistics and Distributions

Contents

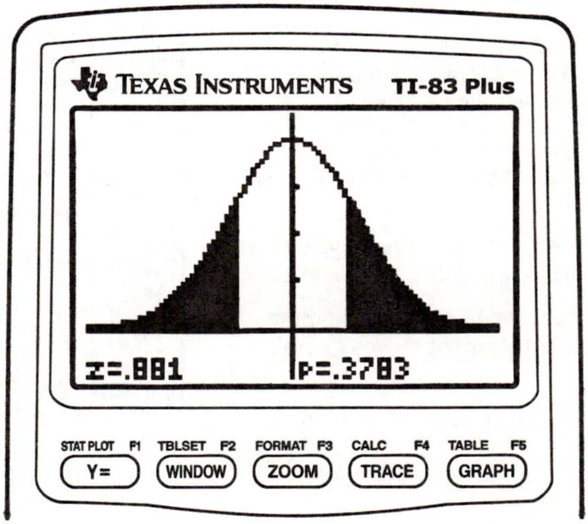

Getting Started is a fast-paced introduction. Read the chapter for details.

Suppose you want to estimate the mean height of a population of women given the random sample below. Because heights among a biological population tend to be normally distributed, a *t* distribution confidence interval can be used when estimating the mean. The 10 height values below are the first 10 of 90 values, randomly generated from a normally distributed population with an assumed mean of 165.1 centimetres and a standard deviation of 6.35 centimetres (randNorm(165.1,6.35,90) with a seed of 789).

Height (in centimetres) of Each of 10 Women

169.43 168.33 159.55 169.97 159.79 181.42 171.17 162.04 167.15 159.53

1. Press [STAT] [ENTER] to display the stat list editor.

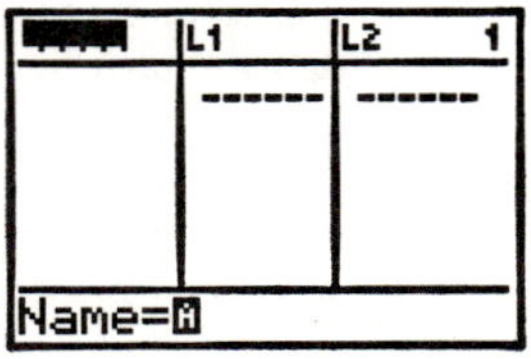

   Press [▲] to move the cursor onto **L1**, and then press [2nd] [INS]. The **Name=** prompt is displayed on the bottom line. The Ⓐ cursor indicates that alpha-lock is on. The existing list name columns shift to the right.

   Note: Your stat editor may not look like the one pictured here, depending on the lists you have already stored.

2. Enter [H] [G] [H] [T] at the **Name=** prompt, and then press [ENTER]. The list to which you will store the women's height data is created.

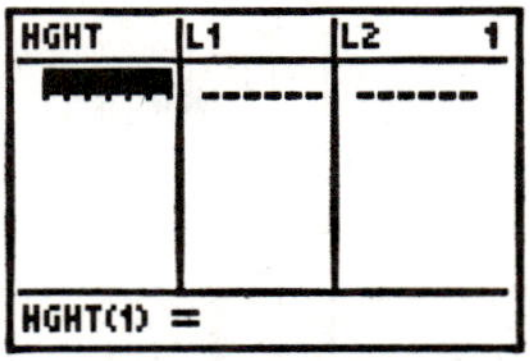

   Press [▼] to move the cursor onto the first row of the list. **HGHT(1)=** is displayed on the bottom line.

3. Press **169** [.] **43** to enter the first height value. As you enter it, it is displayed on the bottom line.

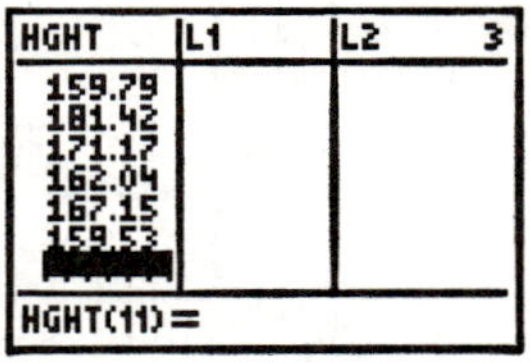

   Press [ENTER]. The value is displayed in the first row, and the rectangular cursor moves to the next row.

   Enter the other nine height values the same way.

4. Press [STAT] [◄] to display the **STAT TESTS** menu, and then press [▼] until **8:TInterval** is highlighted.

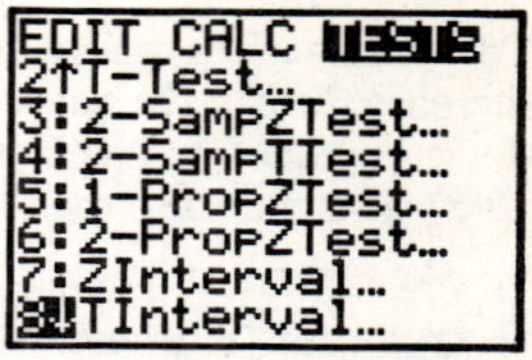

5. Press [ENTER] to select **8:TInterval**. The inferential stat editor for **TInterval** is displayed. If **Data** is not selected for **Inpt:**, press [◄] [ENTER] to select **Data**.

   Press [▼] and [H] [G] [H] [T] at the **List:** prompt (alpha-lock is on).

   Press [▼] [▼] [.] **99** to enter a 99 percent confidence level at the **C-Level:** prompt.

6. Press [▼] to move the cursor onto **Calculate**, and then press [ENTER]. The confidence interval is calculated, and the **TInterval** results are displayed on the home screen.

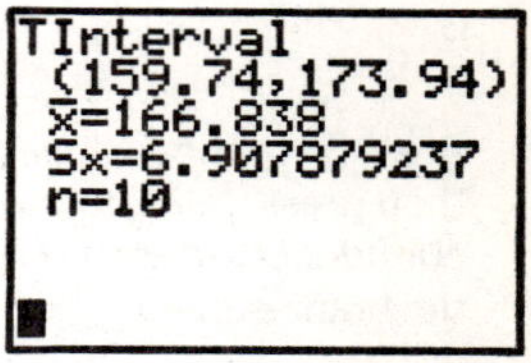

Interpret the results.

The first line, **(159.74,173.94)**, shows that the 99 percent confidence interval for the population mean is between about 159.74 centimetres and 173.94 centimetres. This is about a 14.2 centimetres spread.

The .99 confidence level indicates that in a very large number of samples, we expect 99 percent of the intervals calculated to contain the population mean. The actual mean of the population sampled is 165.1 centimetres (introduction; page 13-2), which is in the calculated interval.

The second line gives the mean height of the sample $\bar{x}$ used to compute this interval. The third line gives the sample standard deviation **Sx**. The bottom line gives the sample size **n**.

To obtain a more precise bound on the population mean μ of women's heights, increase the sample size to 90. Use a sample mean x̄ of 163.8 and sample standard deviation **Sx** of 7.1 calculated from the larger random sample (introduction; page 13-2). This time, use the **Stats** (summary statistics) input option.

7. Press [STAT] [◄] **8** to display the inferential stat editor for **TInterval**.

   Press [►] [ENTER] to select **Inpt:Stats**. The editor changes so that you can enter summary statistics as input.

8. Press [▼] **163** [.] **8** [ENTER] to store 163.8 to **x̄**.

   Press **7** [.] **1** [ENTER] to store 7.1 to **Sx**.

   Press **90** [ENTER] to store 90 to **n**.

9. Press [▼] to move the cursor onto **Calculate**, and then press [ENTER] to calculate the new 99 percent confidence interval. The results are displayed on the home screen.

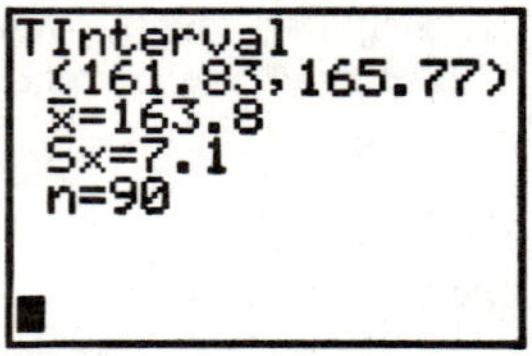

If the height distribution among a population of women is normally distributed with a mean μ of 165.1 centimetres and a standard deviation σ of 6.35 centimetres, what height is exceeded by only 5 percent of the women (the 95th percentile)?

10. Press [CLEAR] to clear the home screen.

    Press [2nd] [DISTR] to display the DISTR (distributions) menu.

11. Press **3** to paste **invNorm(** to the home screen.

Press [.] **95** [,] **165** [.] **1** [,] **6** [.] **35** [)] [ENTER].

.95 is the area, **165.1** is µ, and **6.35** is σ.

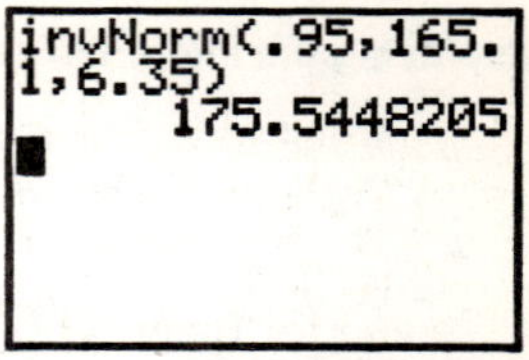

The result is displayed on the home screen; it shows that five percent of the women are taller than 175.5 centimetres.

Now graph and shade the top 5 percent of the population.

12. Press [WINDOW] and set the window variables to these values.

Xmin=145	**Ymin=⁻.02**	**Xres=1**
Xmax=185	**Ymax=.08**	
Xscl=5	**Yscl=0**	

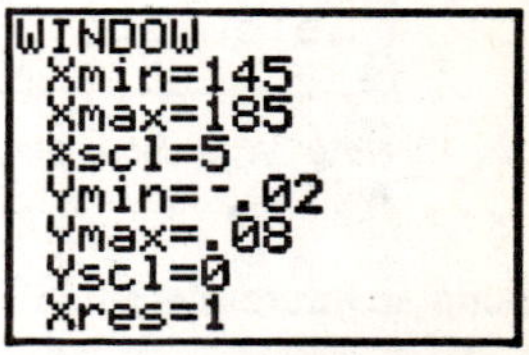

13. Press [2nd] [DISTR] [▸] to display the DISTR DRAW menu.

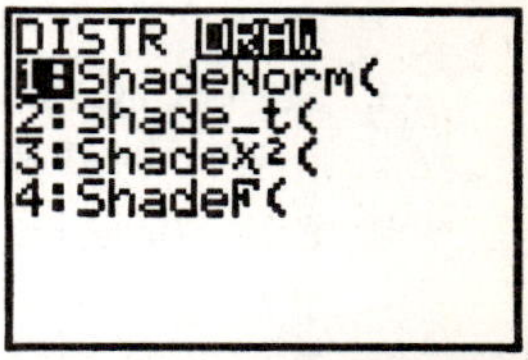

14. Press [ENTER] to paste **ShadeNorm(** to the home screen.

Press [2nd] [ANS] [,] **1** [2nd] [EE] **99** [,] **165** [.] **1** [,] **6** [.] **35** [)].

Ans (175.5448205 from step 11) is the lower bound. 1E99 is the upper bound. The normal curve is defined by a mean µ of 165.1 and a standard deviation σ of 6.35.

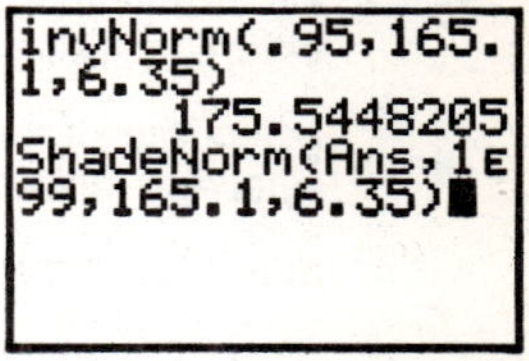

15. Press [ENTER] to plot and shade the normal curve.

Area is the area above the 95th percentile. **low** is the lower bound. **up** is the upper bound.

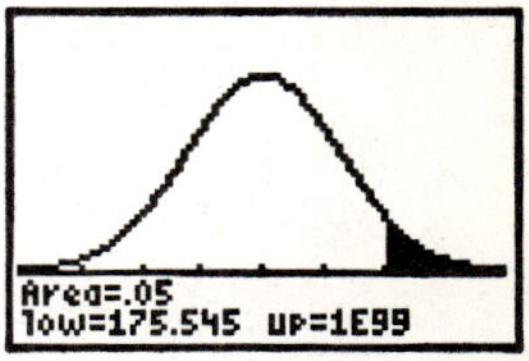

Displaying the Inferential Stat Editors

When you select a hypothesis test or confidence interval instruction from the home screen, the appropriate inferential statistics editor is displayed. The editors vary according to each test or interval's input requirements. Below is the inferential stat editor for **T-Test**.

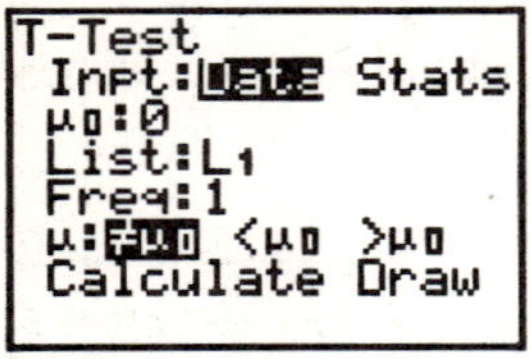

Note: When you select the **ANOVA(** instruction, it is pasted to the home screen. **ANOVA(** does not have an editor screen.

Using an Inferential Stat Editor

To use an inferential stat editor, follow these steps.

1. Select a hypothesis test or confidence interval from the STAT TESTS menu. The appropriate editor is displayed.
2. Select **Data** or **Stats** input, if the selection is available. The appropriate editor is displayed.
3. Enter real numbers, list names, or expressions for each argument in the editor.
4. Select the alternative hypothesis (≠, **<**, or **>**) against which to test, if the selection is available.
5. Select **No** or **Yes** for the **Pooled** option, if the selection is available.
6. Select **Calculate** or **Draw** (when **Draw** is available) to execute the instruction.
   - When you select **Calculate**, the results are displayed on the home screen.
   - When you select **Draw**, the results are displayed in a graph.

This chapter describes the selections in the above steps for each hypothesis test and confidence interval instruction.

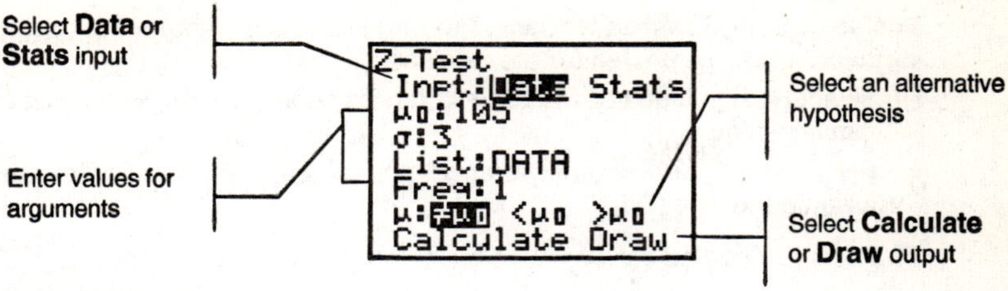

Selecting Data or Stats

Most inferential stat editors prompt you to select one of two types of input. (**1-PropZInt** and **2-PropZTest**, **1-PropZInt** and **2-PropZInt**, χ^2**-Test**, and **LinRegTTest** do not.)

- Select **Data** to enter the data lists as input.
- Select **Stats** to enter summary statistics, such as $\bar{x}$, **Sx**, and **n**, as input.

To select **Data** or **Stats**, move the cursor to either **Data** or **Stats**, and then press [ENTER].

Entering the Values for Arguments

Inferential stat editors require a value for every argument. If you do not know what a particular argument symbol represents, see the tables on pages 13-26 and 13-27.

When you enter values in any inferential stat editor, the TI-83 Plus stores them in memory so that you can run many tests or intervals without having to reenter every value.

Selecting an Alternative Hypothesis (≠ < >)

Most of the inferential stat editors for the hypothesis tests prompt you to select one of three alternative hypotheses.

- The first is a ≠ alternative hypothesis, such as μ≠μ**0** for the **Z-Test**.
- The second is a **<** alternative hypothesis, such as μ**1**<μ**2** for the **2-SampTTest**.
- The third is a **>** alternative hypothesis, such as **p1>p2** for the **2-PropZTest**.

To select an alternative hypothesis, move the cursor to the appropriate alternative, and then press [ENTER].

Selecting the Pooled Option

Pooled (**2-SampTTest** and **2-SampTInt** only) specifies whether the variances are to be pooled for the calculation.

- Select **No** if you do not want the variances pooled. Population variances can be unequal.
- Select **Yes** if you want the variances pooled. Population variances are assumed to be equal.

To select the **Pooled** option, move the cursor to **Yes**, and then press [ENTER].

Selecting Calculate or Draw for a Hypothesis Test

After you have entered all arguments in an inferential stat editor for a hypothesis test, you must select whether you want to see the calculated results on the home screen (**Calculate**) or on the graph screen (**Draw**).

- **Calculate** calculates the test results and displays the outputs on the home screen.
- **Draw** draws a graph of the test results and displays the test statistic and p-value with the graph. The window variables are adjusted automatically to fit the graph.

To select **Calculate** or **Draw**, move the cursor to either **Calculate** or **Draw**, and then press [ENTER]. The instruction is immediately executed.

Selecting Calculate for a Confidence Interval

After you have entered all arguments in an inferential stat editor for a confidence interval, select **Calculate** to display the results. The **Draw** option is not available.

When you press [ENTER], **Calculate** calculates the confidence interval results and displays the outputs on the home screen.

Bypassing the Inferential Stat Editors

To paste a hypothesis test or confidence interval instruction to the home screen without displaying the corresponding inferential stat editor, select the instruction you want from the CATALOG menu. Appendix A describes the input syntax for each hypothesis test and confidence interval instruction.

```
2-SampZTest(
```

Note: You can paste a hypothesis test or confidence interval instruction to a command line in a program. From within the program editor, select the instruction from either the CATALOG (Chapter 15) or the STAT TESTS menu.

STAT TESTS Menu

To display the STAT TESTS menu, press [STAT] [◄]. When you select an inferential statistics instruction, the appropriate inferential stat editor is displayed.

Most STAT TESTS instructions store some output variables to memory. Most of these output variables are in the TEST secondary menu (VARS menu; **5:Statistics**). For a list of these variables, see page 13-28.

EDIT CALC TESTS	
1: Z-Test...	Test for 1 μ, known σ
2: T-Test...	Test for 1 μ, unknown σ
3: 2-SampZTest...	Test comparing 2 μ's, known σ's
4: 2-SampTTest...	Test comparing 2 μ's, unknown σ's
5: 1-PropZTest...	Test for 1 proportion
6: 2-PropZTest...	Test comparing 2 proportions
7: ZInterval...	Confidence interval for 1 μ, known σ
8: TInterval...	Confidence interval for 1 μ, unknown σ
9: 2-SampZInt...	Confidence interval for difference of 2 μ's, known σ's
0: 2-SampTInt...	Confidence interval for difference of 2 μ's, unknown σ's
A: 1-PropZInt...	Confidence interval for 1 proportion
B: 2-PropZInt...	Confidence interval for difference of 2 proportions
C: χ2-Test...	Chi-square test for 2-way tables
D: 2-SampFTest...	Test comparing 2 σ's
E: LinRegTTest...	*t* test for regression slope and ρ
F: ANOVA(	One-way analysis of variance

Note: When a new test or interval is computed, all previous output variables are invalidated.

Inferential Stat Editors for the STAT TESTS Instructions

In this chapter, the description of each STAT TESTS instruction shows the unique inferential stat editor for that instruction with example arguments.

- Descriptions of instructions that offer the **Data/Stats** input choice show both types of input screens.
- Descriptions of instructions that do not offer the **Data/Stats** input choice show only one input screen.

The description then shows the unique output screen for that instruction with the example results.

- Descriptions of instructions that offer the **Calculate/Draw** output choice show both types of screens: calculated and graphic results.
- Descriptions of instructions that offer only the **Calculate** output choice show the calculated results on the home screen.

Z-Test

Z-Test (one-sample *z* test; item **1**) performs a hypothesis test for a single unknown population mean μ when the population standard deviation σ is known. It tests the null hypothesis H_0: $\mu=\mu_0$ against one of the alternatives below.

- H_a: $\mu\neq\mu_0$ (**μ:≠μ0**)
- H_a: $\mu<\mu_0$ (**μ:<μ0**)
- H_a: $\mu>\mu_0$ (**μ:>μ0**)

In the example:

L1={299.4 297.7 301 298.9 300.2 297}

Data | Stats

Input:

Calculated results:

Drawn results:

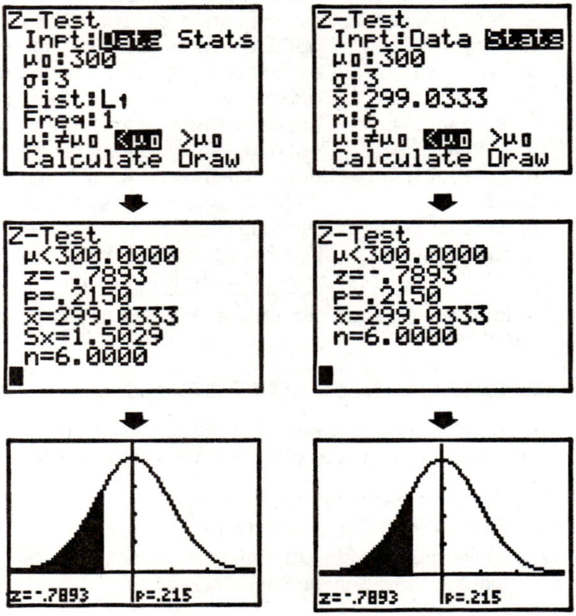

Note: All examples on pages13-10 through 13-25 assume a fixed-decimal mode setting of **4** (Chapter 1). If you set the decimal mode to **Float** or a different fixed-decimal setting, your output may differ from the output in the examples.

T-Test

T-Test (one-sample *t* test; item **2**) performs a hypothesis test for a single unknown population mean μ when the population standard deviation **σ is** unknown. It tests the null hypothesis H_0: $\mu=\mu_0$ against one of the alternatives below.

- H_a: $\mu\neq\mu_0$ (**μ:≠μ0**)
- H_a: $\mu<\mu_0$ (**μ:<μ0**)
- H_a: $\mu>\mu_0$ (**μ:>μ0**)

In the example:

TEST={91.9 97.8 111.4 122.3 105.4 95}

Data | **Stats**

Input: / Calculated results: / Drawn results:

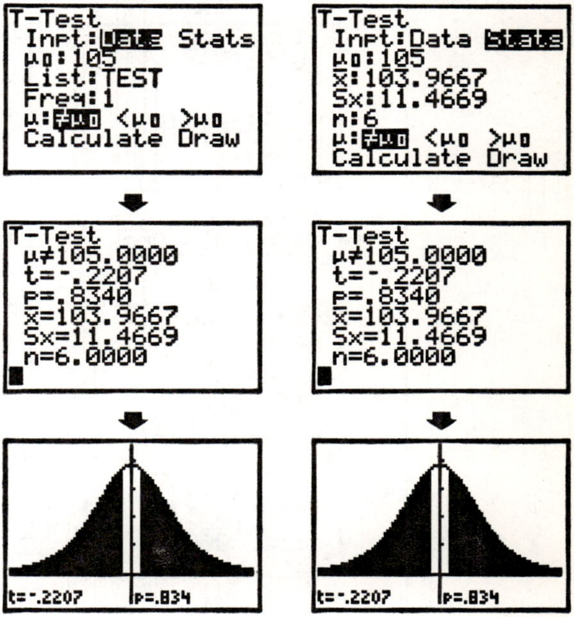

2-SampZTest

2-SampZTest (two-sample *z* test; item **3**) tests the equality of the means of two populations (μ_1 and μ_2) based on independent samples when both population standard deviations (σ_1 and σ_2) are known. The null hypothesis H_0: $\mu_1=\mu_2$ is tested against one of the alternatives below.

- H_a: $\mu_1\neq\mu_2$ (**μ1:≠μ2**)
- H_a: $\mu_1<\mu_2$ (**μ1:<μ2**)
- H_a: $\mu_1>\mu_2$ (**μ1:>μ2**)

In the example:

LISTA={154 109 137 115 140}
LISTB={108 115 126 92 146}

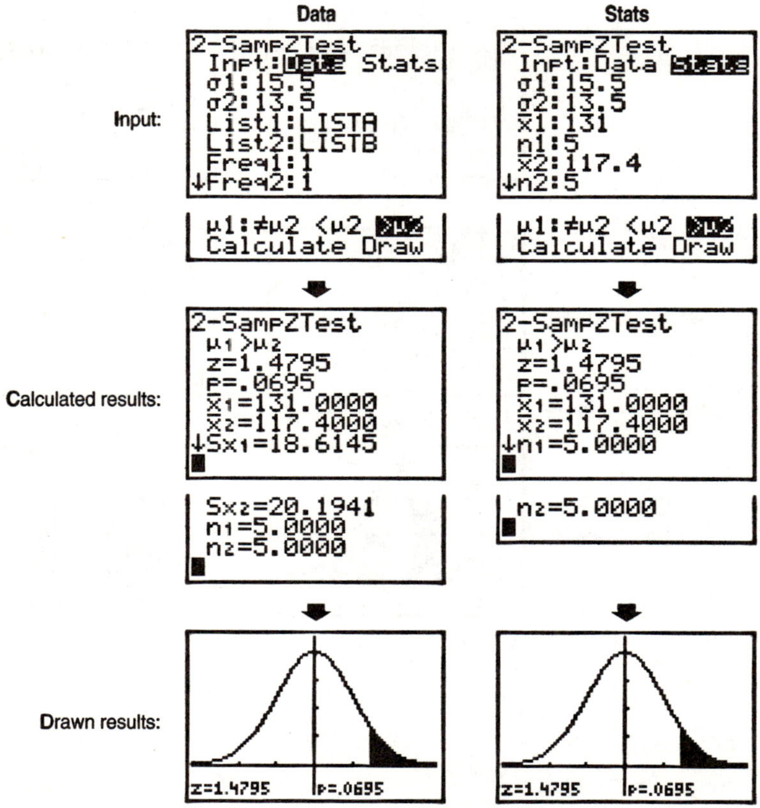

2-SampTTest

2-SampTTest (two-sample *t* test; item **4**) tests the equality of the means of two populations (μ_1 and μ_2) based on independent samples when neither population standard deviation (σ_1 or σ_2) is known. The null hypothesis H_0: $\mu_1=\mu_2$ is tested against one of the alternatives below.

- H_a: $\mu_1 \neq \mu_2$ (**µ1:≠µ2**)
- H_a: $\mu_1 < \mu_2$ (**µ1:<µ2**)
- H_a: $\mu_1 > \mu_2$ (**µ1:>µ2**)

In the example:

SAMP1={12.207 16.869 25.05 22.429 8.456 10.589}
SAMP2={11.074 9.686 12.064 9.351 8.182 6.642}

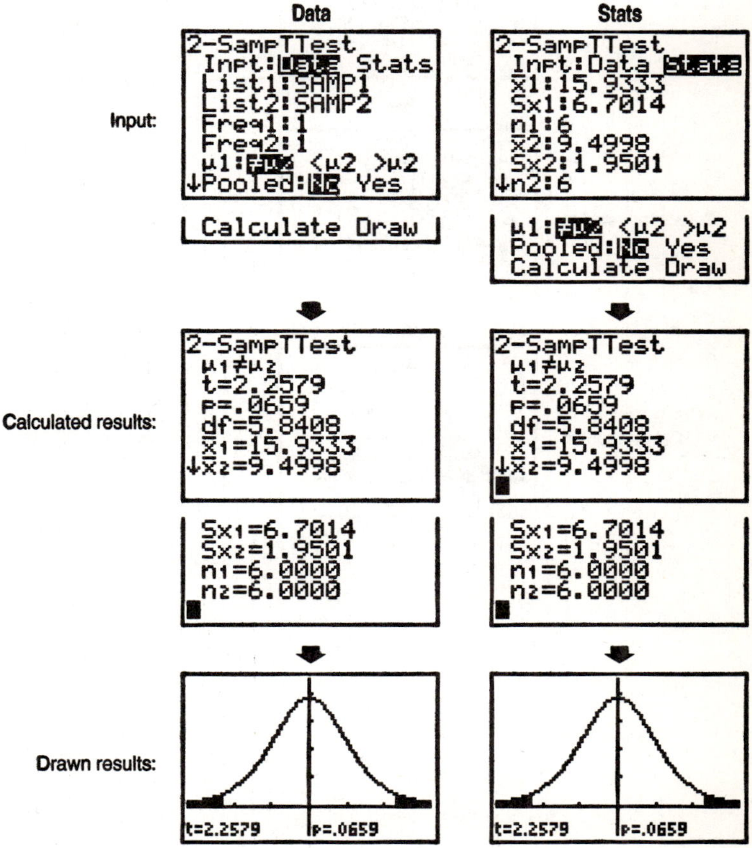

1-PropZTest

1-PropZTest (one-proportion *z* test; item **5**) computes a test for an unknown proportion of successes (prop). It takes as input the count of successes in the sample *x* and the count of observations in the sample ***n***. **1-PropZTest** tests the null hypothesis H_0: prop=p_0 against one of the alternatives below.

- H_a: prop≠p_0 (**prop:≠p0**)
- H_a: prop<p_0 (**prop:<p0**)
- H_a: prop>p_0 (**prop:>p0**)

Input:

Calculated results:

Drawn results:

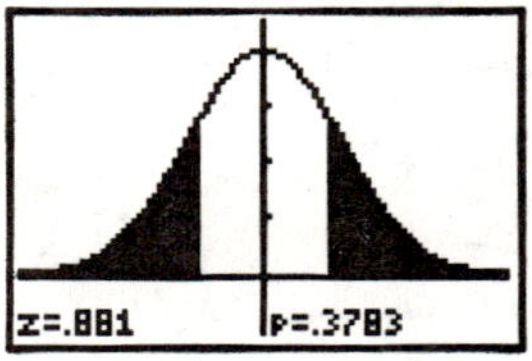

2-PropZTest

2-PropZTest (two-proportion z test; item **6**) computes a test to compare the proportion of successes (p_1 and p_2) from two populations. It takes as input the count of successes in each sample (x_1 and x_2) and the count of observations in each sample (n_1 and n_2). **2-PropZTest** tests the null hypothesis H_0: $p_1=p_2$ (using the pooled sample proportion $\hat{p}$) against one of the alternatives below.

- H_a: $p_1 \neq p_2$ (**p1:≠p2**)
- H_a: $p_1 < p_2$ (**p1:<p2**)
- H_a: $p_1 > p_2$ (**p1:>p2**)

Input:

Calculated results:

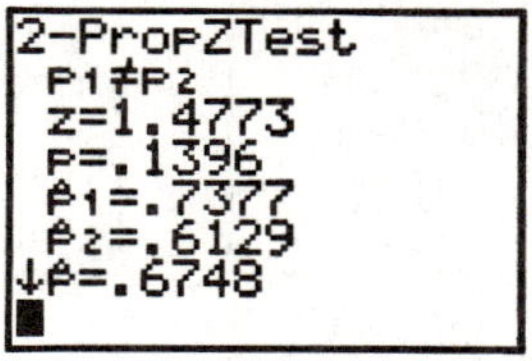

Drawn results:

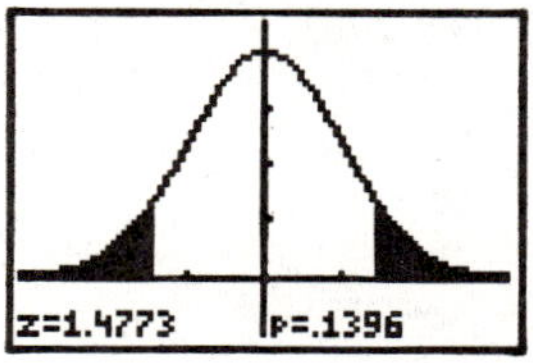

ZInterval

ZInterval (one-sample *z* confidence interval; item **7**) computes a confidence interval for an unknown population mean μ when the population standard deviation σ is known. The computed confidence interval depends on the user-specified confidence level.

In the example:

L1={299.4 297.7 301 298.9 300.2 297}

Data | **Stats**

Input:

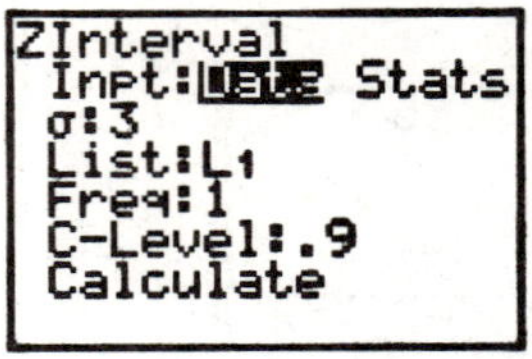

Calculated results:

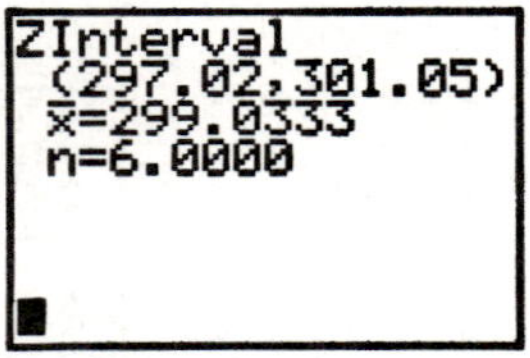

TInterval

TInterval (one-sample *t* confidence interval; item **8**) computes a confidence interval for an unknown population mean μ when the population standard deviation σ is unknown. The computed confidence interval depends on the user-specified confidence level.

In the example:

L6={1.6 1.7 1.8 1.9}

Data | Stats

Input:

Calculated results:

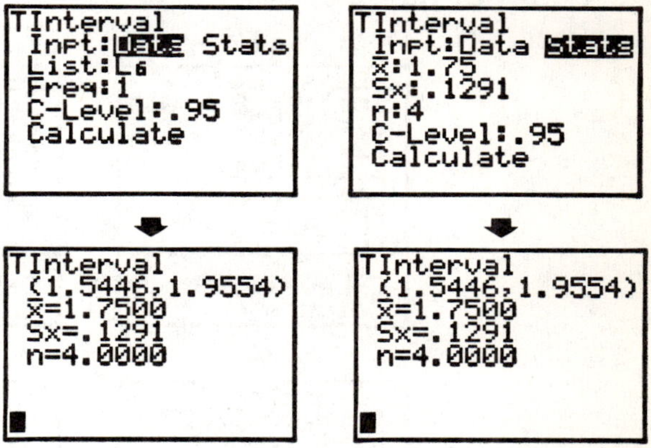

2-SampZInt

2-SampZInt (two-sample *z* confidence interval; item **9**) computes a confidence interval for the difference between two population means ($\mu_1-\mu_2$) when both population standard deviations (σ_1 and σ_2) are known. The computed confidence interval depends on the user-specified confidence level.

In the example:

LISTC={154 109 137 115 140}
LISTD={108 115 126 92 146}

Data **Stats**

Input:

Calculated results:

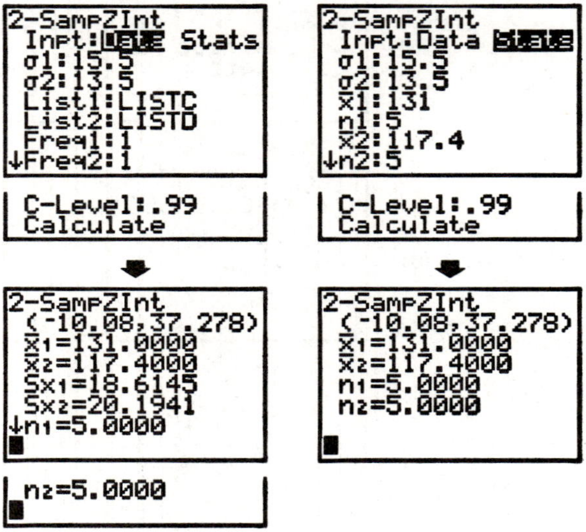

2-SampTInt

2-SampTInt (two-sample *t* confidence interval; item **0**) computes a confidence interval for the difference between two population means ($\mu_1 - \mu_2$) when both population standard deviations (σ_1 and σ_2) are unknown. The computed confidence interval depends on the user-specified confidence level.

In the example:

SAMP1={12.207 16.869 25.05 22.429 8.456 10.589}
SAMP2={11.074 9.686 12.064 9.351 8.182 6.642}

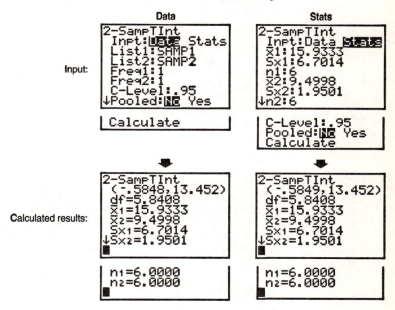

Input:

Data

```
2-SampTInt
 Inpt:Data Stats
 List1:SAMP1
 List2:SAMP2
 Freq1:1
 Freq2:1
 C-Level:.95
↓Pooled:No Yes
```

```
 Calculate
```

Stats

```
2-SampTInt
 Inpt:Data Stats
 x̄1:15.9333
 Sx1:6.7014
 n1:6
 x̄2:9.4998
 Sx2:1.9501
↓n2:6
```

```
 C-Level:.95
 Pooled:No Yes
 Calculate
```

Calculated results:

Data

```
2-SampTInt
 (-.5848,13.452)
 df=5.8408
 x̄1=15.9333
 x̄2=9.4998
 Sx1=6.7014
↓Sx2=1.9501
```

```
 n1=6.0000
 n2=6.0000
```

Stats

```
2-SampTInt
 (-.5849,13.452)
 df=5.8408
 x̄1=15.9333
 x̄2=9.4998
 Sx1=6.7014
↓Sx2=1.9501
```

```
 n1=6.0000
 n2=6.0000
```

1-PropZInt

1-PropZInt (one-proportion z confidence interval; item **A**) computes a confidence interval for an unknown proportion of successes. It takes as input the count of successes in the sample x and the count of observations in the sample n. The computed confidence interval depends on the user-specified confidence level.

Input:

Calculated results:

2-PropZInt

2-PropZInt (two-proportion z confidence interval; item **B**) computes a confidence interval for the difference between the proportion of successes in two populations ($p_1 - p_2$). It takes as input the count of successes in each sample (x_1 and x_2) and the count of observations in each sample (n_1 and n_2). The computed confidence interval depends on the user-specified confidence level.

Input:

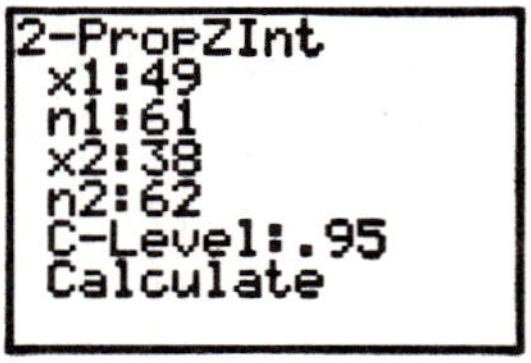

Calculated results:

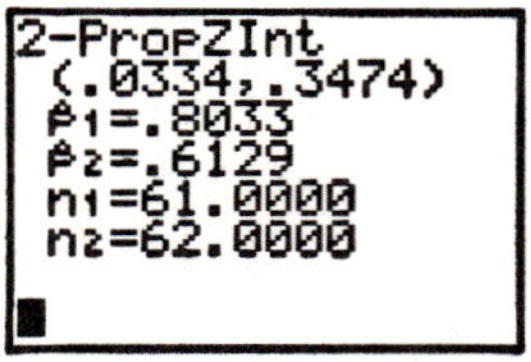

χ^2-Test

χ^2-Test (chi-square test; item **C**) computes a chi-square test for association on the two-way table of counts in the specified *Observed* matrix. The null hypothesis H_0 for a two-way table is: no association exists between row variables and column variables. The alternative hypothesis is: the variables are related.

Before computing a χ^2-Test, enter the observed counts in a matrix. Enter that matrix variable name at the **Observed:** prompt in the χ^2-Test editor; default=**[A]**. At the **Expected:** prompt, enter the matrix variable name to which you want the computed expected counts to be stored; default=**[B]**.

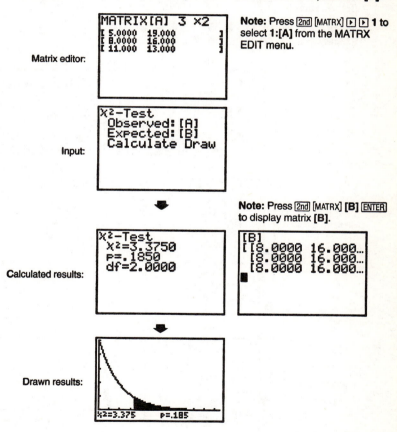

Matrix editor:

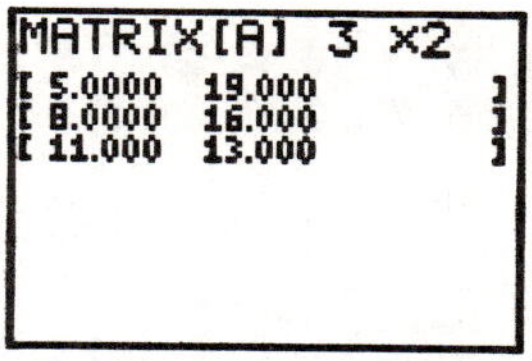

Note: Press [2nd] [MATRX] [▶] [▶] **1** to select **1:[A]** from the MATRX EDIT menu.

Input:

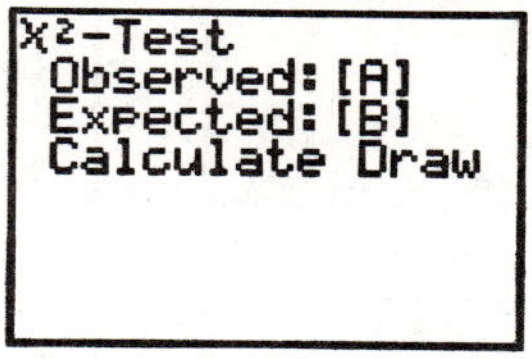

Note: Press [2nd] [MATRX] **[B]** [ENTER] to display matrix **[B]**.

Calculated results:

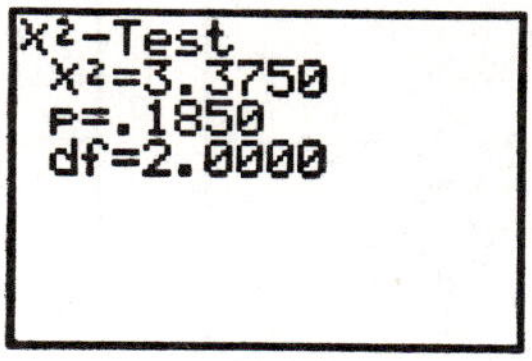

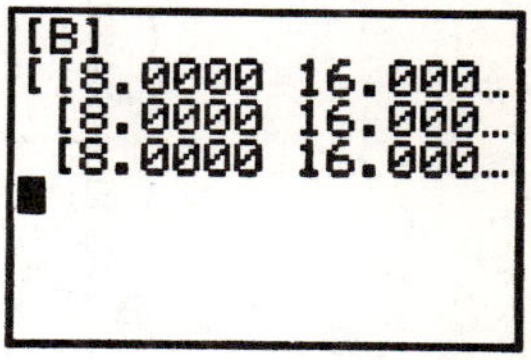

Drawn results:

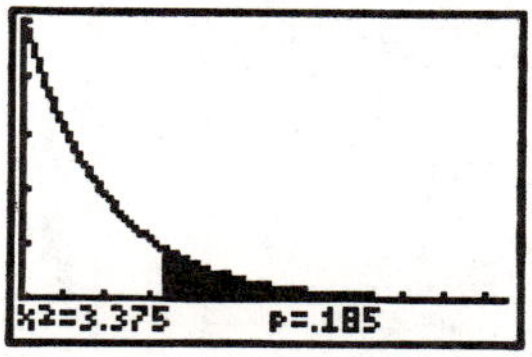

2-SampFTest

2-SampFTest (two-sample F-test; item **D**) computes an F-test to compare two normal population standard deviations (σ_1 and σ_2). The population means and standard deviations are all unknown. **2-SampFTest**, which uses the ratio of sample variances $Sx1^2/Sx2^2$, tests the null hypothesis H_0: $\sigma_1=\sigma_2$ against one of the alternatives below.

- H_a: $\sigma_1\neq\sigma_2$ (**σ1:≠σ2**)
- H_a: $\sigma_1<\sigma_2$ (**σ1:<σ2**)
- H_a: $\sigma_1>\sigma_2$ (**σ1:>σ2**)

In the example:

SAMP4={ 7 -4 18 17 -3 -5 1 10 11 -2}
SAMP5={-1 12 -1 -3 3 -5 5 2-11 -1 -3}

Data | Stats

Input:

Calculated results:

Drawn results:

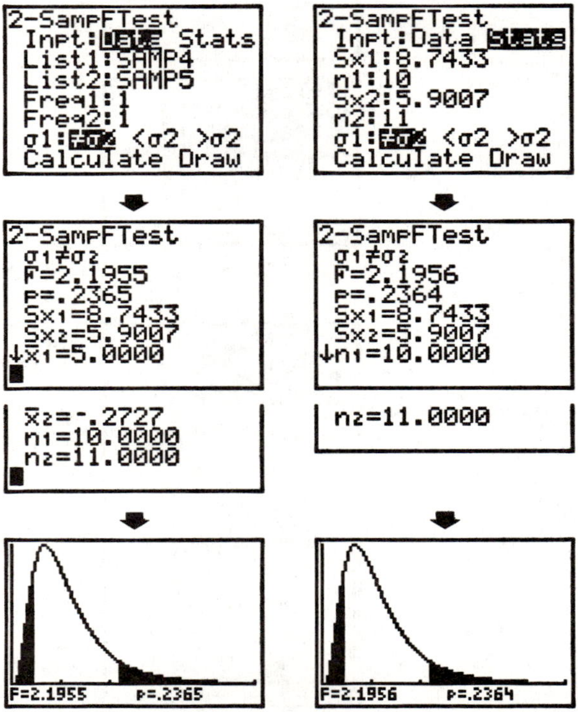

LinRegTTest

LinRegTTest (linear regression t test; item **E**) computes a linear regression on the given data and a t test on the value of slope β and the correlation coefficient ρ for the equation $y=\alpha+\beta x$. It tests the null hypothesis H_0: β=0 (equivalently, ρ=0) against one of the alternatives below.

- H_a: β≠0 and ρ≠0 (β **&** ρ:**≠0**)
- H_a: β<0 and ρ<0 (β **&** ρ:**<0**)
- H_a: β>0 and ρ>0 (β **&** ρ:**>0**)

The regression equation is automatically stored to **RegEQ** (VARS Statistics EQ secondary menu). If you enter a Y= variable name at the **RegEQ:** prompt, the calculated regression equation is automatically stored to the specified Y= equation. In the example below, the regression equation is stored to **Y1**, which is then selected (turned on).

In the example:

L3={ 38 56 59 64 74}
L4={ 41 63 70 72 84}

Input:

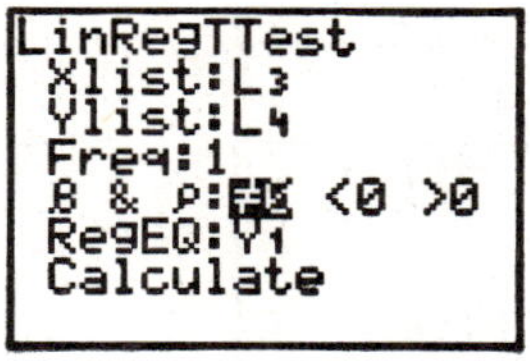

Calculated results:

When **LinRegTTest** is executed, the list of residuals is created and stored to the list name **RESID** automatically. **RESID** is placed on the LIST NAMES menu.

Note: For the regression equation, you can use the fix-decimal mode setting to control the number of digits stored after the decimal point (Chapter 1). However, limiting the number of digits to a small number could affect the accuracy of the fit.

ANOVA(

ANOVA((one-way analysis of variance; item **F**) computes a one-way analysis of variance for comparing the means of two to 20 populations. The ANOVA procedure for comparing these means involves analysis of the variation in the sample data. The null hypothesis H_0: $\mu_1=\mu_2=...=\mu_k$ is tested against the alternative H_a: not all $\mu_1...\mu_k$ are equal.

ANOVA(*list1*,*list2*[,...,*list20*]**)**

In the example:

L1={7 4 6 6 5}
L2={6 5 5 8 7}
L3={4 7 6 7 6}

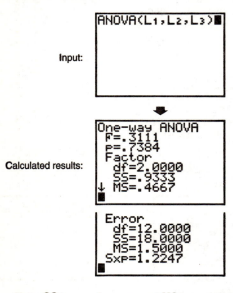

Input:

```
ANOVA(L1,L2,L3)
```

Calculated results:

```
One-way ANOVA
 F=.3111
 p=.7384
 Factor
  df=2.0000
  SS=.9333
↓ MS=.4667
```

```
 Error
  df=12.0000
  SS=18.0000
  MS=1.5000
 Sxp=1.2247
```

Note: SS is sum of squares and **MS** is mean square.

Inferential Statistics Input Descriptions

The tables in this section describe the inferential statistics inputs discussed in this chapter. You enter values for these inputs in the inferential stat editors. The tables present the inputs in the same order that they appear in this chapter.

Input	Description
μ_0	Hypothesized value of the population mean that you are testing.
σ	The known population standard deviation; must be a real number > 0.
List	The name of the list containing the data you are testing.
Freq	The name of the list containing the frequency values for the data in **List**. Default=1. All elements must be integers ≥ 0.
Calculate/Draw	Determines the type of output to generate for tests and intervals. **Calculate** displays the output on the home screen. In tests, **Draw** draws a graph of the results.
$\bar{x}$, Sx, n	Summary statistics (mean, standard deviation, and sample size) for the one-sample tests and intervals.
σ1	The known population standard deviation from the first population for the two-sample tests and intervals. Must be a real number > 0.
σ2	The known population standard deviation from the second population for the two-sample tests and intervals. Must be a real number > 0.
List1, List2	The names of the lists containing the data you are testing for the two-sample tests and intervals. Defaults are **L1** and **L2**, respectively.
Freq1, Freq2	The names of the lists containing the frequencies for the data in **List1** and **List2** for the two-sample tests and intervals. Defaults=1. All elements must be integers ≥ 0.
$\bar{x}$1, Sx1, *n*1, $\bar{x}$2, Sx2, *n*2	Summary statistics (mean, standard deviation, and sample size) for sample one and sample two in the two-sample tests and intervals.
Pooled	Specifies whether variances are to be pooled for **2-SampTTest** and **2-SampTInt**. **No** instructs the TI-83 not to pool the variances. **Yes** instructs the TI-83 to pool the variances.

Input	Description
p_0	The expected sample proportion for **1-PropZTest**. Must be a real number, such that $0 < p_0 < 1$.
x	The count of successes in the sample for the **1-PropZTest** and **1-PropZInt**. Must be an integer ≥ 0.
n	The count of observations in the sample for the **1-PropZTest** and **1-PropZInt**. Must be an integer > 0.
x1	The count of successes from sample one for the **2-PropZTest** and **2-PropZInt**. Must be an integer ≥ 0.
x2	The count of successes from sample two for the **2-PropZTest** and **2-PropZInt**. Must be an integer ≥ 0.
n1	The count of observations in sample one for the **2-PropZTest** and **2-PropZInt**. Must be an integer > 0.
n2	The count of observations in sample two for the **2-PropZTest** and **2-PropZInt**. Must be an integer > 0.
C-Level	The confidence level for the interval instructions. Must be ≥ 0 and <100. If it is ≥ 1, it is assumed to be given as a percent and is divided by 100. Default=0.95.
Observed (Matrix)	The matrix name that represents the columns and rows for the observed values of a two-way table of counts for the **χ^2-Test**. **Observed** must contain all integers ≥ 0. Matrix dimensions must be at least 2×2.
Expected (Matrix)	The matrix name that specifies where the expected values should be stored. **Expected** is created upon successful completion of the **χ^2-Test**.
Xlist, Ylist	The names of the lists containing the data for **LinRegTTest**. Defaults are **L1** and **L2**, respectively. The dimensions of **Xlist** and **Ylist** must be the same.
RegEQ	The prompt for the name of the Y= variable where the calculated regression equation is to be stored. If a Y= variable is specified, that equation is automatically selected (turned on). The default is to store the regression equation to the **RegEQ** variable only.

Test and Interval Output Variables

The inferential statistics variables are calculated as indicated below. To access these variables for use in expressions, press [VARS], **5** (**5:Statistics**), and then select the VARS menu listed in the last column below.

Variables	Tests	Intervals	LinRegTTest, ANOVA	VARS Menu
p-value	**p**		**p**	TEST
test statistics	**z, t, χ^2, F**		**t, F**	TEST
degrees of freedom	**df**	**df**	**df**	TEST
sample mean of x values for sample 1 and sample 2	**$\bar{x}$1, $\bar{x}$2**	**$\bar{x}$1, $\bar{x}$2**		TEST
sample standard deviation of x for sample 1 and sample 2	**Sx1, Sx2**	**Sx1, Sx2**		TEST
number of data points for sample 1 and sample 2	**n1, n2**	**n1, n2**		TEST
pooled standard deviation	**SxP**	**SxP**	**SxP**	TEST
estimated sample proportion	**$\hat{p}$**	**$\hat{p}$**		TEST
estimated sample proportion for population 1	**$\hat{p}$1**	**$\hat{p}$1**		TEST
estimated sample proportion for population 2	**$\hat{p}$2**	**$\hat{p}$2**		TEST
confidence interval pair		**lower, upper**		TEST
mean of x values	**$\bar{x}$**	**$\bar{x}$**		XY
sample standard deviation of x	**Sx**	**Sx**		XY
number of data points	**n**	**n**		XY
standard error about the line			**s**	TEST
regression/fit coefficients			**a, b**	EQ
correlation coefficient			**r**	EQ
coefficient of determination			**r^2**	EQ
regression equation			**RegEQ**	EQ

Note: The variables listed above cannot be archived.

DISTR menu

To display the DISTR menu, press [2nd] [DISTR].

DISTR DRAW	
1: normalpdf(	Normal probability density
2: normalcdf(	Normal distribution probability
3: invNorm(	Inverse cumulative normal distribution
4: tpdf(	Student-*t* probability density
5: tcdf(	Student-*t* distribution probability
6: χ^2pdf(	Chi-square probability density
7: χ^2cdf	Chi-square distribution probability
8: Fpdf(	F probability density
9: Fcdf(	F distribution probability
0: binompdf(	Binomial probability
A: binomcdf(	Binomial cumulative density
B: poissonpdf(	Poisson probability
C: poissoncdf(	Poisson cumulative density
D: geometpdf(	Geometric probability
E: geometcdf(	Geometric cumulative density

Note: ⁻1E99 and 1E99 specify infinity. If you want to view the area left of *upperbound*, for example, specify *lowerbound*=⁻1E99.

normalpdf(

normalpdf(computes the probability density function (pdf) for the normal distribution at a specified x value. The defaults are mean μ=0 and standard deviation σ=1. To plot the normal distribution, paste **normalpdf(** to the **Y=** editor. The probability density function (pdf) is:

$$f(x) = \frac{1}{\sqrt{2\pi}\,\sigma} e^{-\frac{(x-\mu)^2}{2\sigma^2}}, \sigma > 0$$

normalpdf(x[,μ,σ]**)**

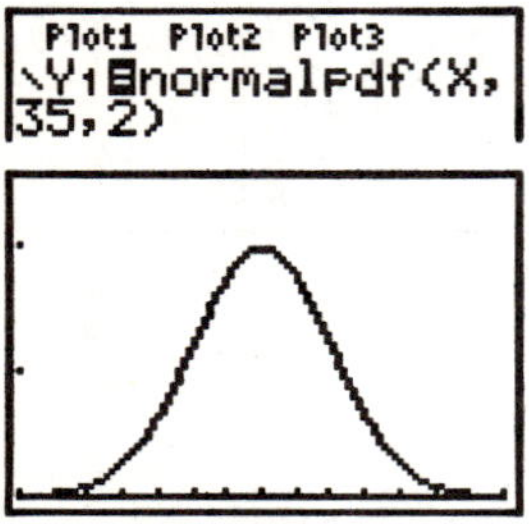

Note: For this example,
Xmin = 28
Xmax = 42
Ymin = 0
Ymax = .25

Tip: For plotting the normal distribution, you can set window variables **Xmin** and **Xmax** so that the mean μ falls between them, and then select **0:ZoomFit** from the ZOOM menu.

normalcdf(

normalcdf(computes the normal distribution probability between *lowerbound* and *upperbound* for the specified mean μ and standard deviation σ. The defaults are μ=0 and σ=1.

normalcdf(*lowerbound*,*upperbound*[,μ,σ]**)**

```
normalcdf(-1E99,
36,35,2)
        .6914624678
```

invNorm(

invNorm(computes the inverse cumulative normal distribution function for a given *area* under the normal distribution curve specified by mean μ and standard deviation σ. It calculates the x value associated with an *area* to the left of the x value. $0 \le area \le 1$ must be true. The defaults are μ=0 and σ=1.

invNorm(*area*[,μ,σ]**)**

```
invNorm(.6914624
678,35,2)
        36.00000004
```

tpdf(

tpdf(computes the probability density function (pdf) for the Student-*t* distribution at a specified x value. *df* (degrees of freedom) must be >0. To plot the Student-*t* distribution, paste **tpdf(** to the Y= editor. The probability density function (pdf) is:

$$f(x) = \frac{\Gamma[(df+1)/2]}{\Gamma(df/2)} \; \frac{(1+x^2/df)^{-(df+1)/2}}{\sqrt{\pi df}}$$

tpdf(*x*,*df***)**

```
 Plot1 Plot2 Plot3
\Y1■tpdf(X,2)
```

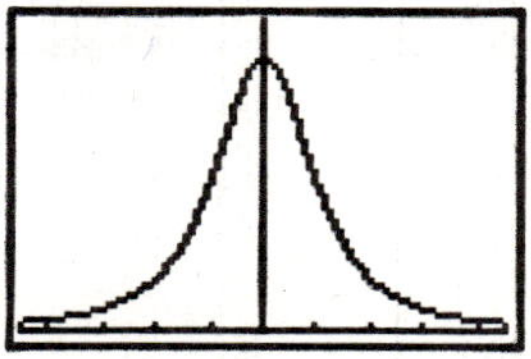

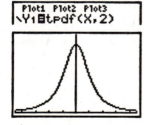

Note: For this example,
Xmin = -4.5
Xmax = 4.5
Ymin = 0
Ymax = .4

tcdf(

tcdf(computes the Student-*t* distribution probability between *lowerbound* and *upperbound* for the specified *df* (degrees of freedom), which must be > 0.

tcdf(*lowerbound*,*upperbound*,*df***)**

```
tcdf(-2,3,18)
         .9657465644
```

χ^2pdf(

χ^2pdf(computes the probability density function (pdf) for the χ^2 (chi-square) distribution at a specified x value. *df* (degrees of freedom) must be an integer > 0. To plot the χ^2 distribution, paste **χ^2pdf(** to the **Y=** editor. The probability density function (pdf) is:

$$f(x) = \frac{1}{\Gamma(df/2)}(1/2)^{df/2}\, x^{df/2-1} e^{-x/2},\ x \ge 0$$

χ^2pdf(*x*,*df***)**

```
 Plot1 Plot2 Plot3
\Y1■X²pdf(X,9)
\Y2■X²pdf(X,7)
\Y3=
\Y4=
\Y5=
\Y6=
\Y7=
```

Note: For this example,
Xmin = 0
Xmax = 30
Ymin = -.02
Ymax = .132

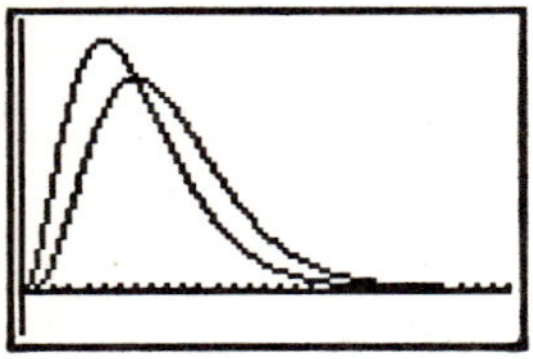

χ^2cdf(

χ^2cdf(computes the χ^2 (chi-square) distribution probability between *lowerbound* and *upperbound* for the specified *df* (degrees of freedom), which must be an integer > 0.

χ^2cdf(*lowerbound*,*upperbound*,*df***)**

```
X²cdf(0,19.023,9
)
         .9750019601
```

Fpdf(

Fpdf(computes the probability density function (pdf) for the **F** distribution at a specified x value. *numerator df* (degrees of freedom) and *denominator df* must be integers > 0. To plot the **F** distribution, paste **Fpdf(** to the **Y=** editor. The probability density function (pdf) is:

$$f(x) = \frac{\Gamma[(n+d)/2]}{\Gamma(n/2)\Gamma(d/2)}\left(\frac{n}{d}\right)^{n/2} x^{n/2-1}(1+nx/d)^{-(n+d)/2}, x \geq 0$$

where n = numerator degrees of freedom
d = denominator degrees of freedom

Fpdf(*x*,*numerator df*,*denominator df***)**

```
 Plot1  Plot2  Plot3
\Y1■Fpdf(X,24,19
)■
```

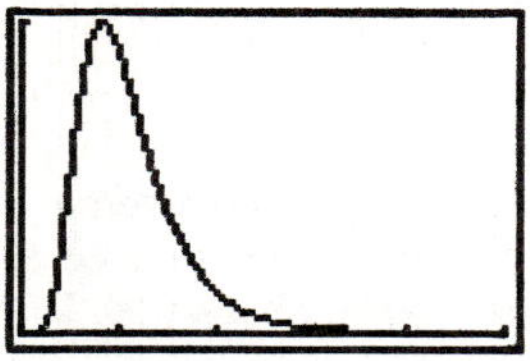

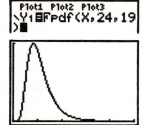

Note: For this example,
Xmin = 0
Xmax = 5
Ymin = 0
Ymax = 1

Fcdf(

Fcdf(computes the **F** distribution probability between *lowerbound* and *upperbound* for the specified *numerator df* (degrees of freedom) and *denominator df*. *numerator df* and *denominator df* must be integers >0.

Fcdf(*lowerbound*,*upperbound*,*numerator df*,*denominator df***)**

```
Fcdf(0,2.4523,24
,19)
      .9749989576
```

binompdf(

binompdf(computes a probability at x for the discrete binomial distribution with the specified *numtrials* and probability of success (p) on each trial. x can be an integer or a list of integers. $0 \le p \le 1$ must be true. *numtrials* must be an integer > 0. If you do not specify x, a list of probabilities from 0 to *numtrials* is returned. The probability density function (pdf) is:

$$f(x) = \binom{n}{x} p^x (1-p)^{n-x} \text{ , } x = 0,1,\ldots,n$$

where n = *numtrials*

binompdf(*numtrials*,*p*[,*x*]**)**

```
binompdf(5,.6,{3
,4,5})
{.3456 .2592 .0…
```

binomcdf(

binomcdf(computes a cumulative probability at x for the discrete binomial distribution with the specified *numtrials* and probability of success (p) on each trial. x can be a real number or a list of real numbers. $0 \le p \le 1$ must be true. *numtrials* must be an integer > 0. If you do not specify x, a list of cumulative probabilities is returned.

binomcdf(*numtrials*,*p*[,*x*]**)**

```
binomcdf(5,.6,{3
,4,5})
{.66304 .92224 …
```

poissonpdf(

poissonpdf(computes a probability at x for the discrete Poisson distribution with the specified mean μ, which must be a real number > 0. x can be an integer or a list of integers. The probability density function (pdf) is:

$$f(x) = e^{-\mu}\mu^x / x! \text{ , } x = 0,1,2,\ldots$$

poissonpdf(*μ*,*x***)**

```
poissonpdf(6,10)
      .0413030934
```

poissoncdf(

poissoncdf(computes a cumulative probability at x for the discrete Poisson distribution with the specified mean μ, which must be a real number > 0. x can be a real number or a list of real numbers.

poissoncdf(μ**,**x**)**

```
poissoncdf(.126,
{0,1,2,3})
{.8816148468 .9...
```

geometpdf(

geometpdf(computes a probability at x, the number of the trial on which the first success occurs, for the discrete geometric distribution with the specified probability of success p. $0 \le p \le 1$ must be true. x can be an integer or a list of integers. The probability density function (pdf) is:

$$f(x) = p(1-p)^{x-1}, x = 1,2,\ldots$$

geometpdf(p**,**x**)**

```
geometpdf(.4,6)
             .031104
```

geometcdf(

geometcdf(computes a cumulative probability at x, the number of the trial on which the first success occurs, for the discrete geometric distribution with the specified probability of success p. $0 \le p \le 1$ must be true. x can be a real number or a list of real numbers.

geometcdf(p**,**x**)**

```
geometcdf(.5,{1,
2,3})
     {.5 .75 .875}
```

DISTR DRAW Menu

To display the DISTR DRAW menu, press [2nd] [DISTR] [▸]. DISTR DRAW instructions draw various types of density functions, shade the area specified by *lowerbound* and *upperbound*, and display the computed area value.

To clear the drawings, select **1:ClrDraw** from the DRAW menu (Chapter 8).

Note: Before you execute a DISTR DRAW instruction, you must set the window variables so that the desired distribution fits the screen.

DISTR DRAW	
1:ShadeNorm(	Shades normal distribution.
2:Shade_t(	Shades Student-*t* distribution.
3:Shadeχ^2(	Shades χ^2 distribution.
4:ShadeF(	Shades F distribution.

Note: ⁻1E99 and 1E99 specify infinity. If you want to view the area left of *upperbound*, for example, specify *lowerbound*=⁻1E99.

ShadeNorm(

ShadeNorm(draws the normal density function specified by mean μ and standard deviation σ and shades the area between *lowerbound* and *upperbound*. The defaults are μ=0 and σ=1.

ShadeNorm(*lowerbound*,*upperbound*[,μ,σ]**)**

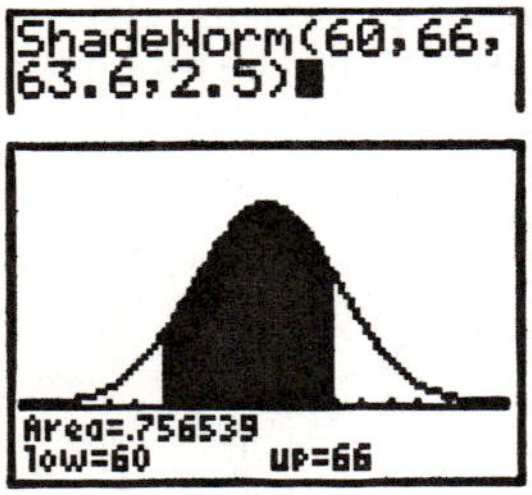

Note: For this example,
Xmin = 55
Xmax = 72
Ymin = ⁻.05
Ymax = .2

Shade_t(

Shade_t(draws the density function for the Student-*t* distribution specified by *df* (degrees of freedom) and shades the area between *lowerbound* and *upperbound.*

Shade_t(*lowerbound*,*upperbound*,*df***)**

```
Shade_t(-1,1E99,
4)
```

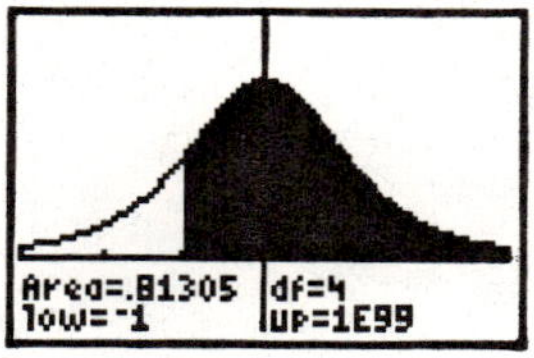

Note: For this example,
Xmin = -3
Xmax = 3
Ymin = -.15
Ymax = .5

Shadeχ^2(

Shadeχ^2(draws the density function for the χ^2 (chi-square) distribution specified by *df* (degrees of freedom) and shades the area between *lowerbound* and *upperbound.*

Shadeχ^2(*lowerbound*,*upperbound*,*df***)**

```
Shadeχ²(0,4,10)
```

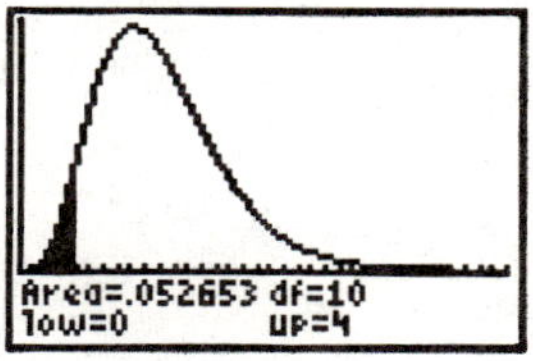

Note: For this example,
Xmin = 0
Xmax = 35
Ymin = -.025
Ymax = .1

ShadeF(

ShadeF(draws the density function for the **F** distribution specified by *numerator df* (degrees of freedom) and *denominator df* and shades the area between *lowerbound* and *upperbound.*

ShadeF(*lowerbound*,*upperbound*,*numerator df*,*denominator df***)**

```
ShadeF(1,2,10,15
)
```

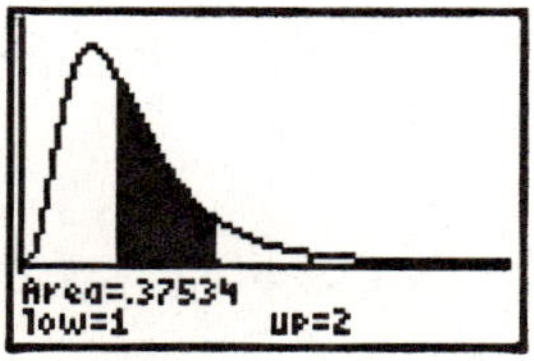

Note: For this example,
Xmin = 0
Xmax = 5
Ymin = -.25
Ymax = .9

14 Applications

Contents

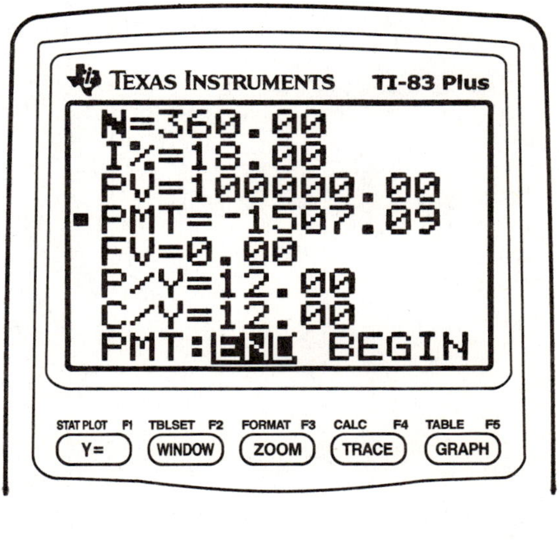

The TI-83 Plus comes with Finance and CBL/CBR applications already listed on the APPLICATIONS menu. Except for the Finance application, you can add and remove applications as space permits. The Finance application is built into the TI-83 Plus code and cannot be deleted.

You can buy additional TI-83 Plus software applications that allow you to customize further your calculator's functionality. The calculator reserves 160 KB of space within ROM memory specifically for applications.

Steps for Running the Finance Application

Follow these basic steps when using the Finance application.

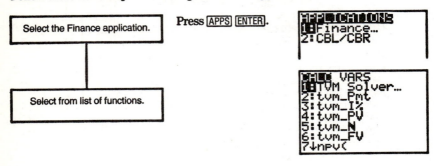

Select the Finance application.

Press [APPS] [ENTER].

Select from list of functions.

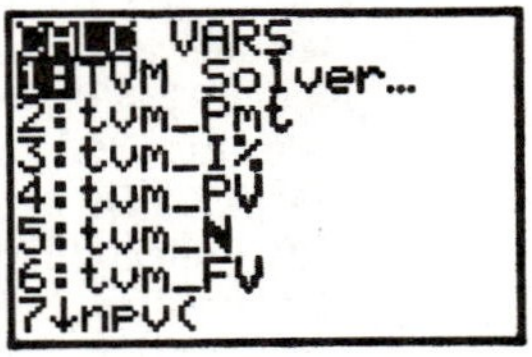

Getting Started is a fast-paced introduction. Read the chapter for details.

You have found a car you would like to buy. The car costs 9,000. You can afford payments of 250 per month for four years. What annual percentage rate (APR) will make it possible for you to afford the car?

1. Press [MODE] [▼] [▶] [▶] [▶] [ENTER] to set the fixed-decimal mode setting to **2**. The TI-83 Plus will display all numbers with two decimal places).

2. Press [APPS] [ENTER] to select **1:Finance** from the APPLICATIONS menu.

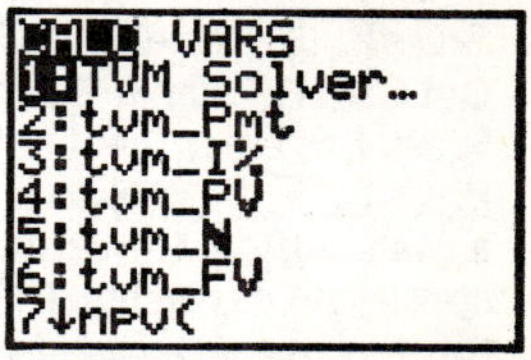

3. Press [ENTER] to select **1:TVM Solver** from the CALC VARS menu. The TVM Solver is displayed.

   Press **48** [ENTER] to store 48 months to **N**. Press [▼] **9000** [ENTER] to store 9,000 to **PV**. Press [(-)] **250** [ENTER] to store ‑250 to **PMT**. (Negation indicates cash outflow.) Press **0** [ENTER] to store 0 to **FV**. Press **12** [ENTER] to store 12 payments per year to **P/Y** and 12 compounding periods per year to **C/Y**. Setting **P/Y** to 12 will compute an annual percentage rate (compounded monthly) for **I%**. Press [▼] [ENTER] to select **PMT:END**, which indicates that payments are due at the end of each period.

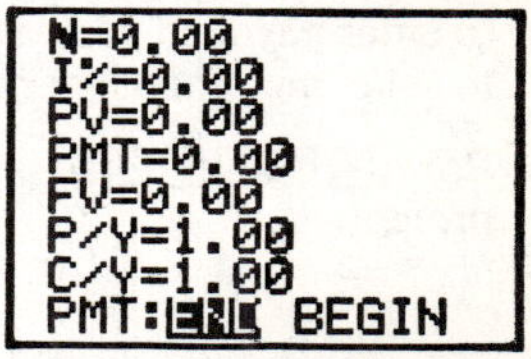

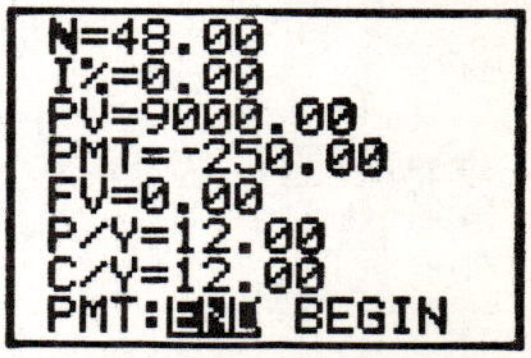

4. Press [▲] [▲] [▲] [▲] [▲] [▲] to move the cursor to the **I%** prompt. Press [ALPHA] [SOLVE] to solve for **I%**. What APR should you look for?

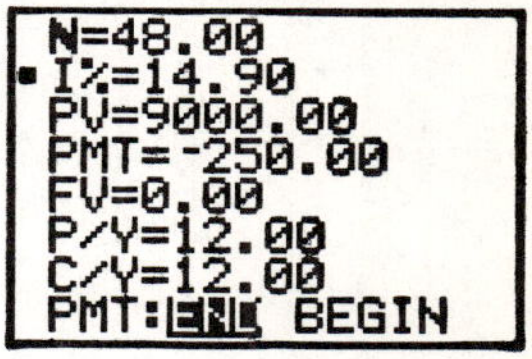

At what annual interest rate, compounded monthly, will 1,250 accumulate to 2,000 in 7 years?

Note: Because there are no payments when you solve compound interest problems, PMT must be set to 0 and P/Y must be set to 1.

1. Press [APPS] [ENTER] to select **1:Finance** from the APPLICATIONS menu.

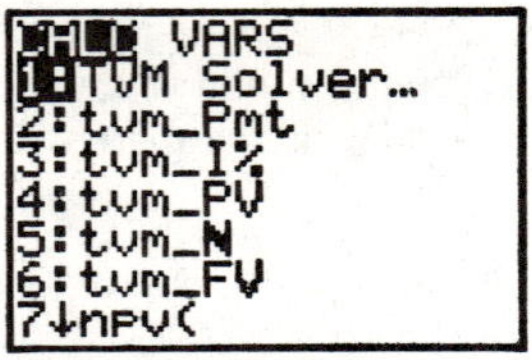

2. Press [ENTER] to select **1:TVM Solver** from the CALC VARS menu. The TVM Solver is displayed. Press **7** to enter the number of periods in years. Press [▼] [▼] [(-)] **1250** to enter the present value as a cash outflow (investment). Press [▼] **0** to specify no payments. Press [▼] **2000** to enter the future value as a cash inflow (return). Press [▼] **1** to enter payment periods per year. Press [▼] **12** to set compounding periods per year to **12**.

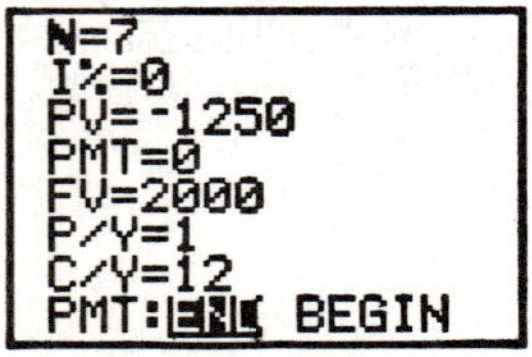

3. Press [▲] [▲] [▲] [▲] [▲] to place the cursor on the I% prompt.

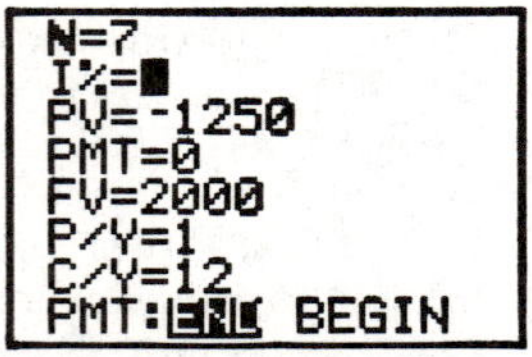

4. Press [ALPHA] [SOLVE] to solve for I%, the annual interest rate.

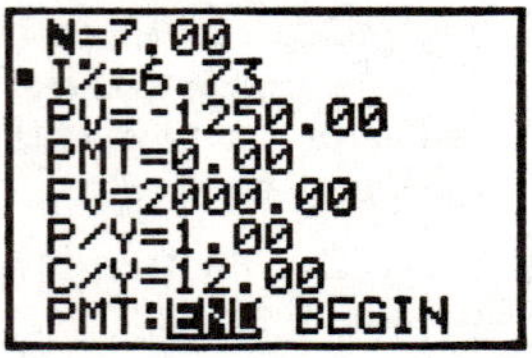

Using the TVM Solver

The TVM Solver displays the time-value-of-money (TVM) variables. Given four variable values, the TVM Solver solves for the fifth variable.

The FINANCE VARS menu section (page 14-15) describes the five TVM variables (**N**, I%, **PV**, **PMT**, and **FV**) and **P/Y** and **C/Y**.

PMT: END BEGIN in the TVM Solver corresponds to the FINANCE CALC menu items **Pmt_End** (payment at the end of each period) and **Pmt_Bgn** (payment at the beginning of each period).

To solve for an unknown TVM variable, follow these steps.

1. Press [APPS] [ENTER] [ENTER] to display the TVM Solver. The screen below shows the default values with the fixed-decimal mode set to two decimal places.

```
N=0.00
I%=0.00
PV=0.00
PMT=0.00
FV=0.00
P/Y=1.00
C/Y=1.00
PMT:END BEGIN
```

2. Enter the known values for four TVM variables.

   Note: Enter cash inflows as positive numbers and cash outflows as negative numbers.

3. Enter a value for **P/Y**, which automatically enters the same value for **C/Y**; if **P/Y** ≠ **C/Y**, enter a unique value for **C/Y**.
4. Select **END** or **BEGIN** to specify the payment method.
5. Place the cursor on the TVM variable for which you want to solve.
6. Press [ALPHA] [SOLVE]. The answer is computed, displayed in the TVM Solver, and stored to the appropriate TVM variable. An indicator square in the left column designates the solution variable.

```
 N=360.00
 I%=18.00
 PV=100000.00
▪PMT=-1507.09
 FV=0.00
 P/Y=12.00
 C/Y=12.00
 PMT:END BEGIN
```

Entering Cash Inflows and Cash Outflows

When using the TI-83 Plus financial functions, you must enter cash inflows (cash received) as positive numbers and cash outflows (cash paid) as negative numbers. The TI-83 Plus follows this convention when computing and displaying answers.

FINANCE CALC Menu

To display the FINANCE CALC menu, press [APPS] [ENTER].

CALC VARS	
1:TVM Solver...	Displays the TVM Solver.
2:tvm_Pmt	Computes the amount of each payment.
3:tvm_I%	Computes the interest rate per year.
4:tvm_PV	Computes the present value.
5:tvm_N	Computes the number of payment periods.
6:tvm_FV	Computes the future value.
7:npv(	Computes the net present value.
8:irr(	Computes the internal rate of return.
9:bal(	Computes the amortization sched. balance.
0:ΣPrn(	Computes the amort. sched. princ. sum.
A:ΣInt(	Computes the amort. sched. interest sum.
B:▸Nom(	Computes the nominal interest rate.
C:▸Eff(	Computes the effective interest rate.
D:dbd(	Calculates the days between two dates.
E:Pmt_End	Selects ordinary annuity (end of period).
F:Pmt_Bgn	Selects annuity due (beginning of period).

Use these functions to set up and perform financial calculations on the home screen.

TVM Solver

TVM Solver displays the TVM Solver (page 14-5).

Calculating Time Value of Money

Use time-value-of-money (TVM) functions (menu items **2** through **6**) to analyze financial instruments such as annuities, loans, mortgages, leases, and savings.

Each TVM function takes zero to six arguments, which must be real numbers. The values that you specify as arguments for these functions are not stored to the TVM variables (page 14-15).

Note: To store a value to a TVM variable, use the TVM Solver (page 14-5) or use [STO▸] and any TVM variable on the FINANCE VARS menu (page 14-15).

If you enter less than six arguments, the TI-83 Plus substitutes a previously stored TVM variable value for each unspecified argument.

If you enter any arguments with a TVM function, you must place the argument or arguments in parentheses.

tvm_Pmt

tvm_Pmt computes the amount of each payment.

tvm_Pmt[(*N*,*I%*,*PV*,*FV*,*P/Y*,*C/Y*)]

Note: In the example above, the values are stored to the TVM variables in the TVM Solver. Then the payment (**tvm_Pmt**) is computed on the home screen using the values in the TVM Solver. Next, the interest rate is changed to 9.5 to illustrate the effect on the payment amount.

tvm_I%

tvm_I% computes the annual interest rate.

tvm_I%[(***N***,*PV*,*PMT*,*FV*,*P/Y*,*C/Y*)]

```
tvm_I%(48,10000,
-250,0,12)
                9.24
Ans→I%
                9.24
```

tvm_PV

tvm_PV computes the present value.

tvm_PV[(***N***,*I%*,*PMT*,*FV*,*P/Y*,*C/Y*)]

```
360→N:11→I%:-100
0→PMT:0→FV:12→P/
Y
               12.00
tvm_PV
           105006.35
```

tvm_N

tvm_N computes the number of payment periods.

tvm_N[(*I%*,*PV*,*PMT*,*FV*,*P/Y*,*C/Y*)]

```
6→I%:9000→PV:-35
0→PMT:0→FV:3→P/Y
                3.00
tvm_N
               36.47
```

tvm_FV

tvm_FV computes the future value.

tvm_FV[(***N***,*I%*,*PV*,*PMT*,*P/Y*,*C/Y*)]

```
6→N:8→I%:-5500→P
V:0→PMT:1→P/Y
                1.00
tvm_FV
             8727.81
```

Calculating a Cash Flow

Use the cash flow functions (menu items **7** and **8**) to analyze the value of money over equal time periods. You can enter unequal cash flows, which can be cash inflows or outflows. The syntax descriptions for **npv(** and **irr(** use these arguments.

- *interest rate* is the rate by which to discount the cash flows (the cost of money) over one period.
- *CF0* is the initial cash flow at time 0; it must be a real number.
- *CFList* is a list of cash flow amounts after the initial cash flow *CF0*.
- *CFFreq* is a list in which each element specifies the frequency of occurrence for a grouped (consecutive) cash flow amount, which is the corresponding element of *CFList*. The default is 1; if you enter values, they must be positive integers < 10,000.

For example, express this uneven cash flow in lists.

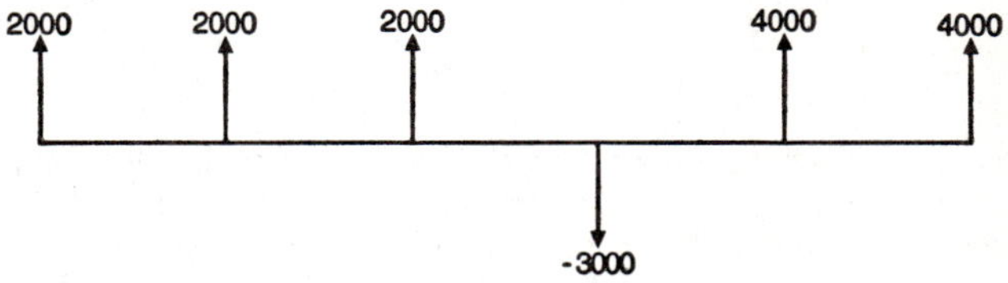

CF0 = **2000**
CFList = **{2000,⁻3000,4000}**
CFFreq = **{2,1,2}**

npv(, irr(

npv((net present value) is the sum of the present values for the cash inflows and outflows. A positive result for **npv** indicates a profitable investment.

npv(*interest rate*,*CF0*,*CFList*[,*CFFreq*]**)**

irr((internal rate of return) is the interest rate at which the net present value of the cash flows is equal to zero.

irr(*CF0*,*CFList*[,*CFFreq*]**)**

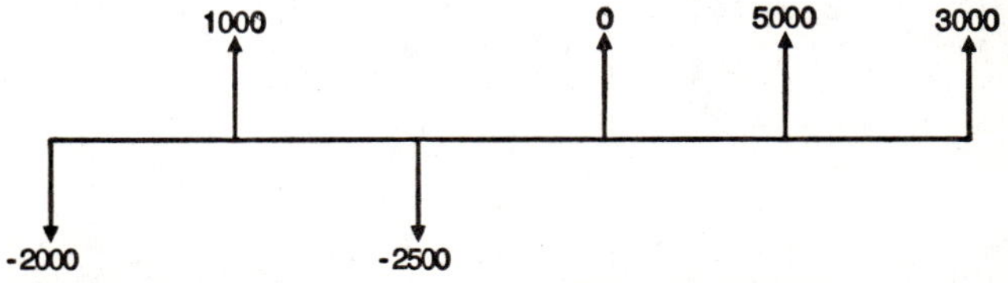


```
{1000,⁻2500,0,50
00,3000}→L1
{1000.00 ⁻2500...
```

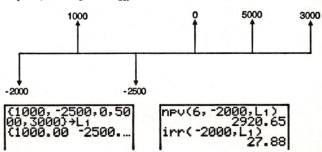

```
npv(6,⁻2000,L1)
         2920.65
irr(⁻2000,L1)
           27.88
```

Calculating an Amortization Schedule

Use the amortization functions (menu items **9**, **0**, and **A**) to calculate balance, sum of principal, and sum of interest for an amortization schedule.

bal(

bal(computes the balance for an amortization schedule using stored values for I%, **PV**, and **PMT**. *npmt* is the number of the payment at which you want to calculate a balance. It must be a positive integer < 10,000. *roundvalue* specifies the internal precision the calculator uses to calculate the balance; if you do not specify *roundvalue*, then the TI-83 Plus uses the current **Float/Fix** decimal-mode setting.

bal(*npmt*[,*roundvalue*]**)**

```
100000→PV:8.5→I%
:-768.91→PMT:12→
P/Y
           12.00
```

```
bal(12)
        99244.07
```

ΣPrn(, ΣInt(

ΣPrn(computes the sum of the principal during a specified period for an amortization schedule using stored values for I%, **PV**, and **PMT**. *pmt1* is the starting payment. *pmt2* is the ending payment in the range. *pmt1* and *pmt2* must be positive integers < 10,000. *roundvalue* specifies the internal precision the calculator uses to calculate the principal; if you do not specify *roundvalue*, the TI-83 Plus uses the current **Float/Fix** decimal-mode setting.

Note: You must enter values for I%, **PV**, **PMT**, and before computing the principal.

ΣPrn(*pmt1*,*pmt2*[,*roundvalue*]**)**

ΣInt(computes the sum of the interest during a specified period for an amortization schedule using stored values for I%, **PV**, and **PMT**. *pmt1* is the starting payment. *pmt2* is the ending payment in the range. *pmt1* and *pmt2* must be positive integers < 10,000. *roundvalue* specifies the internal precision the calculator uses to calculate the interest; if you do not specify *roundvalue*, the TI-83 Plus uses the current **Float/Fix** decimal-mode setting.

ΣInt(*pmt1*,*pmt2*[,*roundvalue*]**)**

```
360→N:100000→PV:
8.5→I%:-768.91→P
MT:12→P/Y
           12.00
```

```
ΣPrn(1,12)
         -755.93
ΣInt(1,12)
        -8470.99
```

Amortization Example: Calculating an Outstanding Loan Balance

You want to buy a home with a 30-year mortgage at 8 percent APR. Monthly payments are 800. Calculate the outstanding loan balance after each payment and display the results in a graph and in the table.

1. Press [MODE]. Press [▼] [▶] [▶] [▶] [ENTER] to set the fixed-decimal mode setting to **2**. Press [▼] [▼] [▶] [ENTER] to select **Par** graphing mode.

2. Press [APPS] [ENTER] [ENTER] to display the **TVM Solver.**

3. Press [ENTER] **360** to enter number of payments. Press [▼] **8** to enter the interest rate. Press [▼] [▼] [(-)] **800** to enter the payment amount. Press [▼] **0** to enter the future value of the mortgage. Press [▼] **12** to enter the payments per year, which also sets the compounding periods per year to **12**. Press [▼] [▼] [ENTER] to select **PMT:END**.

4. Press [▲] [▲] [▲] [▲] [▲] to place the cursor on the **PV** prompt. Press [ALPHA] [SOLVE] to solve for the present value.

5. Press [Y=] to display the parametric Y= editor. Turn off all stat plots. Press [X,T,Θ,n] to define **X1T** as **T**. Press [▼] [APPS] [ENTER] **9** [X,T,Θ,n] [)] to define **Y1T** as **bal(T)**.

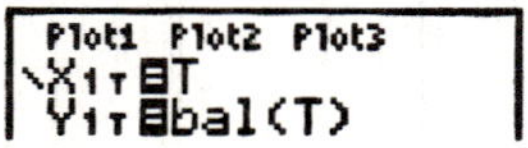

Amortization Example: Calculating an Outstanding Loan Balance (continued)

6. Press [WINDOW] to display the window variables. Enter the values below.

Tmin=0	**Xmin=0**	**Ymin=0**
Tmax=360	**Xmax=360**	**Ymax=125000**
Tstep=12	**Xscl=50**	**Yscl=10000**

7. Press [TRACE] to draw the graph and activate the trace cursor. Press [▶] and [◀] to explore the graph of the outstanding balance over time. Press a number and then press [ENTER] to view the balance at a specific time **T**.

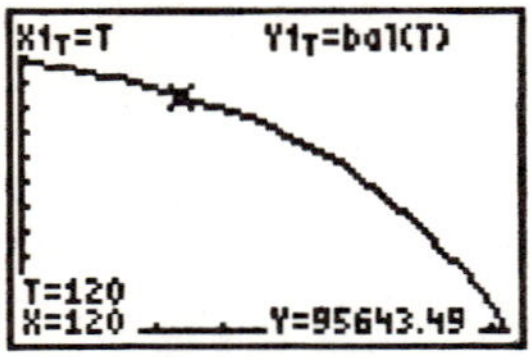

8. Press [2nd] [TBLSET] and enter the values below.
   TblStart=0
   ΔTbl=12

9. Press [2nd] [TABLE] to display the table of outstanding balances (**Y1T**).

T	X1T	Y1T
0.00	0.00	109027
12.00	12.00	108116
24.00	24.00	107130
36.00	36.00	106061
48.00	48.00	104905
60.00	60.00	103652
72.00	72.00	102295

T=0

10. Press [MODE] [▼] [▼] [▼] [▼] [▼] [▼] [▼] [▶] [▶] [ENTER] to select **G-T** split-screen mode, in which the graph and table are displayed simultaneously.

    Press [TRACE] to display **X1T** (time) and **Y1T** (balance) in the table.

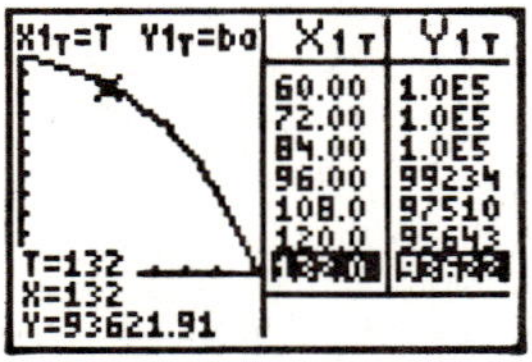

Calculating an Interest Conversion

Use the interest conversion functions (menu items **B** and **C**) to convert interest rates from an annual effective rate to a nominal rate (**▸Nom(**) or from a nominal rate to an annual effective rate (**▸Eff(**).

▸Nom(

▸Nom(computes the nominal interest rate. *effective rate* and *compounding periods* must be real numbers. *compounding periods* must be >0.

▸Nom(*effective rate,compounding periods***)**

```
▸Nom(15.87,4)
           15.00
```

▸Eff(

▸Eff(computes the effective interest rate. *nominal rate* and *compounding periods* must be real numbers. *compounding periods* must be >0.

▸Eff(*nominal rate,compounding periods***)**

```
▸Eff(8,12)
            8.30
```

dbd(

Use the date function **dbd(** (menu item **D**) to calculate the number of days between two dates using the actual-day-count method. *date1* and *date2* can be numbers or lists of numbers within the range of the dates on the standard calendar.

Note: Dates must be between the years 1950 through 2049.

dbd(*date1*,*date2***)**

You can enter *date1* and *date2* in either of two formats.

- MM.DDYY (United States)
- DDMM.YY (Europe)

The decimal placement differentiates the date formats.

```
dbd(12.3190,12.3
192)
           731.00
```

Defining the Payment Method

Pmt_End and **Pmt_Bgn** (menu items **E** and **F**) specify a transaction as an ordinary annuity or an annuity due. When you execute either command, the TVM Solver is updated.

Pmt_End

Pmt_End (payment end) specifies an ordinary annuity, where payments occur at the end of each payment period. Most loans are in this category. **Pmt_End** is the default.

Pmt_End

On the TVM Solver's **PMT:END BEGIN** line, select **END** to set **PMT** to ordinary annuity.

Pmt_Bgn

Pmt_Bgn (payment beginning) specifies an annuity due, where payments occur at the beginning of each payment period. Most leases are in this category.

Pmt_Bgn

On the TVM Solver's **PMT:END BEGIN** line, select **BEGIN** to set **PMT** to annuity due.

FINANCE VARS Menu

To display the FINANCE VARS menu, press [APPS] [ENTER] [▸]. You can use TVM variables in TVM functions and store values to them on the home screen.

CALC VARS	
1: N	Total number of payment periods
2: I%	Annual interest rate
3: PV	Present value
4: PMT	Payment amount
5: FV	Future value
6: P/Y	Number of payment periods per year
7: C/Y	Number of compounding periods/year

N, I%, PV, PMT, FV

N, I%, **PV**, **PMT**, and **FV** are the five TVM variables. They represent the elements of common financial transactions, as described in the table above. I% is an annual interest rate that is converted to a per-period rate based on the values of **P/Y** and **C/Y**.

P/Y and C/Y

P/Y is the number of payment periods per year in a financial transaction.

C/Y is the number of compounding periods per year in the same transaction.

When you store a value to **P/Y**, the value for **C/Y** automatically changes to the same value. To store a unique value to **C/Y**, you must store the value to **C/Y** after you have stored a value to **P/Y**.

The CBL/CBR application allows you to collect real world data. The TI-83 Plus comes with the CBL/CBR application already listed on the APPLICATIONS menu ([APPS] 2).

Steps for Running the CBL/CBR Application

Follow these basic steps when using the CBL/CBR application. You may not have to do all of them each time.

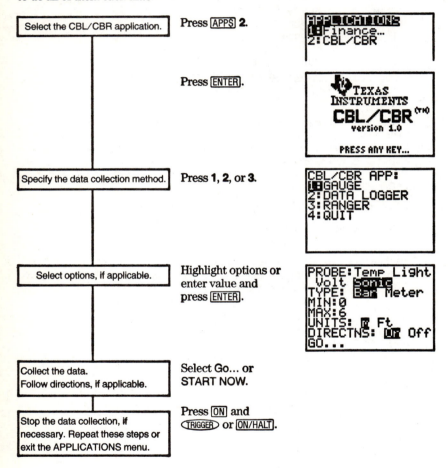

Step	Action	Screen
Select the CBL/CBR application.	Press [APPS] **2**.	
	Press [ENTER].	
Specify the data collection method.	Press **1**, **2**, or **3**.	
Select options, if applicable.	Highlight options or enter value and press [ENTER].	
Collect the data. Follow directions, if applicable.	Select Go... or START NOW.	
Stop the data collection, if necessary. Repeat these steps or exit the APPLICATIONS menu.	Press [ON] and (TRIGGER) or [ON/HALT].	

Selecting the CBL/CBR Application

To use a CBL/CBR application, you need a CBL 2/CBL or CBR (as applicable), a TI-83 Plus, and a unit-to-unit link cable.

1. Press [APPS].

2. Select **2:CBL/CBR** to set up the TI-83 Plus to use either of the applications. An informational screen appears first.

3. Press any key to continue to the next menu.

Specifying the Data Collection Method from the CBL/CBR APP Menu

With a CBL 2/CBL or CBR, you can collect data in one of three ways: GAUGE (bar or meter), DATA LOGGER (a Temp-Time, Light-Time, Volt-Time, or Sonic-Time graph), or RANGER, which runs the RANGER program, the built-in CBR data collection program.

The CBL/CBR APP menu contains the data collection methods.

CBL/CBR APP:	
1:GAUGE	Displays results as either a bar or meter.
2:DATA LOGGER	Displays results as a Temp-Time, Light-Time, Volt-Time, or Sonic-Time graph.
3:RANGER	Sets up and runs the RANGER program.
4:QUIT	Quits the CBL/CBR application.

Note: CBL 2/CBL and CBR differ in that CBL 2/CBL allows you to collect data using one of several different probes including: Temp (Temperature), Light, Volt (Voltage), or Sonic. CBR collects data using only the built-in Sonic probe. You can find more information on CBL 2/CBL and CBR in their user manuals.

Specifying Options for Each Data Collection Method

After you select a data collection method from the CBL/CBR APP menu, a screen showing the options for that method is displayed. The method you choose, as well as the data collection options you choose for that method, determine whether you use the CBR or the CBL 2/CBL. Refer to the charts in the following sections to find the options for the application you are using.

GAUGE

The GAUGE data collection method lets you choose one of four different probes: temp, Light, Volt, or Sonic.

1. Press [APPS] **2** [ENTER].
2. Select **1:GAUGE.**

3. Select options.

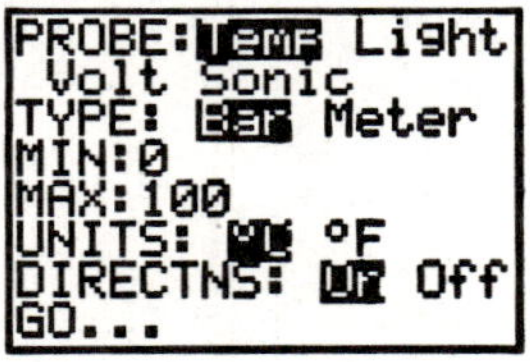

When you select a probe option, all other options change accordingly. Use ▶ and ◀ to move between the probe options. To select a probe, highlight the one you want with the cursor keys, and then press [ENTER].

GAUGE Options (Defaults)				
PROBE:	Temp	Light	Volt	Sonic
Type:	Bar or Meter			
Min:	0	0	-10	0
Max:	100	1	10	6
Units:	°C or °F	mW/cm²	Volt	m or Ft
Directions:	On or Off			

TYPE

The GAUGE data collection results are represented according to TYPE: Bar or Meter. Highlight the one you want with the cursor keys, and then press [ENTER].

Bar

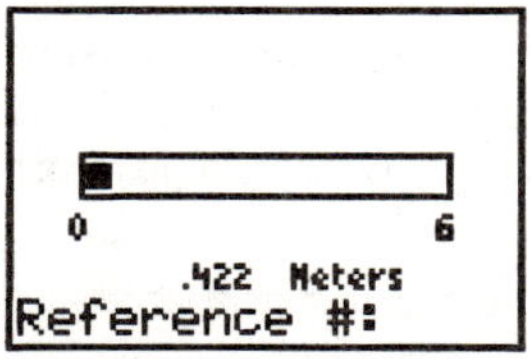

Meter

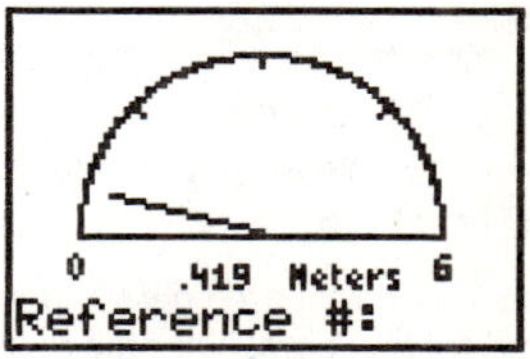

MIN and MAX

MIN and MAX refer to the minimum and maximum UNIT values for the specified probe. Defaults are listed in the table on page 19. See the CBL 2/CBL and CBR guidebook for specific MIN/MAX ranges. Enter values using the number keys.

UNITS

The results are displayed according to the UNITS specified. To specify a unit measurement (Temp or Sonic probes only), highlight the one you want using the cursor keys, enter a value using the number keys, and then press [ENTER].

DIRECTNS (Directions)

If DIRECTNS=On, the calculator displays step-by-step directions on the screen, which help you set up and run the data collection. To select On or Off, highlight the one you want with the cursor keys, and then press [ENTER].

With the Sonic data collection probe, if DIRECTNS=On, the calculator displays a menu screen before starting the application asking you to select **1:CBL** or **2:CBR**. This ensures that you get the appropriate directions. Press **1** to specify CBL 2/CBL or **2** to specify CBR.

Data Collection Comments and Results

To label a specific data point, press [ENTER] to pause the data collection. You will see a Reference#: prompt. Enter a number using the number keys. The calculator automatically converts the reference numbers and the corresponding results into list elements using the following list names (you cannot rename these lists):

Probe	Comment Labels (*X*) Stored to:	Data Results (*Y*) Stored to:
Temp	ʟTREF	ʟTEMP
Light	ʟLREF	ʟLIGHT
Volt	ʟVREF	ʟVOLT
Sonic	ʟDREF	ʟDIST

To see all elements in one of these lists, you can insert these lists into the List editor just as you would any other list. Access list names from the [2nd] [LIST] NAMES menu.

Note: These lists are only temporary placeholders for comment labels and data results for any particular probe. Therefore, every time you collect data and enter comments for one of the four probes, the two lists pertaining to that probe are overwritten with comment labels and data results from the most recently collected data..

If you want to save comment labels and data results from more than one data collection, copy all list elements that you want to save to a list with a different name.

Also, the DATA LOGGER data collection method stores data results to the same list names, overwriting previously-collected data results, even those collected using the GAUGE data collection method.

DATA LOGGER

1. Press [APPS] **2** [ENTER].

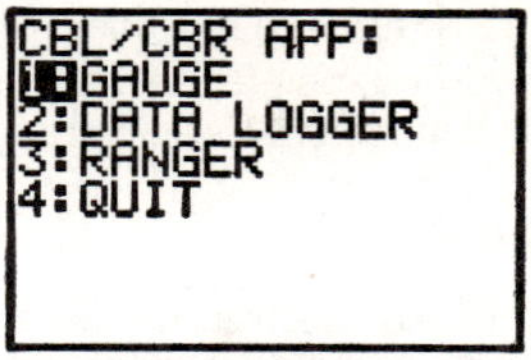

2. Select **2:DATA LOGGER.**

The DATA LOGGER data collection method lets you choose one of four different probes: Temp, Light, Volt, or Sonic. You can use the CBL 2/CBL with all probes; you can use the CBR only with the Sonic probe.

When you select a probe option, all other options change accordingly. Use [▶] and [◀] to move between the probe options. To select a probe, highlight the one you want with the cursor keys, and then press [ENTER].

DATA LOGGER Options (Defaults)				
	Temp	Light	Volt	Sonic
#SAMPLES:	99	99	99	99
INTRVL (SEC):	1	1	1	1
UNITS:	°C or °F	mW/cm^2	Volt	Cm or Ft
PLOT:	RealTme or End			
DIRECTNS:	On or Off			
Ymin ([WINDOW]):	0	0	-10	0
Ymax ([WINDOW]):	100	1	10	6

The DATA LOGGER data collection results are represented as a Temp-Time, Light-Time, Volt-Time, or Distance-Time graph.

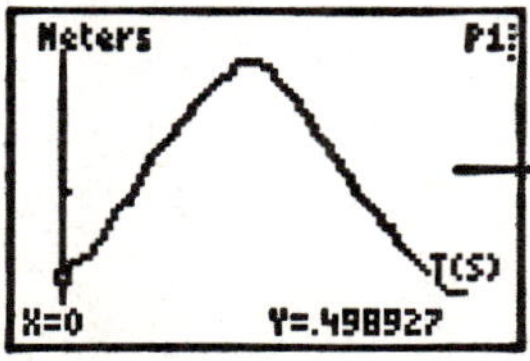

A Distance-Time graph in meters (**Sonic** probe).

#SAMPLES

#SAMPLES refers to how many data samples are collected and then graphed. For example, if #SAMPLES=99, data collection stops after the 99th sample is collected. Enter values using the number keys.

INTRVL (SEC)

INTRVL (SEC) specifies the interval in seconds between each data sample that is collected. For example, if you want to collect 99 samples and INTRVL=1, it takes 99 seconds to finish data collection. Enter values using the number keys. See the CBL 2/CBL or CBR guidebook for more information about interval limits.

UNITS

The results are displayed according to the UNITS specified. To specify a unit measurement (Temp or Sonic only), highlight the one you want using the cursor keys, and then press [ENTER].

PLOT

You can specify whether you want the calculator to collect realtime (RealTme) samples, which means that the calculator graphs data points immediately as they are being collected, or you can wait and show the graph only after all data points have been collected (End). Highlight the option you want with the cursor keys, and then press [ENTER].

Ymin and Ymax

To specify Ymin and Ymax values for the final graph, press [WINDOW] to view the PLOT WINDOW screen. Use [▲] and [▼] to move between options. Enter Ymin and Ymax using the number keys. Press [2nd] [QUIT] to return to the DATA LOGGER options screen.

DIRECTNS (Directions)

If DIRECTNS=On, the calculator displays step-by-step directions on the screen, which help you set up and run the data collection. To select On or Off, highlight the one you want with the cursor keys, and then press [ENTER].

With the Sonic data collection probe, if DIRECTNS=On, the calculator displays a menu screen before starting the application asking you to select **1:CBL** or **2:CBR**. This ensures that you get the appropriate directions. Press **1** to specify CBL 2/CBL or **2** to specify CBR.

Data Collection Results

The calculator automatically converts all collected data points into list elements using the following list names (you cannot rename the lists):

Probe	Time Values (*X*) stored to:	Data Results (*Y*) Stored to:
Temp	ʟTTEMP	ʟTEMP
Light	ʟTLGHT	ʟLIGHT
Volt	ʟTVOLT	ʟVOLT
Sonic	ʟTDIST	ʟDIST

To see all elements in one of these lists, you can insert these lists into the List editor just as you would any other list. Access list names from the [2nd] [LIST] NAMES menu.

Note: These lists are only temporary placeholders for data results for any particular probe. Therefore, every time you collect data for one of the four probes, the list pertaining to that probe is overwritten with data results from the most recently collected data.

If you want to save data results from more than one data collection, copy all list elements that you want to save to a list with a different name.

Also, the GAUGE data collection method stores data results to the same list names, overwriting previously-collected data results, even those collected using the DATA LOGGER data collection method.

RANGER

Selecting the RANGER data collection method runs the CBR RANGER program, a customized program especially for the TI-83 Plus that makes it compatible with the CBR. When the collection process is halted, the CBR RANGER is deleted from RAM. To run the CBR RANGER program again, press [APPS] and select the CBL/CBR application.

Note: The Ranger data collection method only uses the Sonic probe.

1. Press [APPS] **2** [ENTER].
2. Select **3:RANGER.**

3. Press [ENTER].

4. Select options.

For detailed information about the RANGER program as well as option explanations, see the Getting Started with CBR™ guidebook.

Collecting the Data

After you specify all of the options for your data collection method, select the Go option from the GAUGE or DATA LOGGER options screen. If you are using the RANGER data collection method, select 1:SETUP/SAMPLE from the MAIN menu, and then START NOW.

- If DIRECTNS=Off, GAUGE and DATA LOGGER data collection begin immediately.
- If DIRECTNS=On, the calculator displays step-by-step directions.

If PROBE=Sonic, the calculator first displays a menu screen asking you to select **1:CBL** or **2:CBR**. This ensures that you get the appropriate directions. Press **1** to specify CBL 2/CBL or **2** to specify CBR.

If you select START NOW from the MAIN menu of the RANGER data collection method, the calculator displays one directions screen. Press [ENTER] to begin data collection.

To stop the GAUGE data collection method, press [CLEAR] on the TI-83 Plus.

The DATA LOGGER and RANGER data collection methods stop after the specified number of samples have been collected. To stop them before this happens:

1. Press [ON] on the TI-83 Plus.
2. Press (TRIGGER) on the CBR, (START/STOP) on the CBL 2, or [ON/HALT] on the CBL.

To exit from the GAUGE or DATA LOGGER option menus without beginning data collection, press [2nd] [QUIT].

To exit from the RANGER option menu without beginning data collection, select MAIN menu. Select 6:QUIT to return to the CBL/CBR APP menu.

Press **4:QUIT** from the CBL/CBR APP menu to return to the TI-83 Plus Home screen.

15 CATALOG, Strings, Hyperbolic Functions

Contents

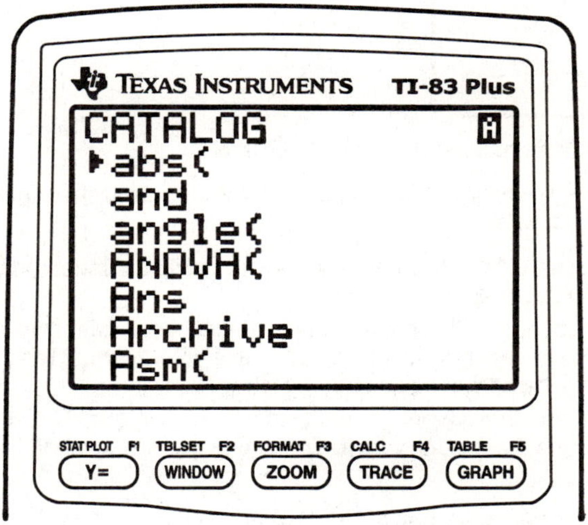

What Is the CATALOG?

The CATALOG is an alphabetical list of all functions and instructions on the TI-83 Plus. You also can access each CATALOG item from a menu or the keyboard, except:

- The six string functions (page 15-6)
- The six hyperbolic functions (page 15-10)
- The **solve(** instruction without the equation solver editor (Chapter 2)
- The inferential stat functions without the inferential stat editors (Chapter 13)

Note: The only CATALOG programming commands you can execute from the home screen are **GetCalc(**, **Get(**, and **Send(**.

Selecting an Item from the CATALOG

To select a CATALOG item, follow these steps.

1. Press [2nd] [CATALOG] to display the CATALOG.

The ▸ in the first column is the selection cursor.

2. Press [▼] or [▲] to scroll the CATALOG until the selection cursor points to the item you want.

   - To jump to the first item beginning with a particular letter, press that letter; alpha-lock is on.
   - Items that begin with a number are in alphabetical order according to the first letter after the number. For example, **2-PropZTest(** is among the items that begin with the letter **P**.
   - Functions that appear as symbols, such as **+**, **$^{-1}$**, **<**, and **√(**, follow the last item that begins with **Z**. To jump to the first symbol, **!**, press [θ].

3. Press [ENTER] to paste the item to the current screen.

Tip: From the top of the CATALOG menu, press [▲] to move to the bottom. From the bottom, press [▼] to move to the top.

What Is a String?

A string is a sequence of characters that you enclose within quotation marks. On the TI-83 Plus, a string has two primary applications.

- It defines text to be displayed in a program.
- It accepts input from the keyboard in a program.

Characters are the units that you combine to form a string.

- Count each number, letter, and space as one character.
- Count each instruction or function name, such as **sin(** or **cos(**, as one character; the TI-83 Plus interprets each instruction or function name as one character.

Entering a String

To enter a string on a blank line on the home screen or in a program, follow these steps.

1. Press [ALPHA] ["] to indicate the beginning of the string.
2. Enter the characters that comprise the string.
   - Use any combination of numbers, letters, function names, or instruction names to create the string.
   - To enter a blank space, press [ALPHA] [␣].
   - To enter several alpha characters in a row, press [2nd] [A-LOCK] to activate alpha-lock.
3. Press [ALPHA] ["] to indicate the end of the string.

   "*string*"
4. Press [ENTER]. On the home screen, the string is displayed on the next line without quotations. An ellipsis (...) indicates that the string continues beyond the screen. To scroll the entire string, press [▶] and [◀].

```
"ABCD 1234 EFGH
5678"
ABCD 1234 EFGH ...
```

Note: Quotation marks do not count as string characters.

String Variables

The TI-83 Plus has 10 variables to which you can store strings. You can use string variables with string functions and instructions.

To display the VARS STRING menu, follow these steps.

1. Press [VARS] to display the VARS menu. Move the cursor to **7:String.**

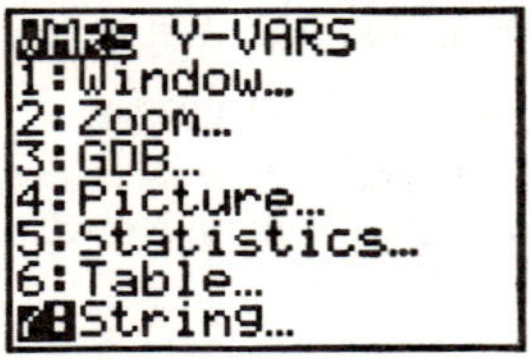

2. Press [ENTER] to display the STRING secondary menu.

Storing a String to a String Variable

To store a string to a string variable, follow these steps.

1. Press [ALPHA] ["], enter the string, and press [ALPHA] ["].
2. Press [STO▸].
3. Press [VARS] **7** to display the VARS STRING menu.
4. Select the string variable (from **Str1** to **Str9**, or **Str0**) to which you want to store the string.

   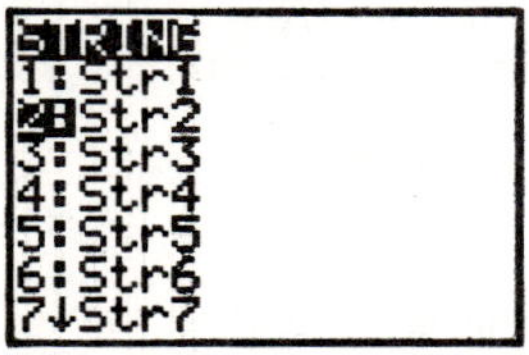

   The string variable is pasted to the current cursor location, next to the store symbol (→).

5. Press [ENTER] to store the string to the string variable. On the home screen, the stored string is displayed on the next line without quotation marks.

   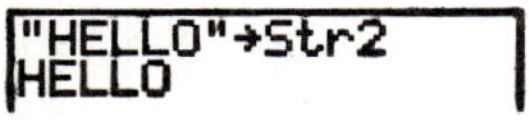

Displaying the Contents of a String Variable

To display the contents of a string variable on the home screen, select the string variable from the VARS STRING menu, and then press [ENTER]. The string is displayed.

Displaying String Functions and Instructions in the CATALOG

String functions and instructions are available only from the CATALOG. The table below lists the string functions and instructions in the order in which they appear among the other CATALOG menu items. The ellipses in the table indicate the presence of additional CATALOG items.

CATALOG	
...	
`Equ▸String(`	Converts an equation to a string.
`expr(`	Converts a string to an expression.
...	
`inString(`	Returns a character's place number.
...	
`length(`	Returns a string's character length.
...	
`String▸Equ(`	Converts a string to an equation.
`sub(`	Returns a string subset as a string.
...	

+ (Concatenation)

To concatenate two or more strings, follow these steps.

1. Enter *string1*, which can be a string or string name.
2. Press [+].
3. Enter *string2*, which can be a string or string name. If necessary, press [+] and enter *string3*, and so on.

   string1+*string2*+*string3*. . .

4. Press [ENTER] to display the strings as a single string.

```
"HIJK "→Str1:Str
1+"LMNOP"
HIJK LMNOP
```

Selecting a String Function from the CATALOG

To select a string function or instruction and paste it to the current screen, follow the steps on page 15-2.

Equ▸String(

Equ▸String(converts to a string an equation that is stored to any VARS Y-VARS variable. **Y***n* contains the equation. **Str***n* (from **Str1** to **Str9**, or **Str0**) is the string variable to which you want the equation to be stored as a string.

Equ▸String(Y*n***,Str***n***)**

```
"3X"→Y1
                Done
Equ▸String(Y1,St
r1)
                Done
Str1
3X
```

expr(

expr(converts the character string contained in *string* to an expression and executes it. *string* can be a string or a string variable.

expr(*string***)**

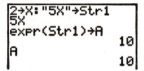

```
2→X:"5X"→Str1
5X
expr(Str1)→A
                  10
A
                  10
```

```
expr("1+2+X²")
                   7
```

inString(

inString(returns the character position in *string* of the first character of *substring*. *string* can be a string or a string variable. *start* is an optional character position at which to start the search; the default is 1.

inString(*string***,***substring*[**,***start*]**)**

```
inString("PQRSTU
V","STU")
                   4
inString("ABCABC
","ABC",4)
                   4
```

Note: If *string* does not contain *substring*, or *start* is greater than the length of *string*, **inString(** returns **0**.

length(

length(returns the number of characters in *string*. *string* can be a string or string variable.

Note: An instruction or function name, such as **sin(** or **cos(**, counts as one character.

length(*string***)**

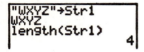

```
"WXYZ"→Str1
WXYZ
length(Str1)
                4
```

String▸Equu(

String▸Equ(converts *string* into an equation and stores the equation to **Y*n***. *string* can be a string or string variable. **String▸Equ(** is the inverse of **Equ▸String(**.

String▸Equ(*string*,**Y***n***)**

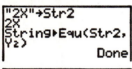

```
"2X"→Str2
2X
String▸Equ(Str2,
Y2)
            Done
```

```
 Plot1 Plot2 Plot3
\Y1=
\Y2=2X
```

sub(

sub(returns a string that is a subset of an existing *string*. *string* can be a string or a string variable. *begin* is the position number of the first character of the subset. *length* is the number of characters in the subset.

sub(*string*,*begin*,*length***)**

```
"ABCDEFG"→Str5
ABCDEFG
sub(Str5,4,2)
DE
```

Entering a Function to Graph during Program Execution

In a program, you can enter a function to graph during program execution using these commands.

```
PROGRAM:INPUT
:Input "ENTRY=",
Str3
:String▸Equ(Str3
,Y3)
:DispGraph
```

```
prgmINPUT
ENTRY=3X█
```

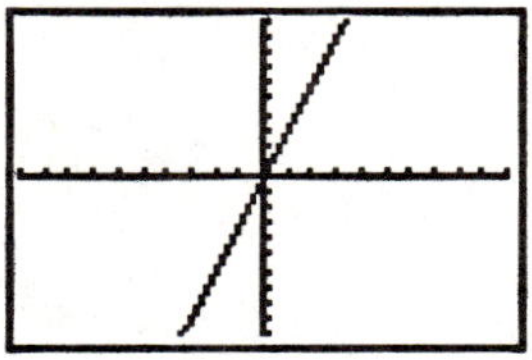

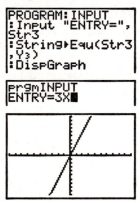

Note: When you execute this program, enter a function to store to **Y3** at the **ENTRY=** prompt.

Hyperbolic Functions

The hyperbolic functions are available only from the CATALOG. The table below lists the hyperbolic functions in the order in which they appear among the other CATALOG menu items. The ellipses in the table indicate the presence of additional CATALOG items.

CATALOG	
...	
cosh(	Hyperbolic cosine
$\cosh^{-1}$(	Hyperbolic arccosine
...	
sinh(	Hyperbolic sine
$\sinh^{-1}$(	Hyperbolic arcsine
...	
tanh(	Hyperbolic tangent
$\tanh^{-1}$(	Hyperbolic arctangent
...	

sinh(, cosh(, tanh(

sinh(, **cosh(**, and **tanh(** are the hyperbolic functions. Each is valid for real numbers, expressions, and lists.

sinh(*value***)**
cosh(*value***)**
tanh(*value***)**

```
sinh(.5)
      .5210953055
cosh({.25,.5,1})

{1.0314131 1.12…
```

$\sinh^{-1}$(, $\cosh^{-1}$(, $\tanh^{-1}$(

$\sinh^{-1}$(is the hyperbolic arcsine function. **$\cosh^{-1}$(** is the hyperbolic arccosine function. **$\tanh^{-1}$(** is the hyperbolic arctangent function. Each is valid for real numbers, expressions, and lists.

$\sinh^{-1}$(*value***)**
$\cosh^{-1}$(*value***)**
$\sinh^{-1}$(*value***)**

```
sinh⁻¹({0,1})
   {0 .881373587}
tanh⁻¹(-.5)
      -.5493061443
```

16 Programming

Contents

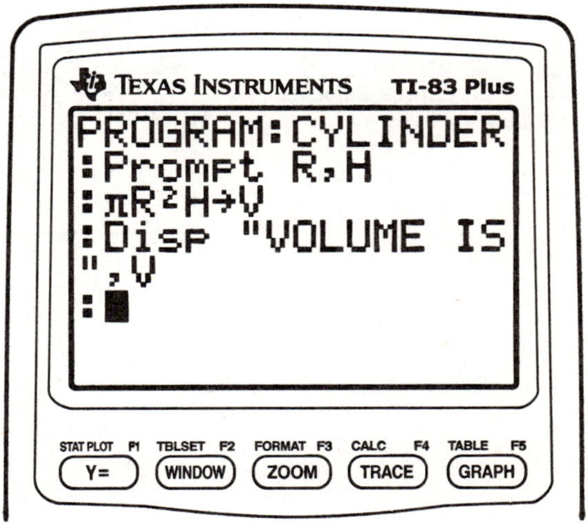

Getting Started is a fast-paced introduction. Read the chapter for details.

A program is a set of commands that the TI-83 Plus executes sequentially, as if you had entered them from the keyboard. Create a program that prompts for the radius R and the height H of a cylinder and then computes its volume.

1. Press [PRGM] [▸] [▸] to display the PRGM NEW menu.

```
EXEC EDIT NEW
1:Create New
```

2. Press [ENTER] to select **1:Create New**. The **Name=** prompt is displayed, and alpha-lock is on. Press [C] [Y] [L] [I] [N] [D] [E] [R], and then press [ENTER] to name the program **CYLINDER**.

   You are now in the program editor. The colon (:) in the first column of the second line indicates the beginning of a command line.

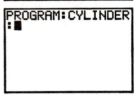

```
PROGRAM:CYLINDER
:█
```

3. Press [PRGM] [▸] **2** to select **2:Prompt** from the PRGM I/O menu. **Prompt** is copied to the command line. Press [ALPHA] [R] [,] [ALPHA] [H] to enter the variable names for radius and height. Press [ENTER].

```
PROGRAM:CYLINDER
:Prompt R,H
:█
```

4. Press [2nd] [π] [ALPHA] [R] [x^2] [ALPHA] [H] [STO▸] [ALPHA] [V] [ENTER] to enter the expression $\pi R^2 H$ and store it to the variable **V**.

```
PROGRAM:CYLINDER
:Prompt R,H
:πR²H→V
:
```

5. Press [PRGM] [▸] **3** to select **3:Disp** from the PRGM I/O menu. **Disp** is pasted to the command line. Press [2nd] [A-LOCK] ["] [V] [O] [L] [U] [M] [E][␣] [I] [S] ["] [ALPHA] [,] [ALPHA] [V] [ENTER] to set up the program to display the text **VOLUME IS** on one line and the calculated value of **V** on the next.

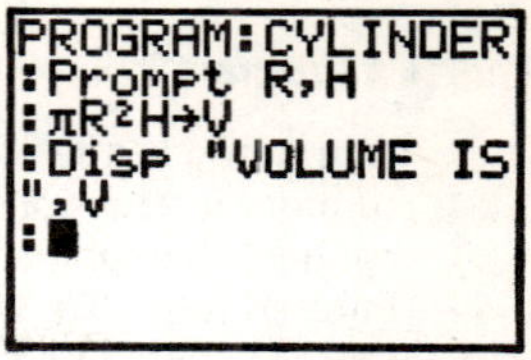
PROGRAM:CYLINDER
:Prompt R,H
:πR²H→V
:Disp "VOLUME IS
",V
:

6. Press [2nd] [QUIT] to display the home screen.

7. Press [PRGM] to display the PRGM EXEC menu. The items on this menu are the names of stored programs.

EXEC EDIT NEW
1:CYLINDER

8. Press [ENTER] to paste **prgmCYLINDER** to the current cursor location. (If **CYLINDER** is not item **1** on your PRGM EXEC menu, move the cursor to **CYLINDER** before you press [ENTER].)

prgmCYLINDER

9. Press [ENTER] to execute the program. Enter **1.5** for the radius, and then press [ENTER]. Enter **3** for the height, and then press [ENTER]. The text **VOLUME IS**, the value of **V**, and **Done** are displayed.

   Repeat steps 7 through 9 and enter different values for **R** and **H**.

prgmCYLINDER
R=?1.5
H=?3
VOLUME IS
21.20575041
Done

What Is a Program?

A program is a set of one or more command lines. Each line contains one or more instructions. When you execute a program, the TI-83 Plus performs each instruction on each command line in the same order in which you entered them. The number and size of programs that the TI-83 Plus can store is limited only by available memory.

Creating a New Program

To create a new program, follow these steps.

1. Press [PRGM] [◄] to display the PRGM NEW menu.

```
EXEC EDIT NEW
1:Create New
```

2. Press [ENTER] to select **1:Create New**. The **Name=** prompt is displayed, and alpha-lock is on.
3. Press a letter from A to Z or θ to enter the first character of the new program name.

   Note: A program name can be one to eight characters long. The first character must be a letter from A to Z or θ. The second through eighth characters can be letters, numbers, or θ.

4. Enter zero to seven letters, numbers, or θ to complete the new program name.
5. Press [ENTER]. The program editor is displayed.
6. Enter one or more program commands (page 16-6).
7. Press [2nd] [QUIT] to leave the program editor and return to the home screen.

Managing Memory and Deleting a Program

To check whether adequate memory is available for a program you want to enter:

1. Press [2nd] [MEM] to display the MEMORY menu.
2. select **2:Mem Mgmt/Del** to display the MEMORY MANAGEMENT/DELETE menu (Chapter 18).
3. Select **7:Prgm** to display the PRGM editor.

```
RAM FREE   24790
ARC FREE  130832
 *PROGRAM1    20
▶ PROGRAM2    22
```

The TI-83 Plus expresses memory quantities in bytes.

You can increase available memory in one of two ways. You can delete one or more programs or you can archive some programs.

To increase available memory by deleting a specific program:

1. Press [2nd] [MEM] and then select **2:Mem Mgmt/Del** from the MEMORY menu.

2. Select **7:Prgm** to display the PRGM editor (Chapter 18).

3. Press [▲] and [▼] to move the selection cursor (▶) next to the program you want to delete, and then press [DEL]. The program is deleted from memory.

   Note: You will receive a message asking you to confirm this delete action. Select **2:yes** to continue.

   To leave the PRGM editor screen without deleting anything, press [2nd] [QUIT], which displays the home screen.

To increase available memory by archiving a program:

1. Press [2nd] [MEM] and then select **2:Mem Mgmt/Del** from the MEMORY menu.

2. Select **2:Mem Mgmt/Del** to display the MEM MGMT/DEL menu.

3. Select **7:Prgm...** to display the PRGM menu.

4. Press [ENTER] to archive the program. An asterisk will appear to the left of the program to indicate it is an archived program.

   To unarchive a program in this screen, put the cursor next to the archived program and press [ENTER]. The asterisk will disappear.

Note: Archive programs cannot be edited or executed. In order to edit or execute an archived program, you must first unarchive it.

Entering a Program Command Line

You can enter on a command line any instruction or expression that you could execute from the home screen. In the program editor, each new command line begins with a colon. To enter more than one instruction or expression on a single command line, separate each with a colon.

Note: A command line can be longer than the screen is wide; long command lines wrap to the next screen line.

While in the program editor, you can display and select from menus. You can return to the program editor from a menu in either of two ways.

- Select a menu item, which pastes the item to the current command line.
- Press [CLEAR].

When you complete a command line, press [ENTER]. The cursor moves to the next command line.

Programs can access variables, lists, matrices, and strings saved in memory. If a program stores a new value to a variable, list, matrix, or string, the program changes the value in memory during execution.

You can call another program as a subroutine (page 16-16 and page 16-23).

Executing a Program

To execute a program, begin on a blank line on the home screen and follow these steps.

1. Press [PRGM] to display the PRGM EXEC menu.
2. Select a program name from the PRGM EXEC menu (page 16-8). **prgm***name* is pasted to the home screen (for example, **prgmCYLINDER**).
3. Press [ENTER] to execute the program. While the program is executing, the busy indicator is on.

Last Answer (**Ans**) is updated during program execution. Last Entry is not updated as each command is executed (Chapter 1).

The TI-83 Plus checks for errors during program execution. It does not check for errors as you enter a program.

Breaking a Program

To stop program execution, press [ON]. The ERR:BREAK menu is displayed.

- To return to the home screen, select **1:Quit**.
- To go where the interruption occurred, select **2:Goto**.

Editing a Program

To edit a stored program, follow these steps.

1. Press [PRGM] [▶] to display the PRGM EDIT menu.

2. Select a program name from the PRGM EDIT menu (page 16-8). Up to the first seven lines of the program are displayed.

   Note: The program editor does not display a ↓ to indicate that a program continues beyond the screen.

3. Edit the program command lines.

   - Move the cursor to the appropriate location, and then delete, overwrite, or insert.
   - Press [CLEAR] to clear all program commands on the command line (the leading colon remains), and then enter a new program command.

Tip: To move the cursor to the beginning of a command line, press [2nd] [◀]; to move to the end, press [2nd] [▶]. To scroll the cursor down seven command lines, press [ALPHA] [▼]. To scroll the cursor up seven command lines, press [ALPHA] [▲].

Inserting and Deleting Command Lines

To insert a new command line anywhere in the program, place the cursor where you want the new line, press [2nd] [INS], and then press [ENTER]. A colon indicates a new line.

To delete a command line, place the cursor on the line, press [CLEAR] to clear all instructions and expressions on the line, and then press [DEL] to delete the command line, including the colon.

Copying and Renaming a Program

To copy all command lines from one program into a new program, follow steps 1 through 5 for Creating a New Program (page 16-4), and then follow these steps.

1. Press [2nd] [RCL]. **Rcl** is displayed on the bottom line of the program editor in the new program (Chapter 1).
2. Press [PRGM] [◄] to display the PRGM EXEC menu.
3. Select a name from the menu. **prgm***name* is pasted to the bottom line of the program editor.
4. Press [ENTER]. All command lines from the selected program are copied into the new program.

Copying programs has at least two convenient applications.

- You can create a template for groups of instructions that you use frequently.
- You can rename a program by copying its contents into a new program.

Note: You also can copy all the command lines from one existing program to another existing program using RCL.

Scrolling the PRGM EXEC and PRGM EDIT Menus

The TI-83 Plus sorts PRGM EXEC and PRGM EDIT menu items automatically into alphanumerical order. Each menu only labels the first 10 items using **1** through **9**, then **0**.

To jump to the first program name that begins with a particular alpha character or θ, press [ALPHA] [*letter from A to Z or θ*].

Tip: From the top of either the PRGM EXEC or PRGM EDIT menu, press [▲] to move to the bottom. From the bottom, press [▼] to move to the top. To scroll the cursor down the menu seven items, press [ALPHA] [▼]. To scroll the cursor up the menu seven items, press [ALPHA] [▲].

PRGM CTL Menu

To display the PRGM CTL (program control) menu, press [PRGM] from the program editor only.

CTL I/O EXEC	
1: If	Creates a conditional test.
2: Then	Executes commands when **If** is true.
3: Else	Executes commands when **If** is false.
4: For(	Creates an incrementing loop.
5: While	Creates a conditional loop.
6: Repeat	Creates a conditional loop.
7: End	Signifies the end of a block.
8: Pause	Pauses program execution.
9: Lbl	Defines a label.
0: Goto	Goes to a label.
A: IS>(	Increments and skips if greater than.
B: DS<(	Decrements and skips if less than.
C: Menu(	Defines menu items and branches.
D: prgm	Executes a program as a subroutine.
E: Return	Returns from a subroutine.
F: Stop	Stops execution.
G: DelVar	Deletes a variable from within program.
H: GraphStyle(	Designates the graph style to be drawn.

These menu items direct the flow of an executing program. They make it easy to repeat or skip a group of commands during program execution. When you select an item from the menu, the name is pasted to the cursor location on a command line in the program.

To return to the program editor without selecting an item, press [CLEAR].

Controlling Program Flow

Program control instructions tell the TI-83 Plus which command to execute next in a program. **If**, **While**, and **Repeat** check a defined condition to determine which command to execute next. Conditions frequently use relational or Boolean tests (Chapter 2), as in:

If A<7:A+1→A

or

If N=1 and M=1:Goto Z

If

Use **If** for testing and branching. If *condition* is false (zero), then the *command* immediately following **If** is skipped. If *condition* is true (nonzero), then the next *command* is executed. **If** instructions can be nested.

:**If** *condition*
:*command* (if true)
:*command*

Program

```
PROGRAM:COUNT
:0→A
:Lbl Z
:A+1→A
:Disp "A IS",A
:If A≥2
:Stop
:Goto Z
```

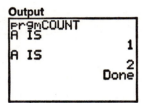

Output

```
prgmCOUNT
A IS
               1
A IS
               2
            Done
```

If-Then

Then following an **If** executes a group of *commands* if *condition* is true (nonzero). **End** identifies the end of the group of *commands*.

:**If** *condition*
:**Then**
:*command* (if true)
:*command* (if true)
:**End**
:*command*

Program

```
PROGRAM:TEST
:1→X:10→Y
:If X<10
:Then
:2X+3→X
:2Y-3→Y
:End
:Disp X,Y
```

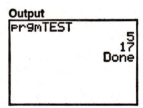

Output

```
prgmTEST
               5
              17
            Done
```

If-Then-Else

Else following **If-Then** executes a group of *commands* if *condition* is false (zero). **End** identifies the end of the group of *commands.*

:If *condition*
:Then
:*command* (if true)
:*command* (if true)
:Else
:*command* (if false)
:*command* (if false)
:End
:*command*

Program

```
PROGRAM:TESTELSE
:Input "X=",X
:If X<0
:Then
:X²→Y
:Else
:X→Y
:End
:Disp {X,Y}
```

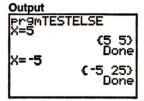

Output

```
prgmTESTELSE
X=5
                {5 5}
                 Done
X=-5
              {-5 25}
                 Done
```

For(

For(loops and increments. It increments *variable* from *begin* to *end* by *increment*. *increment* is optional (default is 1) and can be negative (*end*<*begin*). *end* is a maximum or minimum value not to be exceeded. **End** identifies the end of the loop. **For(** loops can be nested.

:For(*variable*,*begin*,*end*[,*increment*]**)**
:*command* (while *end* not exceeded)
:*command* (while *end* not exceeded)
:End
:*command*

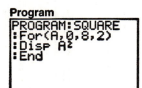

Program

```
PROGRAM:SQUARE
:For(A,0,8,2)
:Disp A²
:End
```

Output

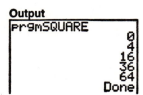

```
prgmSQUARE
                    0
                    4
                   16
                   36
                   64
                 Done
```

While

While performs a group of *commands* while *condition* is true. *condition* is frequently a relational test (Chapter 2). *condition* is tested when **While** is encountered. If *condition* is true (nonzero), the program executes a group of *commands*. **End** signifies the end of the group. When *condition* is false (zero), the program executes each *command* following **End**. **While** instructions can be nested.

:While *condition*
:*command* (while *condition* is true)
:*command* (while *condition* is true)
:End
:*command*

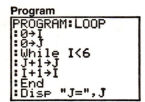

Program

```
PROGRAM:LOOP
:0→I
:0→J
:While I<6
:J+1→J
:I+1→I
:End
:Disp "J=",J
```

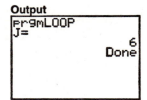

Output

```
prgmLOOP
J=
                 6
              Done
```

Repeat

Repeat repeats a group of *commands* until *condition* is true (nonzero). It is similar to **While**, but *condition* is tested when **End** is encountered; therefore, the group of *commands* is always executed at least once. **Repeat** instructions can be nested.

:Repeat *condition*
:*command* (until *condition* is true)
:*command* (until *condition* is true)
:End
:*command*

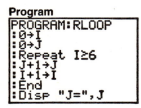

Program

```
PROGRAM:RLOOP
:0→I
:0→J
:Repeat I≥6
:J+1→J
:I+1→I
:End
:Disp "J=",J
```

Output

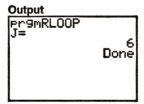

```
prgmRLOOP
J=
                 6
              Done
```

End

End identifies the end of a group of *commands*. You must include an **End** instruction at the end of each **For(**, **While**, or **Repeat** loop. Also, you must paste an **End** instruction at the end of each **If-Then** group and each **If-Then-Else** group.

Pause

Pause suspends execution of the program so that you can see answers or graphs. During the pause, the pause indicator is on in the top-right corner. Press [ENTER] to resume execution.

- **Pause** without a *value* temporarily pauses the program. If the **DispGraph** or **Disp** instruction has been executed, the appropriate screen is displayed.
- **Pause** with *value* displays *value* on the current home screen. *value* can be scrolled.

Pause [*value*]

Program

```
PROGRAM:PAUSE
:10→X
:"X²+2"→Y1
:Disp "X=",X
:Pause
:DispGraph
:Pause
:Disp
```

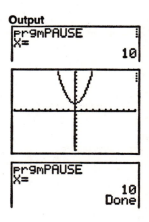

Output

```
prgmPAUSE
X=
                10
```

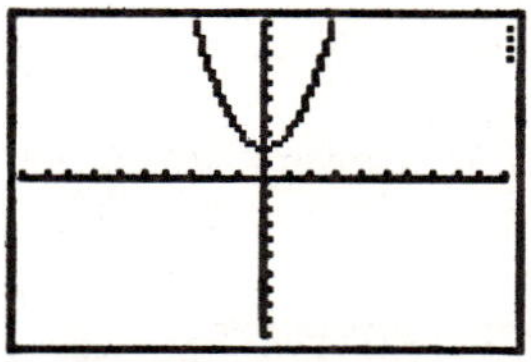

```
prgmPAUSE
X=
                10
              Done
```

Lbl, Goto

Lbl (label) and **Goto** (go to) are used together for branching.

Lbl specifies the *label* for a command. *label* can be one or two characters (**A** through **Z**, **0** through **99**, or θ).

Lbl *label*

Goto causes the program to branch to *label* when **Goto** is encountered.

Goto *label*

Program

```
PROGRAM:CUBE
:Lbl 99
:Input A
:If A≥100
:Stop
:Disp A³
:Pause
:Goto 99
```

Output

```
prgmCUBE
?2
                8
?3
               27
?105
             Done
```

IS>(

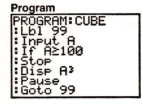

IS>((increment and skip) adds 1 to *variable*. If the answer is > *value* (which can be an expression), the next *command* is skipped; if the answer is ≤ *value*, the next *command* is executed. *variable* cannot be a system variable.

:**IS>(***variable*,*value***)**
:*command* (if answer ≤ *value*)
:*command* (if answer > *value*)

Program

```
PROGRAM:ISKIP
:7→A
:IS>(A,6)
:Disp "NOT > 6"
:Disp "> 6"
```

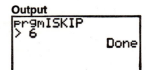

Output

```
prgmISKIP
> 6
             Done
```

Note: IS>(is not a looping instruction.

DS<(

DS<((decrement and skip) subtracts 1 from *variable*. If the answer is < *value* (which can be an expression), the next *command* is skipped; if the answer is ≥ *value*, the next *command* is executed. *variable* cannot be a system variable.

:DS<(variable,value**)**
:*command* (if answer ≥ *value*)
:*command* (if answer < *value*)

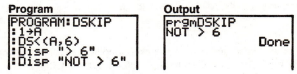

Program

```
PROGRAM:DSKIP
:1→A
:DS<(A,6)
:Disp "> 6"
:Disp "NOT > 6"
```

Output

```
prgmDSKIP
NOT > 6
                Done
```

Note: DS<(is not a looping instruction.

Menu(

Menu(sets up branching within a program. If **Menu(** is encountered during program execution, the menu screen is displayed with the specified menu items, the pause indicator is on, and execution pauses until you select a menu item.

The menu *title* is enclosed in quotation marks ("). Up to seven pairs of menu items follow. Each pair comprises a *text* item (also enclosed in quotation marks) to be displayed as a menu selection, and a *label* item to which to branch if you select the corresponding menu selection.

Menu("*title***","***text1***",***label1***,"***text2***",***label2***, . . .)**

Program

```
PROGRAM:TOSSDICE
:Menu("TOSS DICE
","FAIR DICE",A,
"WEIGHTED DICE",
B)
```

Output

```
TOSS DICE
1:FAIR DICE
2:WEIGHTED DICE
```

The program above pauses until you select **1** or **2**. If you select **2**, for example, the menu disappears and the program continues execution at **Lbl B**.

prgm

Use **prgm** to execute other programs as subroutines (page 16-23). When you select **prgm**, it is pasted to the cursor location. Enter characters to spell a program *name*. Using **prgm** is equivalent to selecting existing programs from the PRGM EXEC menu; however, it allows you to enter the name of a program that you have not yet created.

prgm*name*

Note: You cannot directly enter the subroutine name when using RCL. You must paste the name from the PRGM EXEC menu (page 16-8).

Return

Return quits the subroutine and returns execution to the calling program (page 16-23), even if encountered within nested loops. Any loops are ended. An implied **Return** exists at the end of any program that is called as a subroutine. Within the main program, **Return** stops execution and returns to the home screen.

Stop

Stop stops execution of a program and returns to the home screen. **Stop** is optional at the end of a program.

DelVar

DelVar deletes from memory the contents of *variable*.

DelVar *variable*

```
PROGRAM:DELMATR
:DelVar [A]■
```

GraphStyle(

GraphStyle(designates the style of the graph to be drawn. *function#* is the number of the Y= function name in the current graphing mode. *graphstyle* is a number from **1** to **7** that corresponds to the graph style, as shown below.

1 = ⸜ (line)
2 = ⸜ (thick)
3 = ◥ (shade above)
4 = ◣ (shade below)
5 = -0 (path)
6 = 0 (animate)
7 = ⋱ (dot)

GraphStyle(*function#*,*graphstyle***)**

For example, **GraphStyle(1,5)** in **Func** mode sets the graph style for **Y1** to -0 (path; **5**).

Not all graph styles are available in all graphing modes. For a detailed description of each graph style, see the Graph Styles table in Chapter **3**.

PRGM I/O Menu

To display the PRGM I/O (program input/output) menu, press [PRGM] [▸] from within the program editor only.

CTL **I/O** EXEC	
1: Input	Enters a value or uses the cursor.
2: Prompt	Prompts for entry of variable values.
3: Disp	Displays text, value, or the home screen.
4: DispGraph	Displays the current graph.
5: DispTable	Displays the current table.
6: Output(	Displays text at a specified position.
7: getKey	Checks the keyboard for a keystroke.
8: ClrHome	Clears the display.
9: ClrTable	Clears the current table.
0: GetCalc(	Gets a variable from another TI-83 Plus.
A: Get(	Gets a variable from CBL 2/CBL or CBR.
B: Send(	Sends a variable to CBL 2/CBL or CBR.

These instructions control input to and output from a program during execution. They allow you to enter values and display answers during program execution.

To return to the program editor without selecting an item, press [CLEAR].

Displaying a Graph with Input

Input without a variable displays the current graph. You can move the free-moving cursor, which updates **X** and **Y** (and **R** and θ for **PolarGC** format). The pause indicator is on. Press [ENTER] to resume program execution.

Input

Program

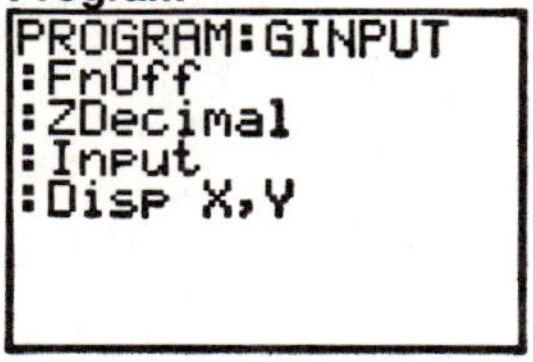

Output

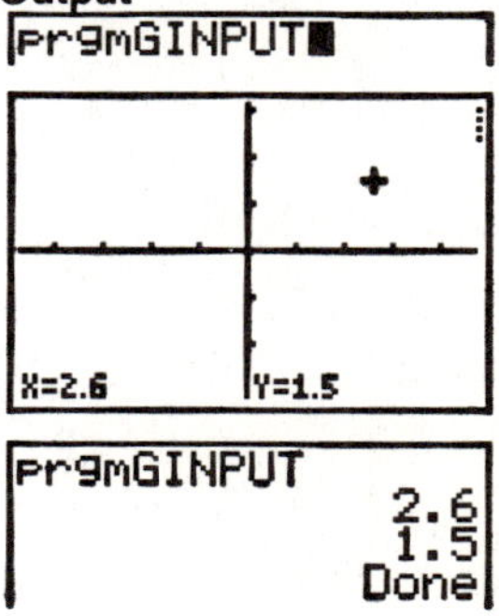

Storing a Variable Value with Input

Input with *variable* displays a **?** (question mark) prompt during execution. *variable* may be a real number, complex number, list, matrix, string, or **Y=** function. During program execution, enter a value, which can be an expression, and then press [ENTER]. The value is evaluated and stored to *variable*, and the program resumes execution.

Input [*variable*]

You can display *text* or the contents of **Str***n* (a string variable) of up to 16 characters as a prompt. During program execution, enter a value after the prompt and then press [ENTER]. The value is stored to *variable*, and the program resumes execution.

Input ["*text*",*variable*]
Input [**Str***n*,*variable*]

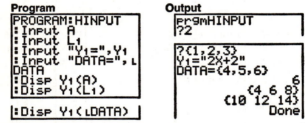

Program

```
PROGRAM:HINPUT
:Input A
:Input L1
:Input "Y1=",Y1
:Input "DATA=",LDATA
:Disp Y1(A)
:Disp Y1(L1)
```

```
:Disp Y1(LDATA)
```

Output

```
prgmHINPUT
?2
```

```
?{1,2,3}
Y1="2X+2"
DATA={4,5,6}
                   6
             {4 6 8}
          {10 12 14}
                Done
```

Note: When a program prompts for input of lists and **Y*n*** functions during execution, you must include the braces (**{ }**) around the list elements and quotation marks (**"**) around the expressions.

Prompt

During program execution, **Prompt** displays each *variable*, one at a time, followed by **=?**. At each prompt, enter a value or expression for each *variable*, and then press [ENTER]. The values are stored, and the program resumes execution.

Prompt *variableA*[,*variableB*,...,*variable n*]

Program	Output
PROGRAM:WINDOW :Prompt Xmin :Prompt Xmax :Prompt Ymin :Prompt Ymax	prgmWINDOW Xmin=?-10 Xmax=?10 Ymin=?-3 Ymax=?3 Done

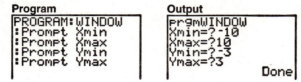

Note: Y= functions are not valid with **Prompt**.

Displaying the Home Screen

Disp (display) without a value displays the home screen. To view the home screen during program execution, follow the **Disp** instruction with a **Pause** instruction.

Disp

Displaying Values and Messages

Disp with one or more *values* displays the value of each.

Disp [*valueA*,*valueB*,*valueC*,...,*value n*]

- If *value* is a variable, the current value is displayed.
- If *value* is an expression, it is evaluated and the result is displayed on the right side of the next line.
- If *value* is text within quotation marks, it is displayed on the left side of the current display line. → is not valid as text.

Program	Output
PROGRAM:A :Disp "THE ANSWE R IS ",π/2	prgmA THE ANSWER IS 1.570796327 Done

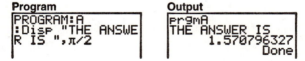

If **Pause** is encountered after **Disp**, the program halts temporarily so you can examine the screen. To resume execution, press [ENTER].

Note: If a matrix or list is too large to display in its entirety, ellipses (...) are displayed in the last column, but the matrix or list cannot be scrolled. To scroll, use **Pause** *value* (page 16-13).

DispGraph

DispGraph (display graph) displays the current graph. If **Pause** is encountered after **DispGraph**, the program halts temporarily so you can examine the screen. Press [ENTER] to resume execution.

DispTable

DispTable (display table) displays the current table. The program halts temporarily so you can examine the screen. Press [ENTER] to resume execution.

Output(

Output(displays *text* or *value* on the current home screen beginning at *row* (**1** through **8**) and *column* (**1** through **16**), overwriting any existing characters.

Tip: You may want to precede **Output(** with **ClrHome** (page 16-21).

Expressions are evaluated and values are displayed according to the current mode settings. Matrices are displayed in entry format and wrap to the next line. → is not valid as text.

Output(*row*,*column*,"*text*"**)**
Output(*row*,*column*,*value***)**

Program

```
PROGRAM:OUTPUT
:3+5→B
:ClrHome
:Output(5,4,"ANS
WER:"
:Output(5,12,B)
```

Output

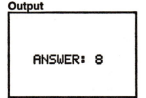

```
ANSWER: 8
```

For **Output(** on a **Horiz** split screen, the maximum value for *row* is 4.

getKey

getKey returns a number corresponding to the last key pressed, according to the key code diagram below. If no key has been pressed, **getKey** returns 0. Use **getKey** inside loops to transfer control, for example, when creating video games.

Program

```
PROGRAM:GETKEY
:While 1
:getKey→K
:While K=0
:getKey→K
:End
:Disp K
:If K=105
:Stop
:End
```

Output

```
prgmGETKEY
                41
                42
                43
               105
              Done
```

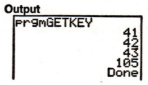

Note: [MATH], [MATRX], [PRGM], and [ENTER] were pressed during program execution.

Note: You can press [ON] at any time during execution to break the program (page 16-6).

TI-83 Plus Key Code Diagram

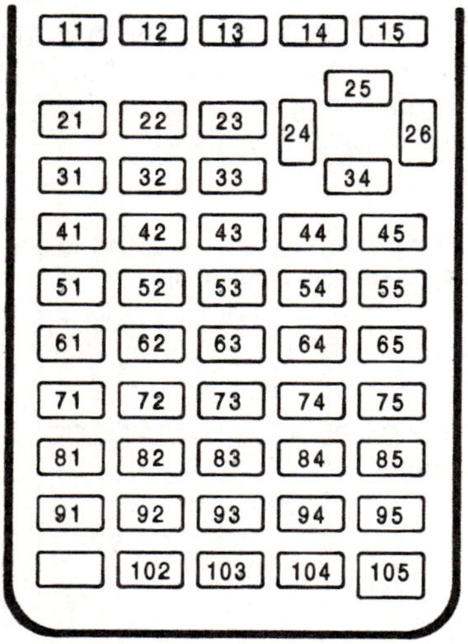

ClrHome, ClrTable

ClrHome (clear home screen) clears the home screen during program execution.

ClrTable (clear table) clears the values in the table during program execution.

GetCalc(

GetCalc(gets the contents of *variable* on another TI-83 Plus and stores it to *variable* on the receiving TI-83 Plus. *variable* can be a real or complex number, list element, list name, matrix element, matrix name, string, Y= variable, graph database, or picture.

GetCalc(*variable***)**

Note: GetCalc(does not work between TI-82 and TI-83 Plus calculators.

Get(, Send(

Get(gets data from the Calculator-Based Laboratory™ (CBL 2™, CBL™) System or Calculator-Based Ranger™ (CBR™) and stores it to *variable* on the receiving TI-83 Plus. *variable* can be a real number, list element, list name, matrix element, matrix name, string, Y= variable, graph database, or picture.

Get(*variable***)**

Note: If you transfer a program that references the **Get(** command to the TI-83 Plus from a TI-82, the TI-83 Plus will interpret it as the **Get(** described above. Use **GetCalc(** to get data from another TI-83 Plus.

Send(sends the contents of *variable* to the CBL 2/CBL or CBR. You cannot use it to send to another TI-83 Plus. *variable* can be a real number, list element, list name, matrix element, matrix name, string, Y= variable, graph database, or picture. *variable* can be a list of elements.

Send(*variable***)**

```
PROGRAM:GETSOUND
:Send({3,.00025,
99,1,0,0,0,0,1})

:Get(L1)
:Get(L2)
```

Note: This program gets sound data and time in seconds from CBL 2/CBL.

Note: You can access **Get(**, **Send(**, and **GetCalc(** from the CATALOG to execute them from the home screen (Chapter 15).

Calling a Program from Another Program

On the TI-83 Plus, any stored program can be called from another program as a subroutine. Enter the name of the program to use as a subroutine on a line by itself.

You can enter a program name on a command line in either of two ways.

- Press [PRGM] [◄] to display the PRGM EXEC menu and select the name of the program (page 16-8). **prgm***name* is pasted to the current cursor location on a command line.
- Select **prgm** from the PRGM CTL menu, and then enter the program name (page 16-9).

prgm*name*

When **prgm***name* is encountered during execution, the next command that the program executes is the first command in the second program. It returns to the subsequent command in the first program when it encounters either **Return** or the implied **Return** at the end of the second program.

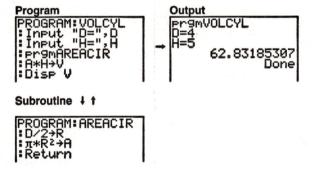

Program

```
PROGRAM:VOLCYL
:Input "D=",D
:Input "H=",H
:prgmAREACIR
:A*H→V
:Disp V
```

→

Output

```
prgmVOLCYL
D=4
H=5
         62.83185307
                Done
```

Subroutine ↓ ↑

```
PROGRAM:AREACIR
:D/2→R
:π*R²→A
:Return
```

Notes about Calling Programs

Variables are global.

label used with **Goto** and **Lbl** is local to the program where it is located. *label* in one program is not recognized by another program. You cannot use **Goto** to branch to a *label* in another program.

Return exits a subroutine and returns to the calling program, even if it is encountered within nested loops.

You can run programs written for the TI-83 Plus in assembly language. Typically, assembly language programs run much faster and provide greater control than than the keystroke programs that you write with the built-in program editor.

Note: Because an assembly langauge program has greater control over the calculator, if your assembly language program has error(s), it may cause your calculator to reset and lose all data, programs, and applications stored in memory.

When you download an assembly language program, it is stored among the other programs as a PRGM menu item. You can:

- Transmit it using the TI-83 Plus communication link (Chapter 19).
- Delete it using the MEM MGMT DEL screen (Chapter 18).

To run an assembly Program, the syntax is: **Asm(***assemblyprgmname***)**

If you write an assembly language program, use the two instructions below from the CATALOG to identify and compile the program.

Instructions	Comments
AsmComp(*prgmASM1*, *prgmASM2***)**	Compiles an assembly language program written in ASCII and stores the hex version
AsmPrgm	Identifies an assembly language program; must be entered as the first line of an assembly language program

To compile an assembly program that you have written:

1. Follow the steps for writing a program (16-4) but be sure to include **AsmPrgm** as the first line of your program.
2. From the home screen, press [2nd] [CATALOG] and then select **AsmComp(** to paste it to the screen
3. Press [PRGM] to display the PRGM EXEC menu.
4. Select the program you want to compile. It will be pasted to the home screen.
5. Press [,] and then select **prgm** from the CATALOG
6. Key in the name you have chosen for the output program.

   Note: This name must be unique – not a copy of an existing program name.
7. Press [)] to complete the sequence.

   The sequence of the arguments should be as follows:

   AsmComp(*prgmASM1*, *prgmASM2***)**
8. Press [ENTER] to compile your program and generate the output program.

17 Activities

Contents

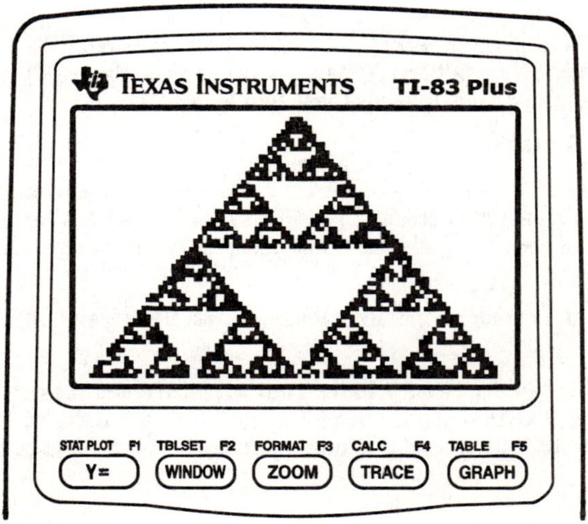

Problem

An experiment found a significant difference between boys and girls pertaining to their ability to identify objects held in their left hands, which are controlled by the right side of their brains, versus their right hands, which are controlled by the left side of their brains. The TI Graphics team conducted a similar test for adult men and women.

The test involved 30 small objects, which participants were not allowed to see. First, they held 15 of the objects one by one in their left hands and guessed what they were. Then they held the other 15 objects one by one in their right hands and guessed what they were. Use box plots to compare visually the correct-guess data from this table.

Correct Guesses

Women Left	Women Right	Men Left	Men Right
8	4	7	12
9	1	8	6
12	8	7	12
11	12	5	12
10	11	7	7
8	11	8	11
12	13	11	12
7	12	4	8
9	11	10	12
11	12	14	11
		13	9
		5	9

Procedure

1. Press [STAT] **5** to select **5:SetUpEditor**. Enter list names **WLEFT**, **WRGHT**, **MLEFT**, and **MRGHT**, separated by commas. Press [ENTER]. The stat list editor now contains only these four lists.
2. Press [STAT] **1** to select **1:Edit**.
3. Enter into **WLEFT** the number of correct guesses each woman made using her left hand (Women Left). Press [▶] to move to **WRGHT** and enter the number of correct guesses each woman made using her right hand (Women Right).
4. Likewise, enter each man's correct guesses in **MLEFT** (Men Left) and **MRGHT** (Men Right).
5. Press [2nd] [STAT PLOT]. Select **1:Plot1**. Turn on plot 1; define it as a modified box plot that uses **WLEFT**. Move the cursor to the top line and select **Plot2**. Turn on plot 2; define it as a modified box plot that uses **WRGHT**.

Procedure (continued)

6. Press [Y=]. Turn off all functions.
7. Press [WINDOW]. Set **Xscl=1** and **Yscl=0**. Press [ZOOM] **9** to select **9:ZoomStat**. This adjusts the viewing window and displays the box plots for the women's results.
8. Press [TRACE].

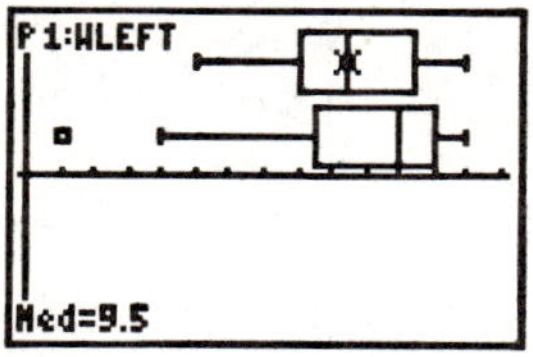

← Women's left-hand data

← Women's right-hand data

Use [◄] and [►] to examine **minX**, **Q1**, **Med**, **Q3**, and **maxX** for each plot. Notice the outlier to the women's right-hand data. What is the median for the left hand? For the right hand? With which hand were the women more accurate guessers, according to the box plots?

9. Examine the men's results. Redefine plot 1 to use **MLEFT**, redefine plot 2 to use **MRGHT**. Press [TRACE].

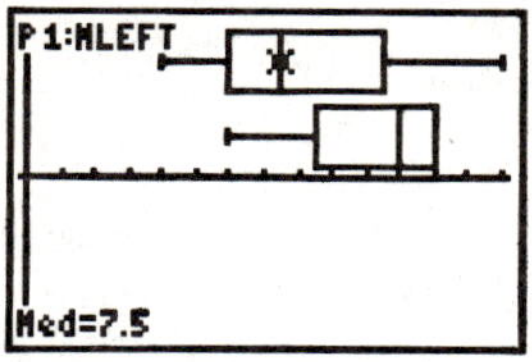

← Men's left-hand data

← Men's right-hand data

Press [◄] and [►] to examine **minX**, **Q1**, **Med**, **Q3**, and **maxX** for each plot. What difference do you see between the plots?

10. Compare the left-hand results. Redefine plot 1 to use **WLEFT**, redefine plot 2 to use **MLEFT**, and then press [TRACE] to examine **minX**, **Q1**, **Med**, **Q3**, and **maxX** for each plot. Who were the better left-hand guessers, men or women?
11. Compare the right-hand results. Define plot 1 to use **WRGHT**, define plot 2 to use **MRGHT**, and then press [TRACE] to examine **minX**, **Q1**, **Med**, **Q3**, and **maxX** for each plot. Who were the better right-hand guessers?

In the original experiment boys did not guess as well with right hands, while girls guessed equally well with either hand. This is not what our box plots show for adults. Do you think that this is because adults have learned to adapt or because our sample was not large enough?

Problem

The fine for speeding on a road with a speed limit of 45 kilometers per hour (kph) is 50; plus 5 for each kph from 46 to 55 kph; plus 10 for each kph from 56 to 65 kph; plus 20 for each kph from 66 kph and above. Graph the piecewise function that describes the cost of the ticket.

The fine (Y) as a function of kilometers per hour (X) is:

$Y = 0$	$0 < X \le 45$
$Y = 50 + 5\,(X - 45)$	$45 < X \le 55$
$Y = 50 + 5 * 10 + 10\,(X - 55)$	$55 < X \le 65$
$Y = 50 + 5 * 10 + 10 * 10 + 20\,(X - 65)$	$65 < X$

Procedure

1. Press [MODE]. Select **Func** and the default settings.
2. Press [Y=]. Turn off all functions and stat plots. Enter the **Y=** function to describe the fine. Use the TEST menu operations to define the piecewise function. Set the graph style for **Y1** to ·. (dot).

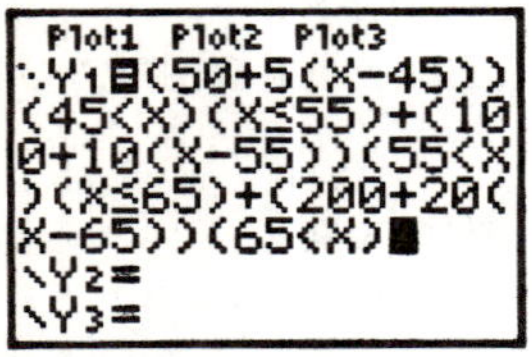

3. Press [WINDOW] and set **Xmin=‾2**, **Xscl=10**, **Ymin=‾5**, and **Yscl=10**. Ignore **Xmax** and **Ymax**; they are set by **ΔX** and **ΔY** in step **4**.
4. Press [2nd] [QUIT] to return to the home screen. Store **1** to **ΔX**, and then store **5** to **ΔY**. **ΔX** and **ΔY** are on the VARS Window X/Y secondary menu. **ΔX** and **ΔY** specify the horizontal and vertical distance between the centers of adjacent pixels. Integer values for **ΔX** and **ΔY** produce nice values for tracing.
5. Press [TRACE] to plot the function. At what speed does the ticket exceed 250?

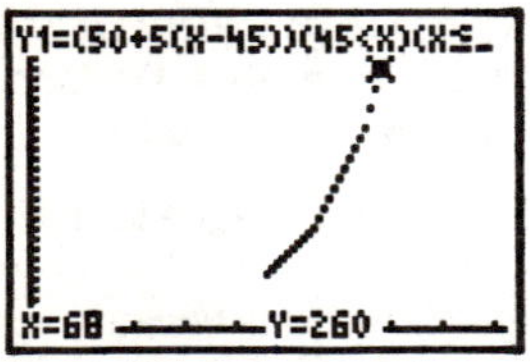

Problem

Graph the inequality $0.4X^3 - 3X + 5 < 0.2X + 4$. Use the TEST menu operations to explore the values of X where the inequality is true and where it is false.

Procedure

1. Press [MODE]. Select **Dot**, **Simul**, and the default settings. Setting **Dot** mode changes all graph style icons to ⠑ (dot) in the Y= editor.
2. Press [Y=]. Turn off all functions and stat plots. Enter the left side of the inequality as **Y4** and the right side as **Y5**.

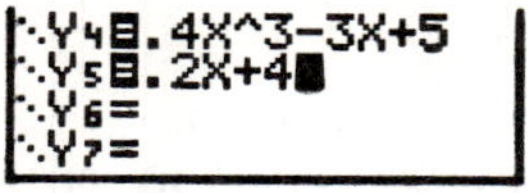

3. Enter the statement of the inequality as **Y6**. This function evaluates to **1** if true or **0** if false.

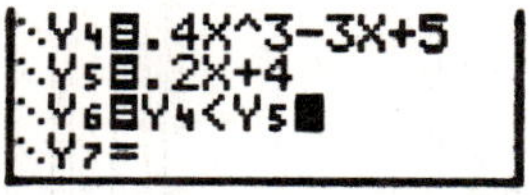

4. Press [ZOOM] **6** to graph the inequality in the standard window.
5. Press [TRACE] [▼] [▼] to move to **Y6**. Then press [◄] and [►] to trace the inequality, observing the value of **Y**.

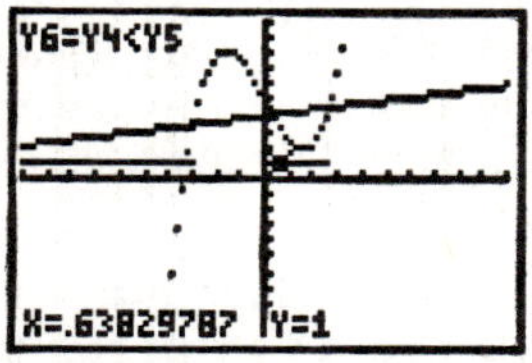

6. Press [Y=]. Turn off **Y4**, **Y5**, and **Y6**. Enter equations to graph only the inequality.

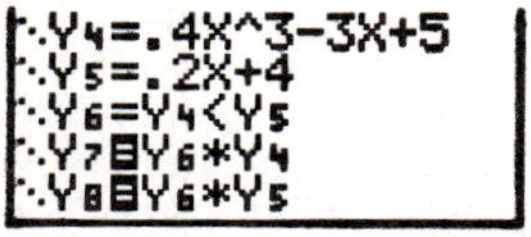

7. Press [TRACE]. Notice that the values of **Y7** and **Y8** are zero where the inequality is false.

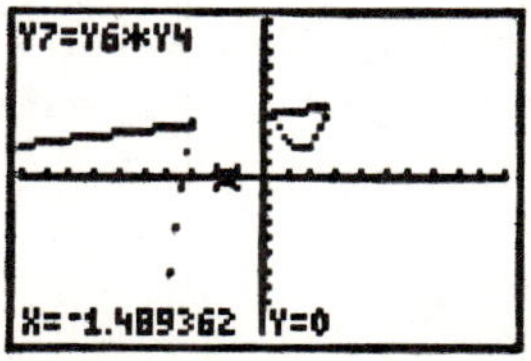

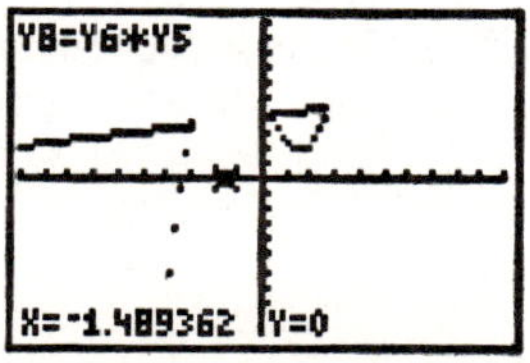

Problem

Using a graph, solve the equation $X^3 - 2X = 2\cos(X)$. Stated another way, solve the system of two equations and two unknowns: $Y = X^3-2X$ and $Y = 2\cos(X)$. Use ZOOM factors to control the decimal places displayed on the graph.

Procedure

1. Press [MODE]. Select the default mode settings. Press [Y=]. Turn off all functions and stat plots. Enter the functions.

```
\Y9=X³-2X
\Y0=2cos(X)
```

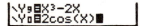

2. Press [ZOOM] **4** to select **4:ZDecimal**. The display shows that two solutions may exist (points where the two functions appear to intersect).

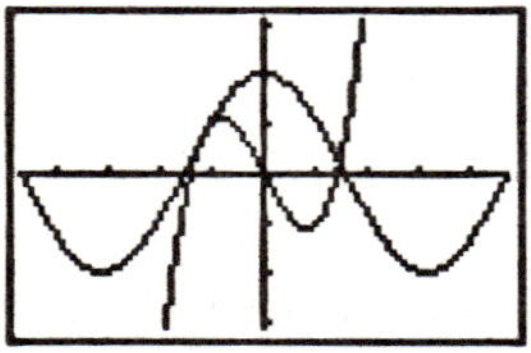

3. Press [ZOOM] [▸] **4** to select **4:SetFactors** from the ZOOM MEMORY menu. Set **XFact=10** and **YFact=10.**
4. Press [ZOOM] **2** to select **2:Zoom In**. Use [◂], [▸], [▴], and [▾] to move the free-moving cursor onto the apparent intersection of the functions on the right side of the display. As you move the cursor, notice that the **X** and **Y** values have one decimal place.
5. Press [ENTER] to zoom in. Move the cursor over the intersection. As you move the cursor, notice that now the **X** and **Y** values have two decimal places.
6. Press [ENTER] to zoom in again. Move the free-moving cursor onto a point exactly on the intersection. Notice the number of decimal places.
7. Press [2nd] [CALC] **5** to select **5:Intersect**. Press [ENTER] to select the first curve and [ENTER] to select the second curve. To guess, move the trace cursor near the intersection. Press [ENTER]. What are the coordinates of the intersection point?
8. Press [ZOOM] **4** to select **4:ZDecimal** to redisplay the original graph.
9. Press [ZOOM]. Select **2:Zoom In** and repeat steps 4 through 8 to explore the apparent function intersection on the left side of the display.

Setting up the Program

This program creates a drawing of a famous fractal, the Sierpinski Triangle, and stores the drawing to a picture. To begin, press [PRGM] [▶] [▶] **1**. Name the program **SIERPINS**, and then press [ENTER]. The program editor is displayed.

Program

```
PROGRAM:SIERPINS
:FnOff :ClrDraw
:PlotsOff
:AxesOff
:0→Xmin:1→Xmax          Set viewing window.
:0→Ymin:1→Ymax
:rand→X:rand→Y
:For(K,1,3000)          Beginning of For group.
:rand→N
:If N≤1/3
:Then
:.5X→X                  If/Then group
:.5Y→Y
:End
:If 1/3<N and N≤2/3
:Then
:.5(.5+X)→X             If/Then group.
:.5(1+Y)→Y
:End
:If 2/3<N
:Then
:.5(1+X)→X              If/Then group.
:.5Y→Y
:End
:Pt-On(X,Y)             Draw point.
:End                    End of For group.
:StorePic 6             Store picture.
```

After you execute the program above, you can recall and display the picture with the instruction **RecallPic 6**.

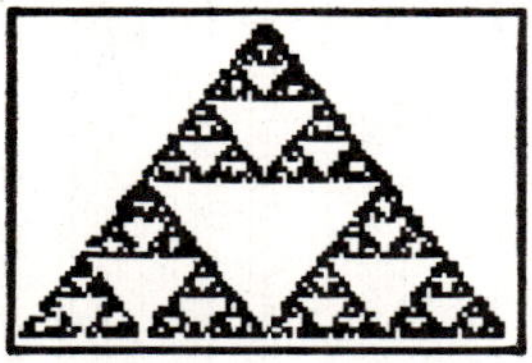

Problem

Using **Web** format, you can identify points with attracting and repelling behavior in sequence graphing.

Procedure

1. Press [MODE]. Select **Seq** and the default mode settings. Press [2nd] [FORMAT]. Select **Web** format and the default format settings.
2. Press [Y=]. Clear all functions and turn off all stat plots. Enter the sequence that corresponds to the expression Y = K X(1−X).

   u(*n*)=Ku(*n*−1)(1−u(*n*−1))
   u(*n*Min)=.01
3. Press [2nd] [QUIT] to return to the home screen, and then store **2.9** to **K**.
4. Press [WINDOW]. Set the window variables.

***n*Min=0**	**Xmin=0**	**Ymin=⁻.26**
***n*Max=10**	**Xmax=1**	**Ymax=1.1**
PlotStart=1	**Xscl=1**	**Yscl=1**
PlotStep=1		
5. Press [TRACE] to display the graph, and then press [▶] to trace the cobweb. This is a cobweb with one attractor.

   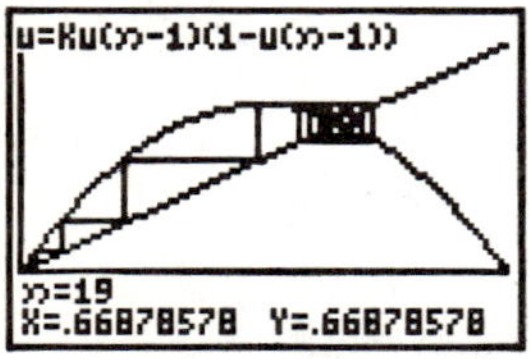

6. Change **K** to **3.44** and trace the graph to show a cobweb with two attractors.
7. Change **K** to **3.54** and trace the graph to show a cobweb with four attractors.

   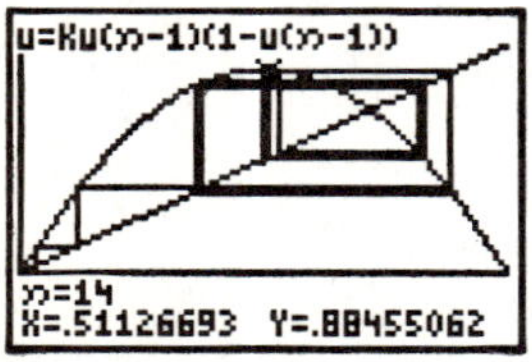

Using a Program to Guess the Coefficients

Setting Up the Program

This program graphs the function A sin(BX) with random integer coefficients between 1 and 10. Try to guess the coefficients and graph your guess as C sin(DX). The program continues until your guess is correct.

Program

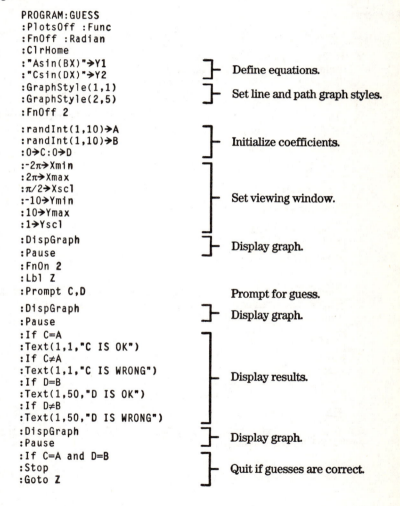

Program	

`PROGRAM:GUESS`
`:PlotsOff :Func`
`:FnOff :Radian`
`:ClrHome`	

`:"Asin(BX)"→Y1` `:"Csin(DX)"→Y2`	Define equations.
`:GraphStyle(1,1)` `:GraphStyle(2,5)`	Set line and path graph styles.
`:FnOff 2`	

`:randInt(1,10)→A`
`:randInt(1,10)→B`
`:0→C:0→D`
	Initialize coefficients.

`:-2π→Xmin`
`:2π→Xmax`
`:π/2→Xscl`
`:-10→Ymin`
`:10→Ymax`
`:1→Yscl`
	Set viewing window.

`:DispGraph` `:Pause`	Display graph.
`:FnOn 2` `:Lbl Z`	
`:Prompt C,D`	Prompt for guess.
`:DispGraph` `:Pause`	Display graph.

`:If C=A`
`:Text(1,1,"C IS OK")`
`:If C≠A`
`:Text(1,1,"C IS WRONG")`
`:If D=B`
`:Text(1,50,"D IS OK")`
`:If D≠B`
`:Text(1,50,"D IS WRONG")`
	Display results.

`:DispGraph` `:Pause`	Display graph.

`:If C=A and D=B`
`:Stop`
`:Goto Z`
	Quit if guesses are correct.

Problem

Using parametric graphing mode, graph the unit circle and the sine curve to show the relationship between them.

Any function that can be plotted in **Func** mode can be plotted in **Par** mode by defining the **X** component as **T** and the **Y** component as **F(T)**.

Procedure

1. Press [MODE]. Select **Par**, **Simul**, and the default settings.
2. Press [WINDOW]. Set the viewing window.

Tmin=0	**Xmin=⁻2**	**Ymin=⁻3**
Tmax=2π	**Xmax=7.4**	**Ymax=3**
Tstep=.1	**Xscl=π/2**	**Yscl=1**

3. Press [Y=]. Turn off all functions and stat plots. Enter the expressions to define the unit circle centered on (0,0).

4. Enter the expressions to define the sine curve.

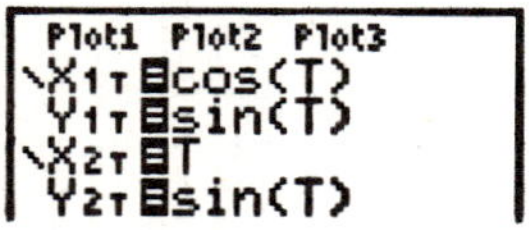

5. Press [TRACE]. As the graph is plotting, you may press [ENTER] to pause and [ENTER] again to resume graphing as you watch the sine function "unwrap" from the unit circle.

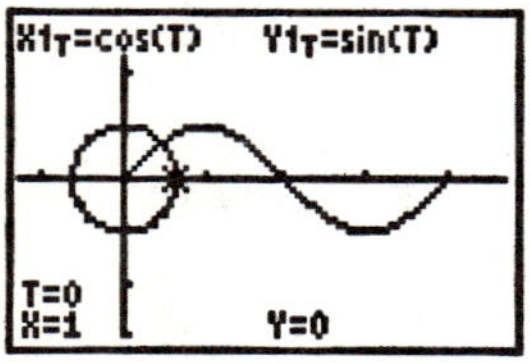

Note: You can generalize the unwrapping. Replace **sin(T)** in **Y2T** with any other trig function to unwrap that function.

Problem

Find the area of the region bounded by

$f(x) = 300x / (x^2 + 625)$
$g(x) = 3\cos(.1x)$
$x \quad = 75$

Procedure

1. Press [MODE]. Select the default mode settings.
2. Press [WINDOW]. Set the viewing window.

Xmin=0	**Ymin=-5**
Xmax=100	**Ymax=10**
Xscl=10	**Yscl=1**
	Xres=1

3. Press [Y=]. Turn off all functions and stat plots. Enter the upper and lower functions.
   Y1=300X/(X²+625)
   Y2=3cos(.1X)
4. Press [2nd] [CALC] **5** to select **5:Intersect**. The graph is displayed. Select a first curve, second curve, and guess for the intersection toward the left side of the display. The solution is displayed, and the value of **X** at the intersection, which is the lower limit of the integral, is stored in **Ans** and **X**.
5. Press [2nd] [QUIT] to go to the home screen. Press [2nd] [DRAW] **7** and use **Shade(** to see the area graphically.

   Shade(Y2,Y1,Ans,75)

   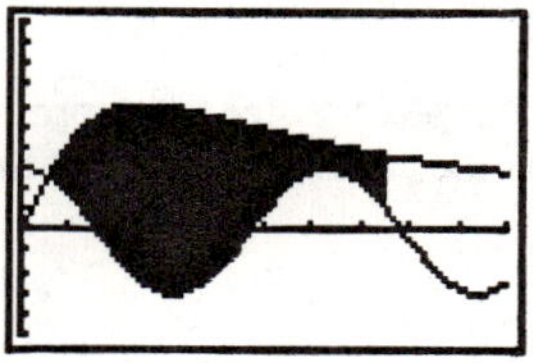

6. Press [2nd] [QUIT] to return to the home screen. Enter the expression to evaluate the integral for the shaded region.

   fnInt(Y1–Y2,X,Ans,75)

   The area is **325.839962.**

Problem

Using two pairs of parametric equations, determine when two objects in motion are closest to each other in the same plane.

A ferris wheel has a diameter (d) of 20 meters and is rotating counterclockwise at a rate (s) of one revolution every 12 seconds. The parametric equations below describe the location of a ferris wheel passenger at time T, where α is the angle of rotation, (0,0) is the bottom center of the ferris wheel, and (10,10) is the passenger's location at the rightmost point, when T=0.

$X(T) = r \cos \alpha$ where $\alpha = 2\pi Ts$ and $r = d/2$
$Y(T) = r + r \sin \alpha$

A person standing on the ground throws a ball to the ferris wheel passenger. The thrower's arm is at the same height as the bottom of the ferris wheel, but 25 meters (b) to the right of the ferris wheel's lowest point (25,0). The person throws the ball with velocity (v_0) of 22 meters per second at an angle (θ) of 66° from the horizontal. The parametric equations below describe the location of the ball at time T.

$X(T) = b - Tv_0 \cos\theta$
$Y(T) = Tv_0 \sin\theta - (g/2)\, T^2$ where $g = 9.8\ m/sec^2$

Procedure

1. Press [MODE]. Select **Par**, **Simul**, and the default settings. **Simul** (simultaneous) mode simulates the two objects in motion over time.

2. Press [WINDOW]. Set the viewing window.

Tmin=0	**Xmin=-13**	**Ymin=0**
Tmax=12	**Xmax=34**	**Ymax=31**
Tstep=.1	**Xscl=10**	**Yscl=10**

3. Press [Y=]. Turn off all functions and stat plots. Enter the expressions to define the path of the ferris wheel and the path of the ball. Set the graph style for **X2T** to ‑0 (path).

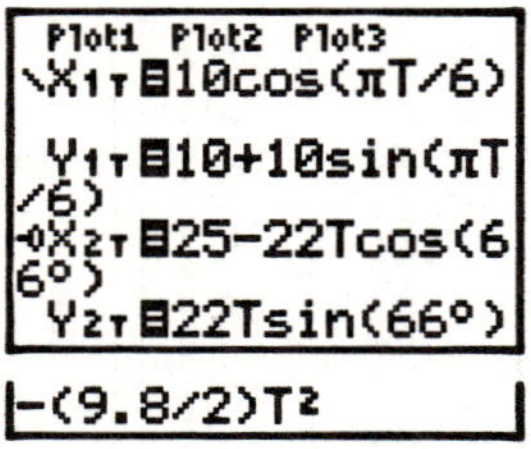

Tip: Try setting the graph styles to ‑0 **X1T** and 0 **X2T**, which simulates a chair on the ferris wheel and the ball flying through the air when you press [GRAPH].

Procedure (continued)

4. Press [GRAPH] to graph the equations. Watch closely as they are plotted. Notice that the ball and the ferris wheel passenger appear to be closest where the paths cross in the top-right quadrant of the ferris wheel.

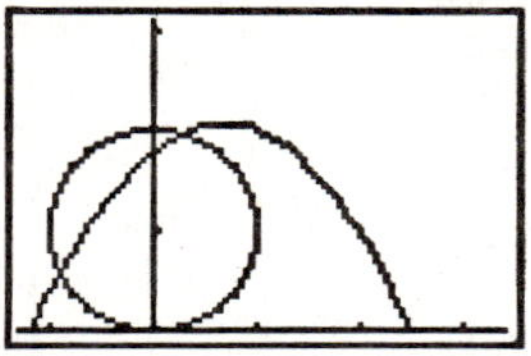

5. Press [WINDOW]. Change the viewing window to concentrate on this portion of the graph.

Tmin=1	**Xmin=0**	**Ymin=10**
Tmax=3	**Xmax=23.5**	**Ymax=25.5**
Tstep=.03	**Xscl=10**	**Yscl=10**

6. Press [TRACE]. After the graph is plotted, press [▶] to move near the point on the ferris wheel where the paths cross. Notice the values of **X**, **Y**, and **T**.

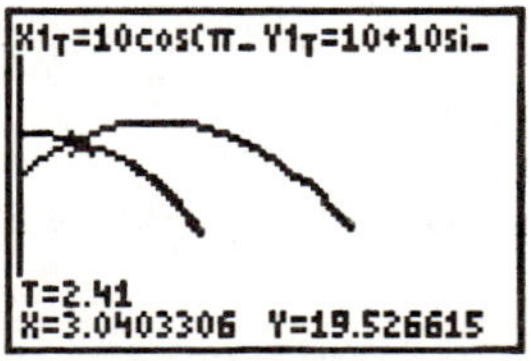

7. Press [▼] to move to the path of the ball. Notice the values of **X** and **Y** (**T** is unchanged). Notice where the cursor is located. This is the position of the ball when the ferris wheel passenger passes the intersection. Did the ball or the passenger reach the intersection first?

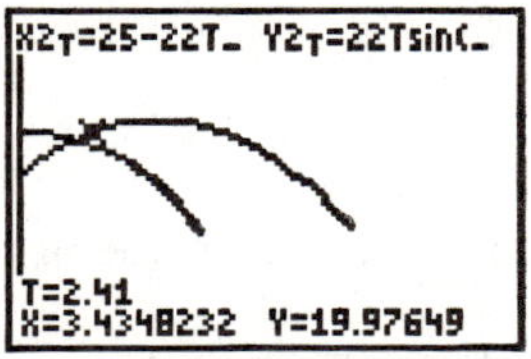

You can use [TRACE] to, in effect, take snapshots in time and explore the relative behavior of two objects in motion.

Problem 1

Using the functions **fnInt(** and **nDeriv(** from the MATH menu to graph functions defined by integrals and derivatives demonstrates graphically that:

$F(x) = \int_1^x 1/t\,dt = \ln(x), x > 0$ and that

$D_x\left[\int_1^x 1/t\,dt\right] = 1/x$

Procedure 1

1. Press [MODE]. Select the default settings.
2. Press [WINDOW]. Set the viewing window.

Xmin=.01	**Ymin=⁻1.5**	**Xres=3**
Xmax=10	**Ymax=2.5**	
Xscl=1	**Yscl=1**	

3. Press [Y=]. Turn off all functions and stat plots. Enter the numerical integral of 1/T from 1 to X and the function ln(X). Set the graph style for **Y1** to ⍀ (line) and **Y2** to -0 (path).

```
Plot1 Plot2 Plot3
\Y1=fnInt(1/T,T,
1,X)
-0Y2=ln(X)
```

4. Press [TRACE]. Press [◄], [▲], [►], and [▼] to compare the values of **Y1** and **Y2**.
5. Press [Y=]. Turn off **Y1** and **Y2**, and then enter the numerical derivative of the integral of 1/X and the function 1/X. Set the graph style for **Y3** to \ (line) and **Y4** to \ (thick).

```
Plot1 Plot2 Plot3
\Y1=fnInt(1/T,T,
1,X)
-0Y2=ln(X)
\Y3=nDeriv(Y1,X,
X)
\Y4=1/X
```

6. Press [TRACE]. Again, use the cursor keys to compare the values of the two graphed functions, **Y3** and **Y4**.

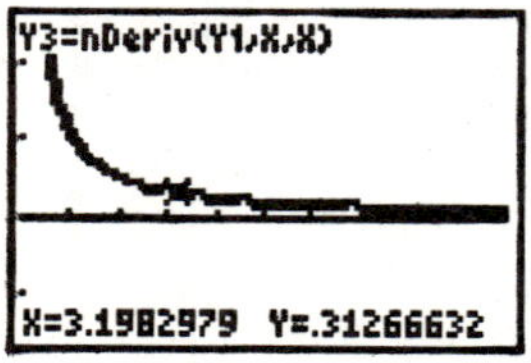

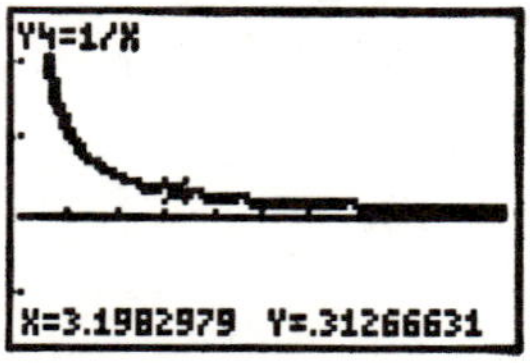

Problem 2

Explore the functions defined by

$$y = \int_{-2}^{x} t^2\,dt, \quad \int_{0}^{x} t^2\,dt, \quad \text{and} \quad \int_{2}^{x} t^2\,dt$$

Procedure 2

1. Press [Y=]. Turn off all functions and stat plots. Use a list to define these three functions simultaneously. Store the function in **Y5**.

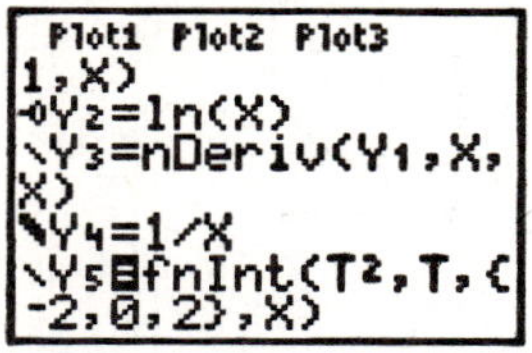

2. Press [ZOOM] **6** to select **6:ZStandard.**

3. Press [TRACE]. Notice that the functions appear identical, only shifted vertically by a constant.

4. Press [Y=]. Enter the numerical derivative of **Y5** in **Y6.**

5. Press [TRACE]. Notice that although the three graphs defined by **Y5** are different, they share the same derivative.

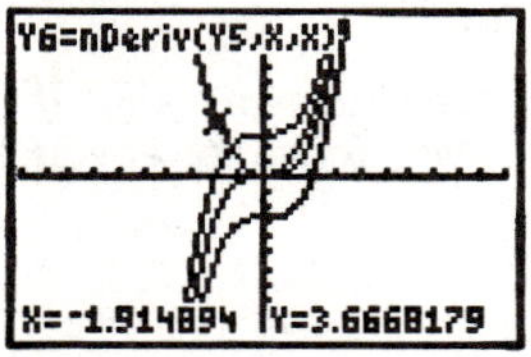

Problem

Use the equation solver to store a formula for the area of a regular N-sided polygon, and then solve for each variable, given the other variables. Explore the fact that the limiting case is the area of a circle, πr^2.

Consider the formula $A = NB^2 \sin(\pi/N) \cos(\pi/N)$ for the area of a regular polygon with N sides of equal length and B distance from the center to a vertex.

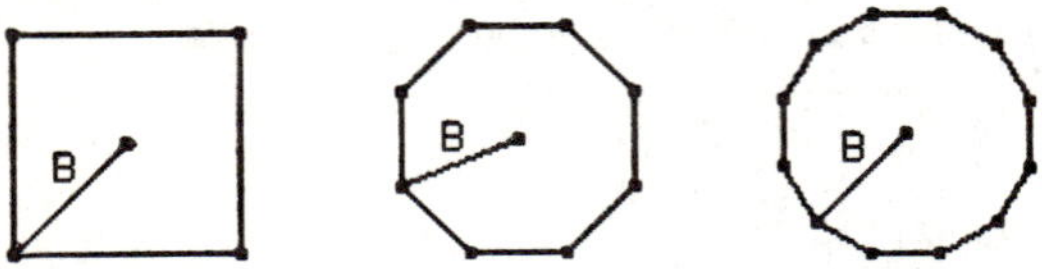

N = 4 sides N = 8 sides N = 12 sides

Procedure

1. Press [MATH] **0** to select **0:Solver** from the MATH menu. Either the equation editor or the interactive solver editor is displayed. If the interactive solver editor is displayed, press [▲] to display the equation editor.

2. Enter the formula as **0=A−NB²sin(π / N)cos(π / N)**, and then press [ENTER]. The interactive solver editor is displayed.

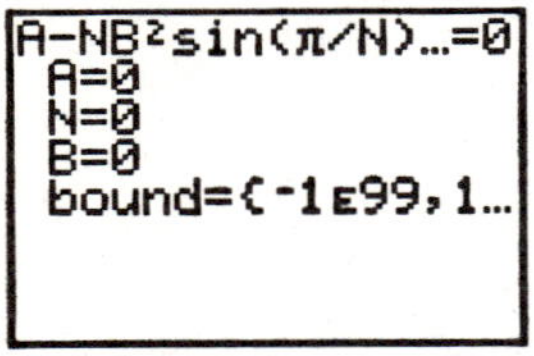

3. Enter **N=4** and **B=6** to find the area (**A**) of a square with a distance (**B**) from center to vertex of 6 centimeters.

4. Press [▲] [▲] to move the cursor onto **A**, and then press [ALPHA] [SOLVE]. The solution for **A** is displayed on the interactive solver editor.

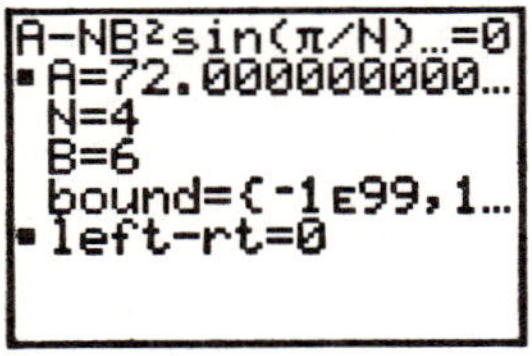

5. Now solve for **B** for a given area with various number of sides. Enter **A=200** and **N=6**. To find the distance **B**, move the cursor onto **B**, and then press [ALPHA] [SOLVE].

6. Enter **N=8**. To find the distance **B**, move the cursor onto **B**, and then press [ALPHA] [SOLVE]. Find **B** for **N=9**, and then for **N=10**.

Procedure (continued)

Find the area given **B=6**, and **N=10**, **100**, **150**, **1000**, and **10000**. Compare your results with $\pi 6^2$ (the area of a circle with radius 6), which is approximately 113.097.

7. Enter **B=6**. To find the area **A**, move the cursor onto **A**, and then press [ALPHA] [SOLVE]. Find **A** for **N=10**, then **N=100**, then **N=150**, then **N=1000**, and finally **N=10000**. Notice that as **N** gets large, the area **A** approaches πB^2.

Now graph the equation to see visually how the area changes as the number of sides gets large.

8. Press [MODE]. Select the default mode settings.

9. Press [WINDOW]. Set the viewing window.

Xmin=0	**Ymin=0**	**Xres=1**
Xmax=200	**Ymax=150**	
Xscl=10	**Yscl=10**	

10. Press [Y=]. Turn off all functions and stat plots. Enter the equation for the area. Use **X** in place of **N**. Set the graph styles as shown.

11. Press [TRACE]. After the graph is plotted, press **100** [ENTER] to trace to **X=100**. Press **150** [ENTER]. Press **188** [ENTER]. Notice that as **X** increases, the value of **Y** converges to $\pi 6^2$, which is approximately 113.097. $Y_2=\pi B^2$ (the area of the circle) is a horizontal asymptote to Y_1. The area of an N-sided regular polygon, with r as the distance from the center to a vertex, approaches the area of a circle with radius r (πr^2) as N gets large.

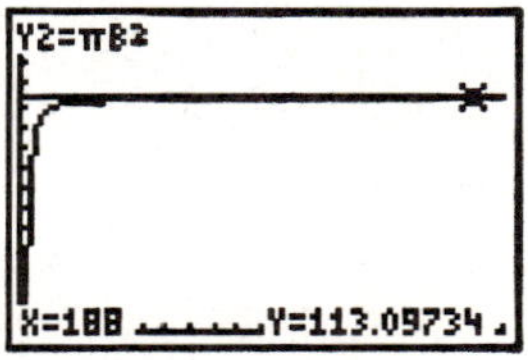

Problem

You are a loan officer at a mortgage company, and you recently closed on a 30-year home mortgage at 8 percent interest with monthly payments of 800. The new home owners want to know how much will be applied to the interest and how much will be applied to the principal when they make the 240th payment 20 years from now.

Procedure

1. Press [MODE] and set the fixed-decimal mode to **2** decimal places. Set the other mode settings to the defaults.

2. Press [APPS] [ENTER] [ENTER] to display the TVM Solver. Enter these values.

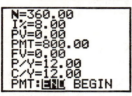

```
N=360.00
I%=8.00
PV=0.00
PMT=800.00
FV=0.00
P/Y=12.00
C/Y=12.00
PMT:END BEGIN
```

   Note: Enter a positive number (**800**) to show **PMT** as a cash inflow. Payment values will be displayed as positive numbers on the graph. Enter **0** for **FV**, since the future value of a loan is 0 once it is paid in full. Enter **PMT: END**, since payment is due at the end of a period.

3. Move the cursor onto the **PV=** prompt, and then press [ALPHA] [SOLVE]. The present value, or mortgage amount, of the house is displayed at the **PV=** prompt.

```
 N=360.00
 I%=8.00
▪PV=-109026.80
 PMT=800.00
 FV=0.00
 P/Y=12.00
 C/Y=12.00
 PMT:END BEGIN
```

Procedure (continued)

Now compare the graph of the amount of interest with the graph of the amount of principal for each payment.

4. Press [MODE]. Set **Par** and **Simul.**

5. Press [Y=]. Turn off all functions and stat plots. Enter these equations and set the graph styles as shown.

```
Plot1 Plot2 Plot3
\X1T=T
 Y1T=ΣPrn(T,T)
\X2T=T
 Y2T=ΣInt(T,T)
\X3T=T
 Y3T=Y1T+Y2T
```

Note: **ΣPrn(** and **ΣInt(** are located on the FINANCE menu (APPS 1:FINANCE).

6. Press [WINDOW]. Set these window variables.

Tmin=1	**Xmin=0**	**Ymin=0**
Tmax=360	**Xmax=360**	**Ymax=1000**
Tstep=12	**Xscl=10**	**Yscl=100**

Tip: To increase the graph speed, change **Tstep** to **24**.

7. Press [TRACE]. After the graph is drawn, press **240** [ENTER] to move the trace cursor to **T=240**, which is equivalent to 20 years of payments.

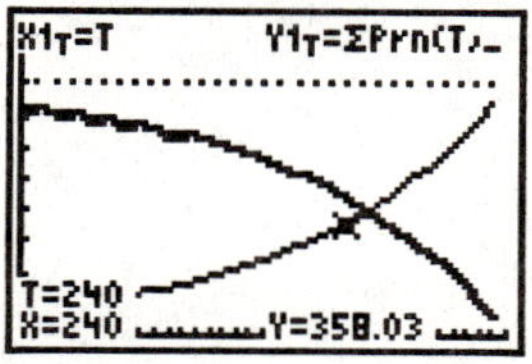

The graph shows that for the 240th payment (**X=240**), 358.03 of the 800 payment is applied to principal (**Y=358.03**).

Note: The sum of the payments (**Y3T=Y1T+Y2T**) is always 800.

Procedure (continued)

8. Press [▼] to move the cursor onto the function for interest defined by **X2T** and **Y2T**. Enter **240**.

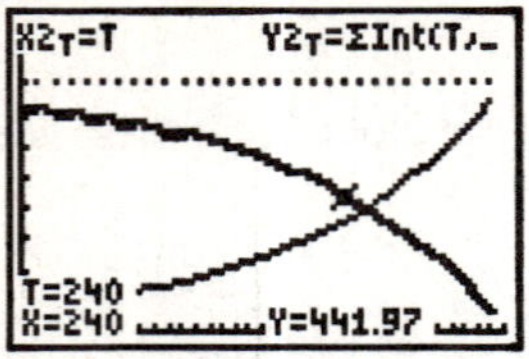

The graph shows that for the 240th payment (**X=240**), 441.97 of the 800 payment is interest (**Y=441.97**).

9. Press [2nd] [QUIT] [APPS] [ENTER] **9** to paste **9:bal(** to the home screen. Check the figures from the graph.

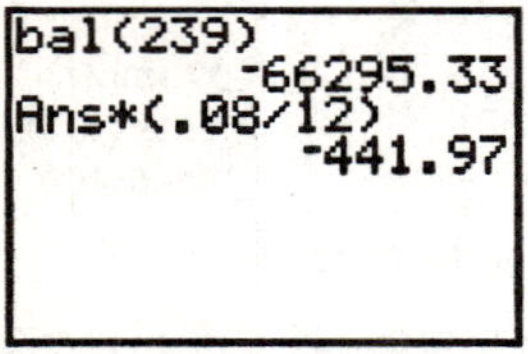

At which monthly payment will the principal allocation surpass the interest allocation?

18 Memory and Variable Management

Contents

MEMORY Menu

At any time you can check available memory or manage existing memory by selecting items from the MEMORY menu. To access this menu, press [2nd] [MEM].

MEMORY	
1: About...	Displays information about the calculator.
2: Mem Mgmt/Del...	Reports memory availability and variable usage.
3: Clear Entries	Clears ENTRY (last-entry storage).
4: ClrAllLists	Clears all lists in memory.
5: Archive...	Archives a selected variable.
6: UnArchive...	UnArchives a selected variable.
7: Reset...	Displays the RAM, ARCHIVE, and ALL menus
8: Group...	Displays GROUP and UNGROUP menus.

To check memory usage, first press [2nd] [MEM] and then press **2:Mem Mgmt/Del.**

Displaying the MEMORY MANAGEMENT/DELETE Menu

Mem Mgmt/Del displays the MEMORY MANAGEMENT/DELETE menu. The two lines at the top report the total amount of available RAM and ARCHIVE memory. By selecting menu items on this screen, you can see the amount of memory each variable type is using. This information can help you determine if some variables need to be deleted from memory to make room for new data, such as programs or applications.

To check memory usage, follow these steps.

1. Press [2nd] [MEM] to display the MEMORY menu.

```
MEMORY
1:About
2:Mem Mgmt/Del...
3:Clear Entries
4:ClrAllLists
5:Archive
6:UnArchive
7↓Reset...
```

Note: The ↑ and ↓ in the top or bottom of the left column indicate that you can scroll up or down to view more variable types.

2. Select **2:Mem Mgmt/Del** to display the MEMORY MANAGEMENT/DELETE menu. The TI-83 Plus expresses memory quantities in bytes.

```
RAM FREE   24347
ARC FREE  163840
1:All...
2:Real...
3:Complex...
4:List...
5:Matrix...
6↓Y-Vars...
```

```
7↑Prgm...
8:Pic...
9:GDB...
0:String...
A:Apps...
B↓AppVars...
```

```
C:Group...
```

3. Select variable types from the list to display memory usage.

   Note: Real, **List**, **Y-Vars**, and **Prgm** variable types never reset to zero, even after memory is cleared.

   Apps are independent applications which are stored in Flash ROM. **AppVars** is a variable holder used to store variables created by independent applications. You cannot edit or change variables in **AppVars** unless you do so through the application which created them.

To leave the MEMORY MANAGEMENT/DELETE menu, press either [2nd] [QUIT] or [CLEAR]. Both options display the home screen.

Deleting an Item

To increase available memory by deleting the contents of any variable (real or complex number, list, matrix, Y= variable, program, Apps, AppVars, picture, graph database, or string), follow these steps.

1. Press [2nd] [MEM] to display the MEMORY menu.
2. Select **2:Mem Mgmt/Del** to display the MEMORY MANAGEMENT/DELETE menu.
3. Select the type of data you want to delete, or select **1:All** for a list of all variables of all types. A screen is displayed listing each variable of the type you selected and the number of bytes each variable is using.

   For example, if you select **4:List**, the LIST editor screen is displayed.

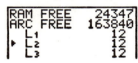

```
RAM FREE    24347
ARC FREE   163840
   L1          12
▸  L2          12
   L3          12
```

4. Press [▲] and [▼] to move the selection cursor (▸) next to the item you want to delete, and then press [DEL]. The variable is deleted from memory. You can delete individual variables one by one from this screen.

   Note: If you are deleting programs or Apps, you will receive a message asking you to confirm this delete action. Select **2:Yes** to continue.

   To leave any variable screen without deleting anything, press [2nd] [QUIT], which displays the home screen.

   Note: You cannot delete some system variables, such as the last-answer variable **Ans** and the statistical variable **RegEQ**.

Clear Entries

Clear Entries clears the contents of the ENTRY (last entry) storage area (Chapter 1). To clear the ENTRY storage area, follow these steps.

1. Press [2nd] [MEM] to display the MEMORY menu.
2. Select **3:Clear Entries** to paste the instruction to the home screen.
3. Press [ENTER] to clear the ENTRY storage area.

```
Clear Entries
         Done
```

To cancel **Clear Entries**, press [CLEAR].

Note: If you select **3:Clear Entries** from within a program, the **Clear Entries** instruction is pasted to the program editor, and the **Entry** (last entry) is cleared when the program is executed.

ClrAllLists

ClrAllLists sets the dimension of each list in RAM only to **0**.

To clear all elements from all lists, follow these steps.

1. Press [2nd] [MEM] to display the MEMORY menu.
2. Select **4:ClrAllLists** to paste the instruction to the home screen.
3. Press [ENTER] to set to **0** the dimension of each list in memory.

```
ClrAllLists
         Done
```

To cancel **ClrAllLists**, press [CLEAR].

ClrAllLists does not delete list names from memory, from the LIST NAMES menu, or from the stat list editor.

Note: If you select **4:ClrAllLists** from within a program, the **ClrAllLists** instruction is pasted to the program editor. The lists are cleared when the program is executed.

RAM ARCHIVE ALL Menu

The RAM ARCHIVE ALL menu gives you the option of resetting all memory (including default settings) or resetting selected portions of memory while preserving other data stored in memory, such as programs and Y= functions. For instance, you can choose to reset all of RAM or just restore the default settings. Be aware that if you choose to reset RAM, all data and programs in RAM will be erased. For archive memory, you can reset variables (Vars), applications (Apps), or both of these. Be aware that if you choose to reset Vars, all data and programs in archive memory will be erased. If you choose to reset Apps, all applications in archive memory will be erased.

When you reset defaults on the TI-83 Plus, all defaults in RAM are restored to the factory settings. Stored data and programs are not changed.

These are some examples of TI-83 Plus defaults that are restored by resetting the defaults.

- Mode settings such as **Normal** (notation); **Func** (graphing); **Real** (numbers); and **Full** (screen)
- Y= functions off
- Window variable values such as **Xmin=⁻10**; **Xmax=10**; **Xscl=1**; **Yscl=1**; and **Xres=1**
- Stat plots off
- Format settings such as **CoordOn** (graphing coordinates on); **AxesOn**; and **ExprOn** (expression on)
- **rand** seed value to 0

Displaying the RAM ARCHIVE ALL Menu

To display the RAM ARCHIVE ALL menu on the TI-83 Plus, follow these steps.

1. Press [2nd] [MEM] to display the MEMORY menu.
2. Select **7:Reset** to display the RAM ARCHIVE ALL menu.

Resetting RAM Memory

Resetting RAM restores RAM system variables to factory settings and deletes all nonsystem variables and all programs. Resetting defaults restores all system variables to default settings without deleting variables and programs in RAM. Resetting RAM or resetting defaults does not affect variables and applications in user data archive.

Tip: Before you reset all RAM memory, consider restoring sufficient available memory by deleting only selected data (page 18-4).

To reset all RAM memory or RAM defaults on the TI-83 Plus, follow these steps.

1. From the RAM ARCHIVE ALL menu, select **1:ALL RAM** to display the RESET RAM menu or **2:Defaults** to display the RESET DEFAULTS menu.

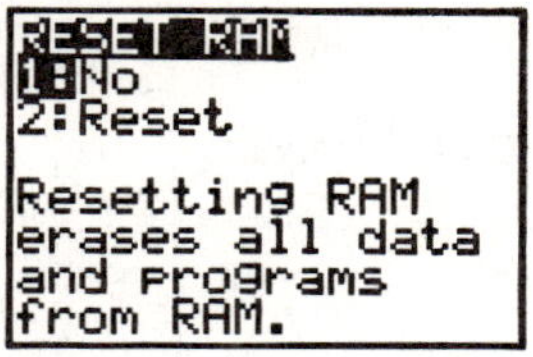

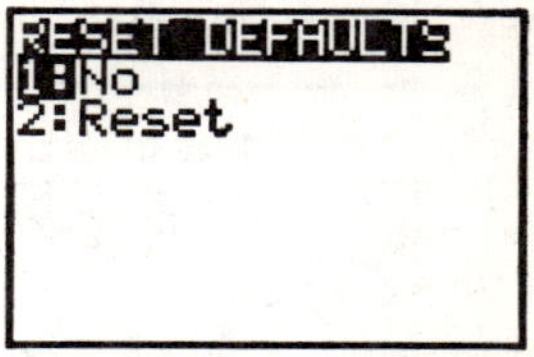

2. If you are resetting RAM, read the message below the RESET RAM menu.

   - To cancel the reset and return to the home screen, press [ENTER].
   - To erase RAM memory or reset defaults, select **2:Reset**. Depending on your choice, the message **RAM cleared** or **Defaults set** is displayed on the home screen.

Resetting Archive Memory

When resetting archive memory on the TI-83 Plus, you can choose to delete from user data archive all variables, all applications, or both variables and applications.

To reset all or part of user data archive memory, follow these steps.

1. From the RAM ARCHIVE ALL menu, press [▸] to display the ARCHIVE menu.

2. Select one of the following:

   - **1:Vars** to display the RESET ARC VAR menu

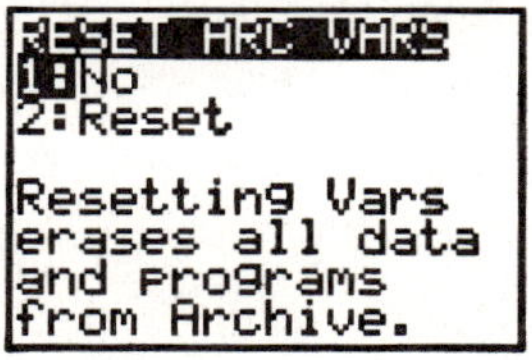

- **2:Apps** to display the RESET ARC APPS menu.

- **3:Both** to display the RESET ARC BOTH menu.

3. Read the message below the menu.
   - To cancel the reset and return to the home screen, press [ENTER].
   - To continue with the reset, select **2:Reset**. A message indicating the type of archive memory cleared will be displayed on the home screen.

Resetting All Memory

When resetting all memory on the TI-83 Plus, RAM and user data archive memory is restored to factory settings. All nonsystem variables, applications, and programs are deleted. All system variables are reset to default settings.

Tip: Before you reset all memory, consider restoring sufficient available memory by deleting only selected data (page 18-4).

To reset all memory on the TI-83 Plus, follow these steps.

1. From the RAM ARCHIVE ALL menu, press [▶] [▶] to display the ALL menu.

2. Select **1:All Memory** to display the RESET MEMORY menu.

3. Read the message below the RESET MEMORY menu.
   - To cancel the reset and return to the home screen, press [ENTER].
   - To continue with the reset, select **2:Reset**. The message **MEM cleared** is displayed on the home screen.

Note: When you clear memory, the contrast sometimes changes. If the screen is faded or blank, adjust the contrast by pressing [2nd] [▲] or [▼].

Archiving and UnArchiving Variables

Archiving allows you to store data, programs, or other variables to the user data archive where they cannot be edited or deleted inadvertently. Archiving also allows you to free up RAM for variables that may require additional memory.

Archived variables cannot be edited or executed. They can only be seen and unarchived. For example, if you archive list **L1**, you will see that **L1** exists in memory but if you select it and paste the name **L1** to the home screen, you won't be able to see its contents or edit it.

Note: Not all variables may be archived. Not all archived variables may be unarchived. For example, system variables including r, t, x, y, and θ cannot be archived. Apps and Groups always exist in Flash ROM so there is no need to archive them. Groups cannot be unarchived. However, you can ungroup or delete them.

Variable Type	Names	Archive? (yes/no)	UnArchive? (yes/no)
Real numbers	**A, B, . . . , Z**	yes	yes
Complex numbers	**A, B, . . . , Z**	yes	yes
Matrices	[A], [B], [C], . . . , [J]	yes	yes
Lists	**L1, L2, L3, L4, L5, L6**, and user-defined names	yes	yes
Programs		yes	yes
Functions	**Y1, Y2, . . . , Y9, Y0**	no	not applicable
Parametric equations	**X1T** and **Y1T**, . . . , **X6T** and **Y6T**	no	not applicable
Polar functions	**r1, r2, r3, r4, r5, r6**	no	not applicable
Sequence functions	**u, v, w**	no	not applicable
Stat plots	**Plot1, Plot2, Plot3**	no	not applicable
Graph databases	**GDB1, GDB2,...**	yes	yes
Graph pictures	**Pic1, Pic2, . . . , Pic9, Pic0**	yes	yes
Strings	**Str1, Str2, . . . Str9, Str0**	yes	yes
Tables	**TblStart, Tb1, TblInput**	no	not applicable
Apps	Applications	see NOTE above	no
AppVars	Application variables	yes	yes
Groups		see NOTE above	no
Variables with reserved names	**minX, maxX, RegEQ,** and others	no	not applicable
System variables	**Xmin, Xmax,** and others	no	not applicable

Archiving and unarchiving can be done in two ways:

- Use the **5:Archive** or **6:UnArchive** commands from the MEMORY menu or CATALOG.
- Use a Memory Management editor screen.

Before archiving or unarchiving variables, particularly those with a large byte size (such as large programs) use the MEMORY menu to:

- Find the size of the variable.
- See if there is enough free space.

For:	Sizes must be such that:
Archive	Archive free size > variable size
UnArchive	RAM free size > variable size

Note: If there is not enough space, unarchive or delete variables as necessary. Be aware that when you unarchive a variable, not all the memory associated with that variable in user data archive will be released since the system keeps track of where the variable has been and where it is now in RAM.

Even if there appears to be enough free space, you may see a Garbage Collection message (page 18-16) when you attempt to archive a variable. Depending on the usability of empty blocks in the user data archive, you may need to unarchive existing variables to create more free space.

To archive or unarchive a list variable (L1) using the Archive/UnArchive options from the MEMORY menu:

1. Press [2nd] [MEM] to display the MEMORY menu.

2. Select **5:Archive** or **6:UnArchive** to place the command in the edit screen.

3. Press [2nd] [L1] to place the **L1** variable in the edit screen.

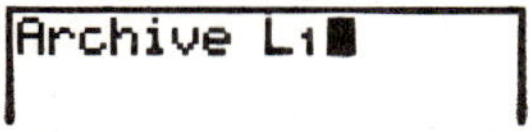

4. Press [ENTER] to complete the archive process.

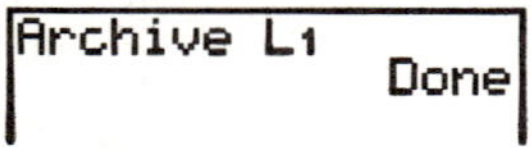

Note: An asterisk will be displayed to the left of the Archived variable name to indicate it is archived.

To archive or unarchive a list variable (L1) using a Memory Management editor:

1. Press [2nd] [MEM] to display the MEMORY menu.

2. Select **2:Mem Mgmt/Del…** to display the MEMORY MANAGEMENT/DELETE menu.

3. Select **4:List…** to display the LIST menu.

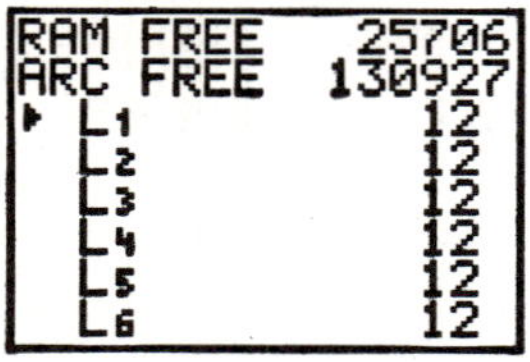

4. Press [ENTER] to archive L1. An asterisk will appear to the left of L1 to indicate it is an archived variable. To unarchive a variable in this screen, put the cursor next to the archived variable and press [ENTER]. The asterisk will disappear.

5. Press [2nd] [QUIT] to leave the LIST menu.

Note: You can access an archived variable for the purpose of linking, deleting, or unarchiving it, but you cannot edit it.

Grouping Variables

Grouping allows you to make a copy of two or more variables residing in RAM and then store them as a group in user data archive. The variables in RAM are not erased. The variables must exist in RAM before they can be grouped. In other words, archived data cannot be included in a group.

To create a group of variables:

1. Press [2nd] [MEM] to display the MEMORY menu.

2. Select **8:Group...** to display GROUP UNGROUP menu.

3. Press [ENTER] to display the GROUP menu.

4. Enter a name for the new group and press [ENTER].

   Note: A group name can be one to eight characters long. The first character must be a letter from A to Z or θ. The second through eighth characters can be letters, numbers, or θ.

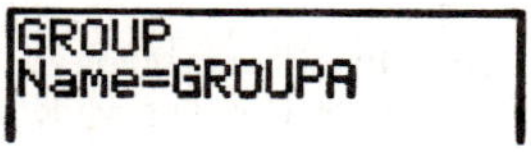

5. Select the type of data you want to group. You can select **1:All+** which shows all variables of all types available and selected. You can also select **1:All-** which shows all variables of all types available but not selected. A screen is displayed listing each variable of the type you selected.

For example, suppose some variables have been created in RAM, and selecting **1:All-** displays the following screen.

6. Press [▲] and [▼] to move the selection cursor (▸) next to the first item you want to copy into a group, and then press [ENTER]. A small square will remain to the left of all variables selected for grouping.

Repeat the selection process until all variables for the new group are selected and then press [▸] to display the DONE menu.

7. Press [ENTER] to complete the grouping process.

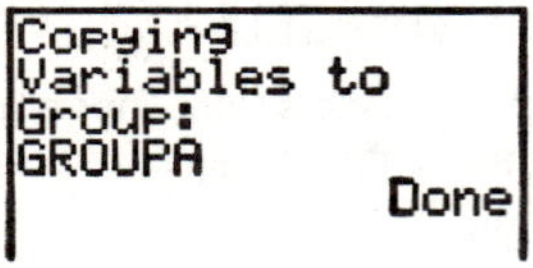

Note: You can only group variables in RAM. You cannot group some system variables, such as the last-answer variable **Ans** and the statistical variable **RegEQ**.

Ungrouping Variables

Ungrouping allows you to make a copy of variables in a group stored in user data archive and place them ungrouped in RAM.

To ungroup, follow the steps on page 18-16.

DuplicateName Menu

During the ungrouping action, if a duplicate variable name is detected in RAM, the DUPLICATE NAME menu is displayed.

DuplicateName	
1: Rename	Prompts to rename receiving variable.
2: Overwrite	Overwrites data in receiving duplicate variable.
3: Overwrite All	Overwrites data in all receiving duplicate variables.
4: Omit	Skips transmission of sending variable.
5: Quit	Stops transmission at duplicate variable.

Notes about Menu Items

- When you select **1:Rename**, the **Name=** prompt is displayed, and alpha-lock is on. Enter a new variable name, and then press [ENTER]. Ungrouping resumes.
- When you select **2:Overwrite**, the unit overwrites the data of the duplicate variable name found in RAM. Ungrouping resumes.
- When you select **3: Overwrite All**, the unit overwrites the data of all duplicate variable names found in RAM. Ungrouping resumes.
- When you select **4:Omit**, the unit does not ungroup the variable in conflict with the duplicated variable name found in RAM. Ungrouping resumes with the next item.
- When you select **5:Quit**, ungrouping stops, and no further changes are made.

To ungroup a group of variables:

1. Press [2nd] [MEM] to display the MEMORY menu.

2. Select **8:Group...** to display the GROUP UNGROUP menu.

3. Press [▸] to display the UNGROUP menu.

4. Press [▴] and [▾] to move the selection cursor (▸) next to the group variable you want to ungroup, and then press [ENTER].

   The ungroup action is completed.

Note: Ungrouping does not remove the group from user data archive. You must delete the group in user data archive to remove it.

Garbage Collection Message

If you use the user data archive extensively, you may see a **Garbage Collect?** message. This occurs if you try to archive a variable when there is not enough free archive memory. The TI-83 Plus will attempt to rearrange the archived variables to make additional room.

Responding to the Garbage Collection Message

- To cancel, select **1:No**.
- If you choose **1:No**, the message **ERR:ARCHIVE FULL** will be displayed.

- To continue archiving, select **2:Yes**.

If you select **2:Yes**, the process message **Garbage Collecting...** or **Defragmenting...** will be displayed.

Note: The process message **Defragmenting...** is displayed whenever an application marked for deletion is encountered.

Garbage collection may take up to 20 minutes, depending on how much of archive memory has been used to store variables.

After garbage collection, depending on how much additional space is freed, the variable may or may not be archived. If not, you can unarchive some variables and try again.

Why Not Perform Garbage Collection Automatically Without a Message?

The message:

- Lets you know why an archive will take longer than usual. It also alerts you that the archive may fail if there is not enough memory.
- Can alert you when a program is caught in a loop that repetitively fills the user data archive. Cancel the archive and determine the reason.

Why Is Garbage Collection Necessary?

The user data archive is divided into sectors. When you first begin archiving, variables are stored consecutively in sector 1. This continues to the end of the sector.

An archived variable is stored in a continuous block within a single sector. Unlike an application stored in user data archive, an archived variable cannot cross a sector boundary. If there is not enough space left in the sector, the next variable is stored at the beginning of the next sector. Typically, this leaves an empty block at the end of the previous sector.

Each variable that you archive is stored in the first empty block large enough to hold it.

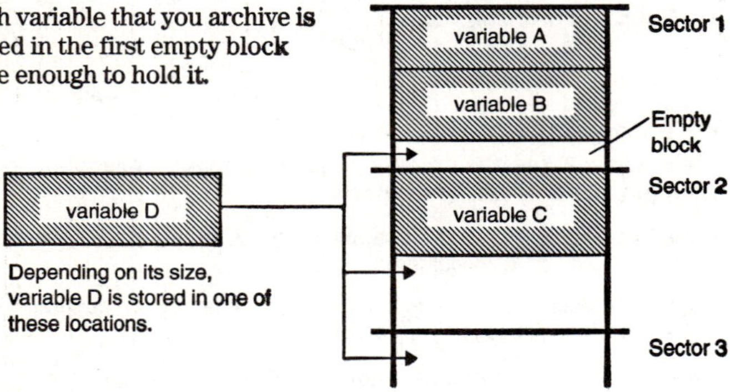

This process continues to the end of the last sector. Depending on the size of individual variables, the empty blocks may account for a significant amount of space. Garbage collection occurs when the variable you are archiving is larger than any empty block.

How Unarchiving a Variable Affects the Process

When you unarchive a variable, it is copied to RAM but it is not actually deleted from user data archive memory.

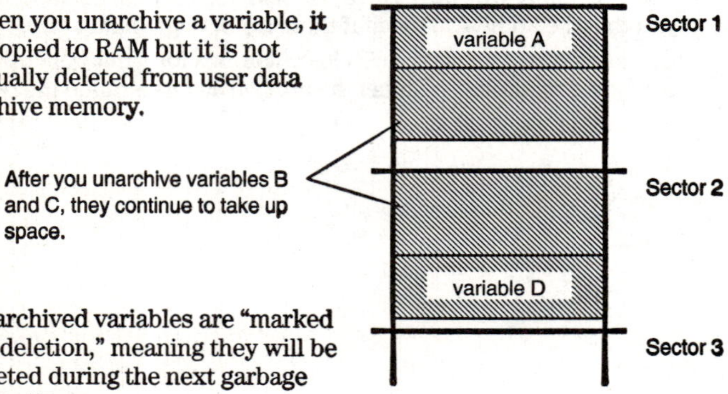

Unarchived variables are “marked for deletion,” meaning they will be deleted during the next garbage collection.

If the MEMORY Screen Shows Enough Free Space

Even if the MEMORY screen shows enough free space to archive a variable or store an application, you may still get a **Garbage Collect?** message or an **ERR: ARCHIVE FULL** message (18-20).

When you unarchive a variable, the Archive free amount increases immediately, but the space is not actually available until after the next garbage collection.

If the Archive free amount shows enough available space for your variable, there probably will be enough space to archive it after garbage collection (depending on the usability of any empty blocks).

The Garbage Collection Process

The garbage collection process:

- Deletes unarchived variables from the user data archive.
- Rearranges the remaining variables into consecutive blocks.

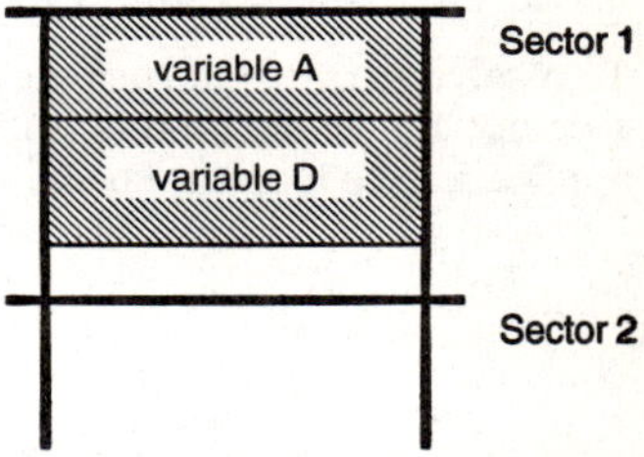

Note: Power loss during garbage collection may cause all memory (RAM and Archive) to be deleted.

Using the GarbageCollect Command

You can reduce the number of automatic garbage collections by periodically optimizing memory. This is done by using the **GarbageCollect** command.

To use the **GarbageCollect** command, follow these steps.

1. Press [2nd] [CATALOG] to display the CATALOG.

2. Press [▼] or [▲] to scroll the CATALOG until the selection cursor points to the **GarbageCollect** command.
3. Press [ENTER] to paste the command to the current screen.
4. Press [ENTER] to display the **Garbage Collect?** message.
5. Select **2:Yes** to begin garbage collection.

Even if the MEMORY screen shows enough free space to archive a variable or store an application, you may still get an **ERR: ARCHIVE FULL** message.

An **ERR:ARCHIVE FULL** message may be displayed:

- When there is insufficient space to archive a variable within a continuous block and within a single sector.
- When there is insufficient space to store an application within a continuous block of memory.

When the message is displayed, it will indicate the largest single space of memory available for storing a variable and an application.

To resolve the problem, use the **GarbageCollect** command to optimize memory. If memory is still insufficient, you must delete variables or applications to increase space.

19 Communication Link

Contents

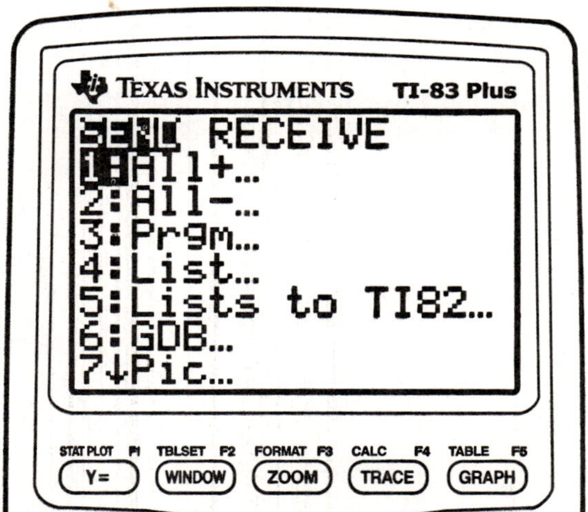

Getting Started is a fast-paced introduction. Read the chapter for details.

Create and store a variable and a matrix, and then transfer them to another TI-83 Plus.

1. On the home screen of the sending unit, press **5** [.] **5** [STO▸] [ALPHA] **Q**. Press [ENTER] to store 5.5 to **Q**.

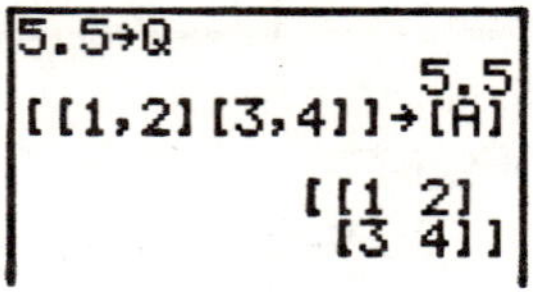

2. Press [2nd] [[ ] [2nd] [[ ] **1** [,] **2** [2nd] [ ]] [2nd] [[ ] **3** [,] **4** [2nd] [ ]] [2nd] [ ]] [STO▸] [2nd] [MATRX] **1**. Press [ENTER] to store the matrix to **[A]**.

3. On the sending unit, press [2nd] [MEM] to display the MEMORY menu.

4. On the sending unit, press **2** to select **2:Mem Mgmt/Del**. The MEMORY MANAGEMENT DELETE menu is displayed.

5. On the sending unit, press **5** to select **5:Matrix**. The MATRIX editor screen is displayed.

6. On the sending unit, press [ENTER] to archive **[A]**. An * will appear, signifying that **[A]** is now archived.

7. Connect the calculators with the link cable. Push both ends in firmly.

8. On the receiving unit, press [2nd] [LINK] [▶] to display the RECEIVE menu. Press **1** to select **1:Receive**. The message **Waiting...** is displayed and the busy indicator is on.

9. On the sending unit, press [2nd] [LINK] to display the SEND menu.

10. Press **2** to select **2:All-**. The All- SELECT screen is displayed.

11. Press [▼] until the selection cursor (▶) is next to **[A] MATRX**. Press [ENTER].

12. Press [▼] until the selection cursor is next to **Q REAL**. Press [ENTER]. A square dot next to **[A]** and **Q** indicates that each is selected to send.

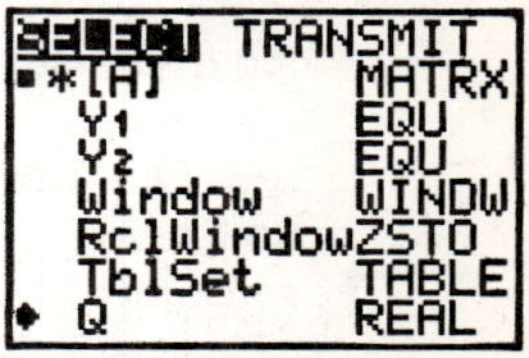

13. On the sending unit, press [▶] to display the TRANSMIT menu.

14. On the sending unit, press **1** to select **1:Transmit** and begin transmission. The receiving unit displays the message **Receiving....** When the items are transmitted, both units display the name and type of each transmitted variable.

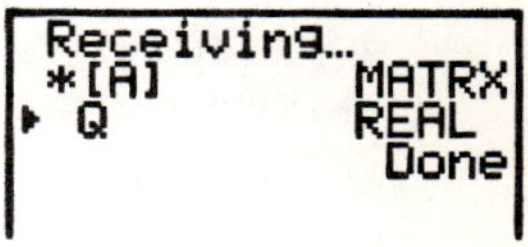

TI-83 Plus Link Capabilities

The TI-83 Plus has a port to connect and communicate with another TI-83 Plus, a TI-83, a TI-82, a TI-73, a Calculator-Based Laboratory™ (CBL 2™, CBL™) System, or a Calculator-Based Ranger™ (CBR™) System. A unit-to-unit link cable is included with the TI-83 Plus for this purpose. With a TI-GRAPH LINK™ (optional accessory), you can also link the TI-83 Plus to a personal computer. This chapter describes how to communicate with another calculator.

Linking Two TI-83 Plus calculators

You can transfer all variables, programs, and applications to another TI-83 Plus or backup the entire memory of a TI-83 Plus.

- Variables stored in RAM on the sending TI-83 Plus will be sent to RAM of the receiving TI-83 Plus.
- Variables and applications stored in user data archive of the sending TI-83 Plus will be sent to the user data archive of the receiving TI-83 Plus.

The software that enables this communication is built into the TI-83 Plus. To transmit from one TI-83 Plus to another, follow the steps on pages 19-6 and 19-7.

Linking a TI-83 and a TI-83 Plus

You can transfer from a TI-83 to a TI-83 Plus all variables and programs if they fit in the RAM of the TI-83 Plus. The RAM of the TI-83 Plus is slightly less than the RAM of the TI-83.

Also, you can transfer from a TI-83 Plus to a TI-83 all variables except for applications, application variables, grouped variables, new variable types, or programs with new features in them such as **Archive, UnArchive, Asm(, AsmComp**, and **AsmPrgm.**

If archived variables on the TI-83 Plus are variable types recognized and used on the TI-83, you can transmit these variables to the TI-83. They will be automatically sent to the TI-83's RAM during the transfer process.

The software that enables this communication is built into the TI-83 Plus. To transmit data from a TI-83 Plus to a TI-83, follow the steps on page 19-11.

Note: You cannot perform a memory backup from a TI-83 to a TI-83 Plus or a TI-83 Plus to a TI-83.

Linking a TI-82 and a TI-83 Plus

You can transfer from a TI-82 to a TI-83 Plus all variables and programs. Also, you can transfer from a TI-83 Plus to a TI-82 real lists **L1** through **L6** with up to 99 elements.

The software that enables this communication is built into the TI-83 Plus. To transmit data from a TI-83 Plus to a TI-82, follow the steps on page 19-12. To transmit data from a TI-82 to a TI-83 Plus, follow the steps on page 19-13.

Note: You cannot perform a memory backup from a TI-82 to a TI-83 Plus or a TI-83 Plus to a TI-82.

Linking a TI-73 and a TI-83 Plus

You can transfer real numbers, pics, real lists **L1** through **L6**, and named lists from a TI-73 to a TI-83 Plus or from a TI-83 Plus to a TI-73. Named lists may also be exchanged. However, θ is not recognized by the TI-73, so you cannot include this symbol in any list names sent to the TI-73.

The software that enables this communication is built into the TI-83 Plus.

To transmit data from a TI-83 Plus to a TI-73, follow the steps on page 19-14. To transmit data from a TI-73 to a TI-83 Plus, follow the steps on page 19-15.

Note: You cannot perform a memory backup from a TI-73 to a TI-83 Plus or a TI-83 Plus to a TI-73.

Connecting Two Calculators with the Cable

The TI-83 Plus link port is located at the center of the bottom edge of the calculator.

1. Insert either end of the cable into the port **very firmly**.
2. Insert the other end of the cable into the other calculator's port.

Linking to the CBL/CBR System

The Calculator-Based Laboratory (CBL 2/CBL) System and the Calculator-Based Ranger (CBR) System are optional accessories that connect to a TI-83 Plus with the unit-to-unit link cable. With a CBL 2/CBL or CBR and a TI-83 Plus, you can collect and analyze real-world data. The software that enables this communication is built into the TI-83 Plus. (Chapter 14).

Linking to a PC or Macintosh®

TI-GRAPH LINK™ (sold separately) is an accessory that links a TI-83 Plus to enable communication with a personal computer.

LINK SEND Menu

To display the LINK SEND menu, press [2nd] [LINK].

SEND RECEIVE	
1: All+...	Displays all items selected.
2: All-...	Displays all items deselected.
3: Prgm...	Displays all program names.
4: List...	Displays all list names.
5: Lists to TI82...	Displays list names **L1** through **L6**.
6: GDB...	Displays all graph databases.
7: Pic...	Displays all picture data types.
8: Matrix...	Displays all matrix data types.
9: Real...	Displays all real variables.
0: Complex...	Displays all complex variables.
A: Y-Vars...	Displays all Y= variables.
B: String...	Displays all string variables.
C: Apps...	Displays all software applications.
D: AppVars...	Displays all software application variables.
E: Group...	Displays all grouped variables.
F: SendId	Sends the Calculator ID number immediately. (You do not need to select SEND.)
G: SendSW	Sends software updates to another TI-83 Plus.
H: Back Up...	Selects all for backup to TI-83 Plus.

When you select an item on the LINK SEND menu, the corresponding SELECT screen is displayed.

Note: Each SELECT screen, except All+ SELECT, is displayed initially with no data selected.

Selecting Items to Send

To select items to send on the sending unit, follow these steps.

1. Press [2nd] [LINK] to display the LINK SEND menu.
2. Select the menu item that describes the data type to send. The corresponding SELECT screen is displayed.
3. Press [▲] and [▼] to move the selection cursor (▸) to an item you want to select or deselect.
4. Press [ENTER] to select or deselect the item. Selected names are marked with a ▪.

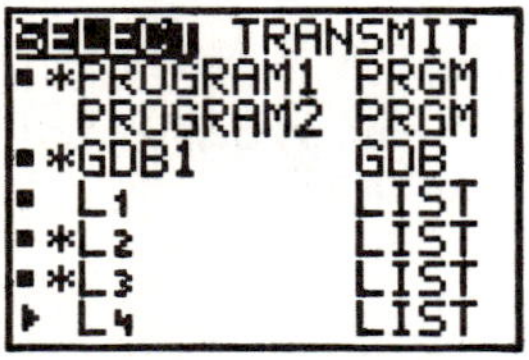

Note: An asterisk (*) to the left of an item indicates the item is archived (Chapter 18).

5. Repeat steps 3 and 4 to select or deselect additional items.

LINK RECEIVE Menu

To display the LINK RECEIVE menu, press [2nd] [LINK] [▶].

SEND RECEIVE	
1: Receive	Sets unit to receive data transmission.

Receiving Unit

When you select **1:Receive** from the LINK RECEIVE menu on the receiving unit, the message **Waiting...** and the busy indicator are displayed. The receiving unit is ready to receive transmitted items. To exit the receive mode without receiving items, press [ON], and then select **1:Quit** from the Error in Xmit menu.

To transmit, follow the steps on page 19-8.

When transmission is complete, the unit exits the receive mode. You can select **1:Receive** again to receive more items. The receiving unit then displays a list of items received. Press [2nd] [QUIT] to exit the receive mode.

DuplicateName Menu

During transmission, if a variable name is duplicated, the DuplicateName menu is displayed on the receiving unit.

DuplicateName	
1: Rename	Prompts to rename receiving variable.
2: Overwrite	Overwrites data in receiving variable.
3: Omit	Skips transmission of sending variable.
4: Quit	Stops transmission at duplicate variable.

When you select **1:Rename**, the **Name=** prompt is displayed, and alpha-lock is on. Enter a new variable name, and then press [ENTER]. Transmission resumes.

When you select **2:Overwrite**, the sending unit's data overwrites the existing data stored on the receiving unit. Transmission resumes.

When you select **3:Omit**, the sending unit does not send the data in the duplicated variable name. Transmission resumes with the next item.

When you select **4:Quit**, transmission stops, and the receiving unit exits receive mode.

Insufficient Memory in Receiving Unit

During transmission, if the receiving unit does not have sufficient memory to receive an item, the Memory Full menu is displayed on the receiving unit.

- To skip this item for the current transmission, select **1:Omit.** Transmission resumes with the next item.
- To cancel the transmission and exit receive mode, select **2:Quit.**

Transmitting Items

After you have selected items to send on the sending unit (page 19-5) and set the receiving unit to receive (page 19-6), follow these steps to transmit the items.

1. Press [▶] on the sending unit to display the TRANSMIT menu.

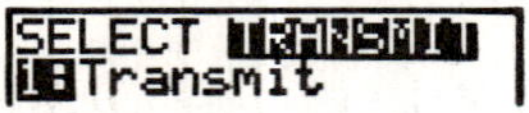

2. Confirm that **Waiting...** is displayed on the receiving unit, which indicates it is set to receive (page 19-6).

3. Press [ENTER] to select **1:Transmit**. The name and type of each item are displayed line by line on the sending unit as the item is queued for transmission, and then on the receiving unit as each item is accepted.

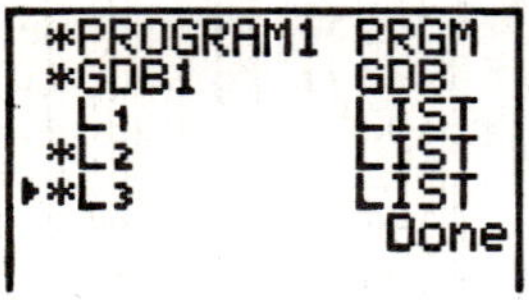

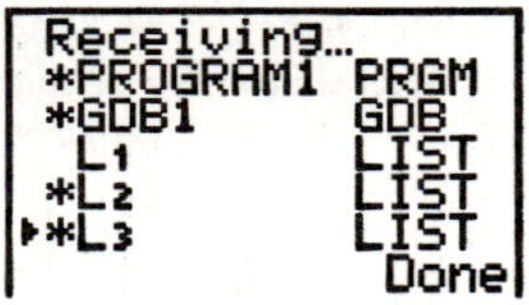

Note: Items sent from the RAM of the sending unit are transmitted to the RAM of the receiving unit. Items sent from user data archive of the sending unit are transmitted to user data archive of the receiving unit.

After all selected items have been transmitted, the message **Done** is displayed on both calculators. Press [▲] and [▼] to scroll through the names.

Stopping a Transmission

To stop a link transmission, press [ON]. The Error in Xmit menu is displayed on both units. To leave the error menu, select **1:Quit**.

Error Conditions

A transmission error occurs after one or two seconds if:

- A cable is not attached to the sending unit.
- A cable is not attached to the receiving unit.

  Note: If the cable is attached, push it in firmly and try again.
- The receiving unit is not set to receive transmission.
- You attempt a backup between a TI-73, a TI-82, or a TI-83 and a TI-83 Plus.
- You attempt a data transfer from a TI-83 Plus to a TI-83 with variables or features not recognized by the TI-83.
- New variable types and features not recognized by the TI-83 include applications, application variables, grouped variables, new variable types, or programs with new features in them such as **Archive**, **UnArchive**, **SendID**, **SendSW**, **Asm(**, **AsmComp(**, and **AsmPrgm**.
- You attempt a data transfer from a TI-83 Plus to a TI-82 with data other than real lists **L1** through **L6** or without using menu item **5:Lists to TI82**.
- You attempt a data transfer from a TI-83 Plus to a TI-73 with data other than real numbers, pics, real lists **L1** through **L6** or named lists with θ as part of the name.

Although a transmission error does not occur, these two conditions may prevent successful transmission.

- You try to use **Get(** with a calculator instead of a CBL 2/CBL or CBR.
- You try to use **GetCalc(** with a TI-83 instead of a TI-83 Plus.

Transmitting Items to an Additional TI-83 Plus

After sending or receiving data, you can repeat the same transmission to additional TI-83 Plus units—from either the sending unit or the receiving unit—without having to reselect data to send. The current items remain selected.

Note: You cannot repeat transmission if you selected All+ or All-.

To transmit to an additional TI-83 Plus, follow these steps.

1. Set the TI-83 Plus to receive (page 19-7).
2. Do not select or deselect any new items to send. If you select or deselect an item, all selections or deselections from the previous transmission are cleared.
3. Disconnect the link cable from one TI-83 Plus and connect it to the additional TI-83 Plus.
4. Set the additional TI-83 Plus to receive (page 19-7).
5. Press [2nd] [LINK] on the sending TI-83 Plus to display the LINK SEND menu.
6. Select the menu item that you used for the last transmission. The data from your last transmission is still selected.
7. Press [▶] to display the LINK TRANSMIT menu.
8. Confirm that the receiving unit is set to receive (page 19-7).
9. Press [ENTER] to select **1:Transmit** and begin transmitting.

Transmitting Items to a TI-83

You can transfer from a TI-83 Plus to a TI-83 all variables except for applications, application variables, grouped variables, new variable types, or programs with new features in them such as archiving, and unarchiving.

If archived variables on the TI-83 Plus are variable types recognized and used on the TI-83, you can transmit these variables to the TI-83. They will be automatically sent to the RAM of the receiving TI-83 during the transfer process.

To transmit to a TI-83 data that is stored in a TI-83 Plus, follow these steps.

1. Set the TI-83 to receive (page 19-7).
2. Press [2nd] [LINK] on the sending TI-83 Plus to display the LINK SEND menu.
3. Select the menu items you want to transmit.
4. Press [▶] to display the LINK TRANSMIT menu.
5. Confirm that the receiving unit is set to receive (page 19-7).
6. Press [ENTER] to select **1:Transmit** and begin transmitting.

Transmitting Lists to a TI-82

The only data type you can transmit from a TI-83 Plus to a TI-82 is real list data stored in **L1** through **L6** with up to 99 elements for each list.

To transmit to a TI-82 the list data that is stored to TI-83 Plus lists **L1**, **L2**, **L3**, **L4**, **L5**, or **L6**, follow these steps.

1. Set the TI-82 to receive (page 19-7).
2. Press [2nd] [LINK] **5** on the sending TI-83 Plus to select **5:Lists to TI82**. The SELECT screen is displayed.
3. Select each list to transmit.
4. Press [▶] to display the LINK TRANSMIT menu.
5. Confirm that the receiving unit is set to receive (page 19-7).
6. Press [ENTER] to select **1:Transmit** and begin transmitting.

Note: If dimension > 99 for a TI-83 Plus list that is selected to send, the receiving TI-82 will truncate the list at the ninety-ninth element during transmission.

Resolved Differences between the TI-82 and TI-83 Plus

Generally, you can transmit items to a TI-83 Plus from a TI-82, but differences between the two products may affect some transmitted data. This table shows differences for which the software built into the TI-83 Plus automatically adjusts when a TI-83 Plus receives TI-82 data.

TI-82	TI-83 Plus
n**Min**	**PlotStart**
n**Start**	***n*****Min**
U***n***	**u**
V***n***	**v**
U***n*****Start**	**u(*****n*****Min)**
V***n*****Start**	**v(*****n*****Min)**
TblMin	**TblStart**

For example, if you transmit from a TI-82 to a TI-83 Plus a program that contains ***n*****Start** on a command line and then display the program on the receiving TI-83 Plus, you will see that ***n*****Min** has automatically replaced ***n*****Start** on the command line.

Unresolved Differences between the TI-82 and TI-83 Plus

The software built into the TI-83 Plus cannot resolve some differences between the TI-82 and TI-83 Plus, which are described below. You must edit the data on the TI-83 Plus after you transmit to account for these differences, or the TI-83 Plus will misinterpret the data.

The TI-83 Plus reinterprets TI-82 prefix functions to include open parentheses, which may add extraneous parentheses to transmitted expressions.

For example, if you transmit **sin X+5** from a TI-82 to a TI-83 Plus, the TI-83 Plus reinterprets it as **sin(X+5**. Without a closing parenthesis after **X**, the TI-83 Plus interprets this as **sin(X+5)**, not the sum of **5** and **sin(X)**.

If a TI-82 instruction that the TI-83 Plus cannot translate is transmitted, the ERR:INVALID menu is displayed when the TI-83 Plus attempts to execute the instruction. For example, on the TI-82, the character group **Un-1** is pasted to the cursor location when you press [2nd] [Un−1]. The TI-83 Plus cannot directly translate **Un-1** to the TI-83 Plus syntax **u(n−1)**, so the ERR:INVALID menu is displayed.

Note: TI-83 Plus implied multiplication rules differ from those of the TI-82. For example, the TI-83 Plus evaluates **1/2X** as **(1/2)∗X**, while the TI-82 evaluates **1/2X** as **1/(2∗X)** (Chapter 2).

Transmitting Lists to a TI-73

You can transmit from a TI-83 Plus to a TI-73 lists, real numbers, pics, real list data stored in **L1** through **L6** and named lists.

Note: Since the TI-73 does not recognize θ, named lists may not contain θ in the list name.

To transmit to a TI-73 list data that is stored to the TI-83 Plus lists **L1**, **L2**, **L3**, **L4**, **L5**, or **L6**, follow these steps.

1. Set the TI-73 to receive (page 19-7).
2. Press [2nd] [LINK] **4** on the sending TI-83 Plus to select **4:Lists....** The SELECT screen is displayed.
3. Select lists you want to send.
4. Press [▶] on the sending TI-83 Plus to display the LINK TRANSMIT menu.
5. Confirm that the receiving unit is set to receive (page 19-7).
6. Press [ENTER] to select **1:Transmit** and begin transmitting.

Transmitting Lists to a TI-83 Plus

You can transmit from a TI-73 to a TI-83 Plus lists, real numbers, pics, and real list data stored in **L1** through **L6** and any named lists.

To transmit to a TI-83 Plus list data that is stored to the TI-73 lists **L1**, **L2**, **L3**, **L4**, **L5**, or **L6**, follow these steps.

1. Set the TI-83 Plus to receive (page 19-7).
2. Press [APPS] on the sending TI-73 to display the APPLICATIONS menu.
3. Press [ENTER] on the sending TI-73 to select **1:Link** and display the LINK SEND menu.
4. Choose **0:Vars to TI83**.and then select the lists you want to send.
5. Press [▶] on the sending TI-73 to display the LINK TRANSMIT menu.
6. Confirm that the receiving unit is set to receive (page 19-7).
7. Press [ENTER] to select **1:Transmit** and begin transmitting.

Memory Backup

To copy the exact contents of memory in the sending TI-83 Plus to the memory of the receiving TI-83 Plus, put the other unit in receive mode. Then, on the receiving unit, select **H:Back Up** from the LINK SEND menu.

- **Warning: H:Back Up** overwrites the memory in the receiving unit; all information in the memory of the receiving unit is lost.

  Note: If you do not want to do a backup, select **2:Quit** to return to the LINK SEND menu.

- Select **1:Transmit** to begin transmission.

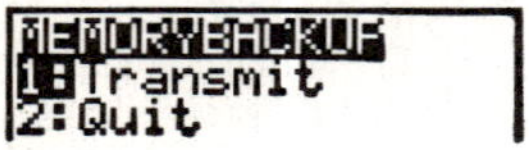

Receiving Unit

As a safety check to prevent accidental loss of memory, the message **WARNING - Backup** is displayed when the receiving unit receives notice of a backup.

- To continue with the backup process, select **1:Continue**. The backup transmission begins.
- To prevent the backup, select **2:Quit**.

Note: If a transmission error is returned during a backup, the receiving unit is reset.

Memory Backup Complete

When the backup is complete, both the sending calculator and receiving calculator display a confirmation screen.

A Tables and Reference Information

Contents

Functions return a value, list, or matrix. You can use functions in an expression. Instructions initiate an action. Some functions and instructions have arguments. Optional arguments and accompanying commas are enclosed in brackets ([ ]). For details about an item, including argument descriptions and restrictions, turn to the page listed on the right side of the table.

From the CATALOG, you can paste any function or instruction to the home screen or to a command line in the program editor. However, some functions and instructions are not valid on the home screen. The items in this table appear in the same order as they appear in the CATALOG.

† indicates either keystrokes that are valid in the program editor only or ones that paste certain instructions when you are in the program editor. Some keystrokes display menus that are available only in the program editor. Others paste mode, format, or table-set instructions only when you are in the program editor.

Function or Instruction/ Arguments	Result	Key or Keys/ Menu or Screen/Item	
abs(*value***)**	Returns the absolute value of a real number, expression, list, or matrix.	[MATH] NUM **1:abs(**	2-13 10-10
abs(*complex value***)**	Returns the magnitude of a complex number or list.	[MATH] CPX **5:abs(**	2-19
valueA **and** *valueB*	Returns 1 if both *valueA* and *valueB* are ≠ 0. *valueA* and *valueB* can be real numbers, expressions, or lists.	[2nd] [TEST] LOGIC **1:and**	2-26
angle(*value***)**	Returns the polar angle of a complex number or list of complex numbers.	[MATH] CPX **4:angle(**	2-19
ANOVA(*list1*,*list2* [,*list3*,...,*list20*]**)**	Performs a one-way analysis of variance for comparing the means of two to 20 populations.	[STAT] TESTS **F:ANOVA(**	13-24
Ans	Returns the last answer.	[2nd] [ANS]	1-18
Archive	Moves the specified variables from RAM to the user data archive memory.	[2nd] [MEM] **5:Archive**	18-10
Asm(*assemblyprgmname***)**	Executes an assembly language program.	[2nd] [CATALOG] **Asm(**	16-24

Function or Instruction/ Arguments	Result	Key or Keys/ Menu or Screen/Item	
AsmComp(*prgmASM1*, *prgmASM2***)**	Compiles an assembly language program written in ASCII and stores the hex version.	[2nd] [CATALOG] **AsmComp(**	16-24
AsmPrgm	Must be used as the first line of an assembly language program.	[2nd] [CATALOG] **AsmPrgm**	16-24
augment(*matrixA*,*matrixB***)**	Returns a matrix, which is *matrixB* appended to *matrixA* as new columns.	[2nd] [MATRX] MATH **7:augment(**	10-14
augment(*listA*,*listB***)**	Returns a list, which is *listB* concatenated to the end of *listA*.	[2nd] [LIST] OPS **9:augment(**	11-15
AxesOff	Turns off the graph axes.	† [2nd] [FORMAT] **AxesOff**	3-14
AxesOn	Turns on the graph axes.	† [2nd] [FORMAT] **AxesOn**	3-14
a+b*i*	Sets the mode to rectangular complex number mode (a+b*i*).	† [MODE] **a+b*i***	1-12
bal(*npmt*[,*roundvalue*]**)**	Computes the balance at *npmt* for an amortization schedule using stored values for **PV**, I%, and **PMT** and rounds the computation to *roundvalue*.	[APPS] **1:Finance** CALC **9:bal(**	14-10
binomcdf(*numtrials*,*p*[,*x*]**)**	Computes a cumulative probability at *x* for the discrete binomial distribution with the specified *numtrials* and probability *p* of success on each trial.	[2nd] [DISTR] DISTR **A:binomcdf(**	13-32

Function or Instruction/ Arguments	Result	Key or Keys/ Menu or Screen/Item	
binompdf(*numtrials,p*[*,x*]**)**	Computes a probability at *x* for the discrete binomial distribution with the specified *numtrials* and probability *p* of success on each trial.	[2nd] [DISTR] DISTR **0:binompdf(**	13-32
χ^2cdf(*lowerbound, upperbound,df***)**	Computes the χ^2 distribution probability between *lowerbound* and *upperbound* for the specified degrees of freedom *df*.	[2nd] [DISTR] DISTR **7:χ^2cdf(**	13-30
χ^2pdf(*x,df***)**	Computes the probability density function (pdf) for the χ^2 distribution at a specified *x* value for the specified degrees of freedom *df*.	[2nd] [DISTR] DISTR **6:χ^2pdf(**	13-30
χ^2-Test(*observedmatrix, expectedmatrix* [*,drawflag*]**)**	Performs a chi-square test. *drawflag*=**1** draws results; *drawflag*=**0** calculates results.	† [STAT] TESTS **C:χ^2-Test(**	13-21
Circle(*X,Y,radius***)**	Draws a circle with center (*X,Y*) and *radius*.	[2nd] [DRAW] DRAW **9:Circle(**	8-11
Clear Entries	Clears the contents of the Last Entry storage area.	[2nd] [MEM] MEMORY **3:Clear Entries**	18-5
ClrAllLists	Sets to **0** the dimension of all lists in memory.	[2nd] [MEM] MEMORY **4:ClrAllLists**	18-5
ClrDraw	Clears all drawn elements from a graph or drawing.	[2nd] [DRAW] DRAW **1:ClrDraw**	8-4
ClrHome	Clears the home screen.	† [PRGM] I/O **8:ClrHome**	16-21
ClrList *listname1* [*,listname2, ..., listname n*]	Sets to **0** the dimension of one or more *listnames*.	[STAT] EDIT **4:ClrList**	12-20
ClrTable	Clears all values from the table.	† [PRGM] I/O **9:ClrTable**	16-21

Function or Instruction/ Arguments	Result	Key or Keys/ Menu or Screen/Item	
conj(*value*)	Returns the complex conjugate of a complex number or list of complex numbers.	[MATH] CPX **1:conj(**	2-18
Connected	Sets connected plotting mode; resets all Y= editor graph-style settings to \\ .	† [MODE] **Connected**	1-11
CoordOff	Turns off cursor coordinate value display.	† [2nd] [FORMAT] **CoordOff**	3-14
CoordOn	Turns on cursor coordinate value display.	† [2nd] [FORMAT] **CoordOn**	3-14
cos(*value*)	Returns cosine of a real number, expression, or list.	[COS]	2-3
cos^{-1}(*value*)	Returns arccosine of a real number, expression, or list.	[2nd] [COS^{-1}]	2-3
cosh(*value*)	Returns hyperbolic cosine of a real number, expression, or list.	[2nd] [CATALOG] **cosh(**	15-10
$cosh^{-1}$(*value*)	Returns hyperbolic arccosine of a real number, expression, or list.	[2nd] [CATALOG] **$cosh^{-1}$(**	15-10
CubicReg [*Xlistname, Ylistname,freqlist, regequ*]	Fits a cubic regression model to *Xlistname* and *Ylistname* with frequency *freqlist*, and stores the regression equation to *regequ*.	[STAT] CALC **6:CubicReg**	12-26
cumSum(*list*)	Returns a list of the cumulative sums of the elements in *list*, starting with the first element.	[2nd] [LIST] OPS **6:cumSum(**	11-12
cumSum(*matrix*)	Returns a matrix of the cumulative sums of *matrix* elements. Each element in the returned matrix is a cumulative sum of a *matrix* column from top to bottom.	[2nd] [MATRX] MATH **0:cumSum(**	10-15
dbd(*date1,date2*)	Calculates the number of days between *date1* and *date2* using the actual-day-count method.	[APPS] **1:Finance** CALC **D:dbd(**	14-14
value▶**Dec**	Displays a real or complex number, expression, list, or matrix in decimal format.	[MATH] MATH **2:▶Dec**	2-5

Function or Instruction/ Arguments	Result	Key or Keys/ Menu or Screen/Item
Degree	Sets degree angle mode.	† [MODE] **Degree** 1-11
DelVar *variable*	Deletes from memory the contents of *variable*.	† [PRGM] CTL **G:DelVar** 16-16
DependAsk	Sets table to ask for dependent-variable values.	† [2nd] [TBLSET] **Depend: Ask** 7-3
DependAuto	Sets table to generate dependent-variable values automatically.	† [2nd] [TBLSET] **Depend: Auto** 7-3
det(*matrix***)**	Returns determinant of *matrix*.	[2nd] [MATRX] MATH **1:det(** 10-12
DiagnosticOff	Sets diagnostics-off mode; **r**, $\mathbf{r^2}$, and $\mathbf{R^2}$ are not displayed as regression model results.	[2nd] [CATALOG] **DiagnosticOff** 12-23
DiagnosticOn	Sets diagnostics-on mode; **r**, $\mathbf{r^2}$, and $\mathbf{R^2}$ are displayed as regression model results.	[2nd] [CATALOG] **DiagnosticOn** 12-23
dim(*listname***)**	Returns the dimension of *listname*.	[2nd] [LIST] OPS **3:dim(** 11-11
dim(*matrixname***)**	Returns the dimension of *matrixname* as a list.	[2nd] [MATRX] MATH **3:dim(** 10-12
length→**dim(***listname***)**	Assigns a new dimension (*length*) to a new or existing *listname*.	[2nd] [LIST] OPS **3:dim(** 11-11
{*rows*,*columns***}**→ **dim(***matrixname***)**	Assigns new dimensions to a new or existing *matrixname*.	[2nd] [MATRX] MATH **3:dim(** 10-13
Disp	Displays the home screen.	† [PRGM] I/O **3:Disp** 16-19
Disp [*valueA*,*valueB*, *valueC*,...,*value n*]	Displays each value.	† [PRGM] I/O **3:Disp** 16-19

Function or Instruction/ Arguments	Result	Key or Keys/ Menu or Screen/Item	
DispGraph	Displays the graph.	† [PRGM] I/O **4:DispGraph**	16-20
DispTable	Displays the table.	† [PRGM] I/O **5:DispTable**	16-20
value▸**DMS**	Displays *value* in DMS format.	[2nd] [ANGLE] ANGLE **4:▸DMS**	2-24
Dot	Sets dot plotting mode; resets all Y= editor graph-style settings to ⋱.	† [MODE] **Dot**	1-11
DrawF *expression*	Draws *expression* (in terms of **X**) on the graph.	[2nd] [DRAW] DRAW **6:DrawF**	8-9
DrawInv *expression*	Draws the inverse of *expression* by plotting **X** values on the y-axis and **Y** values on the x-axis.	[2nd] [DRAW] DRAW **8:DrawInv**	8-9
:DS<(*variable*,*value***)** **:***commandA* **:***commands*	Decrements *variable* by 1; skips *commandA* if *variable* < *value*.	† [PRGM] CTL **B:DS<(**	16-15
e^(*power***)**	Returns **e** raised to *power*.	[2nd] [e^x]	2-4
e^(*list***)**	Returns a list of **e** raised to a *list* of powers.	[2nd] [e^x]	2-4
Exponent: *value*E*exponent*	Returns *value* times 10 to the *exponent*.	[2nd] [EE]	1-7
Exponent: *list*E*exponent*	Returns *list* elements times 10 to the *exponent*.	[2nd] [EE]	1-7
Exponent: *matrix*E*exponent*	Returns *matrix* elements times 10 to the *exponent*.	[2nd] [EE]	1-7
▸**Eff(***nominal rate*, *compounding periods***)**	Computes the effective interest rate.	[APPS] **1:Finance** CALC **C:▸Eff(**	14-13
Else *See* **If:Then:Else**			

Function or Instruction/ Arguments	Result	Key or Keys/ Menu or Screen/Item	
End	Identifies end of **For(**, **If-Then-Else**, **Repeat**, or **While** loop.	† [PRGM] CTL **7:End**	16-11
Eng	Sets engineering display mode.	† [MODE] **Eng**	1-10
Equ▸String(Y= *var*,**Str***n***)**	Converts the contents of a Y= *var* to a string and stores it in **Str***n*.	[2nd] [CATALOG] **Equ▸String(**	15-7
expr(*string***)**	Converts *string* to an expression and executes it.	[2nd] [CATALOG] **expr(**	15-7
ExpReg [*Xlistname*, *Ylistname*,*freqlist*,*regequ*]	Fits an exponential regression model to *Xlistname* and *Ylistname* with frequency *freqlist*, and stores the regression equation to *regequ*.	[STAT] CALC **0:ExpReg**	12-26
ExprOff	Turns off the expression display during TRACE.	† [2nd] [FORMAT] **ExprOff**	3-14
ExprOn	Turns on the expression display during TRACE.	† [2nd] [FORMAT] **ExprOn**	3-14
Fcdf(*lowerbound*, *upperbound*, *numerator df*, *denominator df***)**	Computes the F distribution probability between *lowerbound* and *upperbound* for the specified *numerator df* (degrees of freedom) and *denominator df*.	[2nd] [DISTR] DISTR **9:Fcdf(**	13-31
Fill(*value*,*matrixname***)**	Stores *value* to each element in *matrixname*.	[2nd] [MATRX] MATH **4:Fill(**	10-13
Fill(*value*,*listname***)**	Stores *value* to each element in *listname*.	[2nd] [LIST] OPS **4:Fill(**	11-11
Fix #	Sets fixed-decimal mode for # of decimal places.	† [MODE] **0123456789** (select one)	1-10
Float	Sets floating decimal mode.	† [MODE] **Float**	1-10

Function or Instruction/ Arguments	Result	Key or Keys/ Menu or Screen/Item	
fMax(*expression,variable, lower,upper*[,*tolerance*]**)**	Returns the value of *variable* where the local maximum of *expression* occurs, between *lower* and *upper*, with specified *tolerance*.	[MATH] MATH **7:fMax(**	2-6
fMin(*expression,variable, lower,upper*[,*tolerance*]**)**	Returns the value of *variable* where the local minimum of *expression* occurs, between *lower* and *upper*, with specified *tolerance*.	[MATH] MATH **6:fMin(**	2-6
fnInt(*expression,variable, lower,upper*[,*tolerance*]**)**	Returns the function integral of *expression* with respect to *variable*, between *lower* and *upper*, with specified *tolerance*.	[MATH] MATH **9:fnInt(**	2-7
FnOff [*function#, function#,...,function n*]	Deselects all Y= functions or specified Y= functions.	[VARS] Y-VARS On/Off **2:FnOff**	3-8
FnOn [*function#, function#,...,function n*]	Selects all Y= functions or specified Y= functions.	[VARS] Y-VARS On/Off **1:FnOn**	3-8
:For(*variable,begin,end* [,*increment*]**)** :*commands* **:End** :*commands*	Executes *commands* through **End**, incrementing *variable* from *begin* by *increment* until *variable>end*.	† [PRGM] CTL **4:For(**	16-11
fPart(*value***)**	Returns the fractional part or parts of a real or complex number, expression, list, or matrix.	[MATH] NUM **4:fPart(**	2-14 10-11
Fpdf(*x,numerator df, denominator df***)**	Computes the F distribution probability between *lowerbound* and *upperbound* for the specified *numerator df* (degrees of freedom) and *denominator df*.	[2nd] [DISTR] DISTR **8:Fpdf(**	13-31

Function or Instruction/ Arguments	Result	Key or Keys/ Menu or Screen/Item	
value▶**Frac**	Displays a real or complex number, expression, list, or matrix as a fraction simplified to its simplest terms.	[MATH] MATH **1:▶Frac**	2-5
Full	Sets full screen mode.	† [MODE] **Full**	1-12
Func	Sets function graphing mode.	† [MODE] **Func**	1-11
GarbageCollect	Displays the garbage collection menu to allow cleanup of unused archive memory.	[2nd] [CATALOG] **GarbageCollect**	18-19
gcd(*valueA*,*valueB***)**	Returns the greatest common divisor of *valueA* and *valueB*, which can be real numbers or lists.	[MATH] NUM **9:gcd(**	2-15
geometcdf(*p*,*x***)**	Computes a cumulative probability at *x*, the number of the trial on which the first success occurs, for the discrete geometric distribution with the specified probability of success *p*.	[2nd] [DISTR] DISTR **E:geometcdf(**	13-33
geometpdf(*p*,*x***)**	Computes a probability at *x*, the number of the trial on which the first success occurs, for the discrete geometric distribution with the specified probability of success *p*.	[2nd] [DISTR] DISTR **D:geometpdf(**	13-33
Get(*variable***)**	Gets data from the CBL 2/CBL or CBR System and stores it in *variable*.	† [PRGM] I/O **A:Get(**	16-22
GetCalc(*variable***)**	Gets contents of *variable* on another TI-83 Plus and stores it to *variable* on the receiving TI-83 Plus.	† [PRGM] I/O **0:GetCalc(**	16-22
getKey	Returns the key code for the current keystroke, or **0**, if no key is pressed.	† [PRGM] I/O **7:getKey**	16-21
Goto *label*	Transfers control to *label*.	† [PRGM] CTL **0:Goto**	16-14

Function or Instruction/ Arguments	**Result**	**Key or Keys/ Menu or Screen/Item**	
GraphStyle(*function#*, *graphstyle#***)**	Sets a *graphstyle* for *function#*.	† [PRGM] CTL **H:GraphStyle(**	16-16
GridOff	Turns off grid format.	† [2nd] [FORMAT] **GridOff**	3-14
GridOn	Turns on grid format.	† [2nd] [FORMAT] **GridOn**	3-14
G-T	Sets graph-table vertical split-screen mode.	† [MODE] **G-T**	1-12
Horiz	Sets horizontal split-screen mode.	† [MODE] **Horiz**	1-12
Horizontal *y*	Draws a horizontal line at *y*.	[2nd] [DRAW] DRAW **3:Horizontal**	8-6
identity(*dimension***)**	Returns the identity matrix of *dimension* rows × *dimension* columns.	[2nd] [MATRX] MATH **5:identity(**	10-13
:If *condition* **:***commandA* **:***commands*	If *condition* = 0 (false), skips *commandA*.	† [PRGM] CTL **1:If**	16-10
:If *condition* **:Then** **:***commands* **:End** **:***commands*	Executes *commands* from **Then** to **End** if *condition* = 1 (true).	† [PRGM] CTL **2:Then**	16-10
:If *condition* **:Then** **:***commands* **:Else** **:***commands* **:End** **:***commands*	Executes *commands* from **Then** to **Else** if *condition* = 1 (true); from **Else** to **End** if *condition* = 0 (false).	† [PRGM] CTL **3:Else**	16-11
imag(*value***)**	Returns the imaginary (nonreal) part of a complex number or list of complex numbers.	[MATH] CPX **3:imag(**	2-18

Function or Instruction/ Arguments	Result	Key or Keys/ Menu or Screen/Item	
IndpntAsk	Sets table to ask for independent-variable values.	† [2nd] [TBLSET] **Indpnt: Ask**	7-3
IndpntAuto	Sets table to generate independent-variable values automatically.	† [2nd] [TBLSET] **Indpnt: Auto**	7-3
Input	Displays graph.	† [PRGM] I/O **1:Input**	16-17
Input [*variable*] **Input** ["*text*",*variable*]	Prompts for value to store to *variable*.	† [PRGM] I/O **1:Input**	16-18
Input [**Str***n*,*variable*]	Displays **Str***n* and stores entered value to *variable*.	† [PRGM] I/O **1:Input**	16-18
inString(*string*,*substring* [,*start*]**)**	Returns the character position in *string* of the first character of *substring* beginning at *start*.	[2nd] [CATALOG] **inString(**	15-7
int(*value***)**	Returns the largest integer ≤ a real or complex number, expression, list, or matrix.	[MATH] NUM **5:int(**	2-14 10-11
ΣInt(*pmt1*,*pmt2* [,*roundvalue*]**)**	Computes the sum, rounded to *roundvalue*, of the interest amount between *pmt1* and *pmt2* for an amortization schedule.	[APPS] **1:Finance** CALC **A:ΣInt(**	14-10
invNorm(*area*[,μ,σ]**)**	Computes the inverse cumulative normal distribution function for a given *area* under the normal distribution curve specified by μ and σ.	[2nd] [DISTR] DISTR **3:invNorm(**	13-29
iPart(*value***)**	Returns the integer part of a real or complex number, expression, list, or matrix.	[MATH] NUM **3:iPart(**	2-14 10-11

Function or Instruction/ Arguments	Result	Key or Keys/ Menu or Screen/Item	
irr(*CF0*,*CFList*[,*CFFreq*]**)**	Returns the interest rate at which the net present value of the cash flow is equal to zero.	[APPS] **1:Finance** CALC **8:irr(**	14-8
:IS>(*variable*,*value***)** *:commandA* *:commands*	Increments *variable* by 1; skips *commandA* if *variable*>*value*.	† [PRGM] CTL **A:IS>(**	16-14
ʟlistname	Identifies the next one to five characters as a user-created list name.	[2nd] [LIST] OPS **B:ʟ**	11-16
LabelOff	Turns off axes labels.	† [2nd] [FORMAT] **LabelOff**	3-14
LabelOn	Turns on axes labels.	† [2nd] [FORMAT] **LabelOn**	3-14
Lbl *label*	Creates a *label* of one or two characters.	† [PRGM] CTL **9:Lbl**	16-14
lcm(*valueA*,*valueB***)**	Returns the least common multiple of *valueA* and *valueB*, which can be real numbers or lists.	[MATH] NUM **8:lcm(**	2-15
length(*string***)**	Returns the number of characters in *string*.	[2nd] [CATALOG] **length(**	15-8
Line(*X1*,*Y1*,*X2*,*Y2***)**	Draws a line from (*X1*,*Y1*) to (*X2*,*Y2*).	[2nd] [DRAW] DRAW **2:Line(**	8-5
Line(*X1*,*Y1*,*X2*,*Y2*,**0)**	Erases a line from (*X1*,*Y1*) to (*X2*,*Y2*).	[2nd] [DRAW] DRAW **2:Line(**	8-5

Function or Instruction/ Arguments	Result	Key or Keys/ Menu or Screen/Item	
LinReg(a+bx) [*Xlistname*, *Ylistname*,*freqlist*, *regequ*]	Fits a linear regression model to *Xlistname* and *Ylistname* with frequency *freqlist*, and stores the regression equation to *regequ*.	[STAT] CALC **8:LinReg(a+bx)**	12-26
LinReg(ax+b) [*Xlistname*, *Ylistname*,*freqlist*, *regequ*]	Fits a linear regression model to *Xlistname* and *Ylistname* with frequency *freqlist*, and stores the regression equation to *regequ*.	[STAT] CALC **4:LinReg(ax+b)**	12-25
LinRegTTest [*Xlistname*, *Ylistname*,*freqlist*, *alternative*,*regequ*]	Performs a linear regression and a *t*-test. *alternative*=**⁻1** is <; *alternative*=**0** is ≠; *alternative*=**1** is >.	† [STAT] TESTS **E:LinRegTTest**	13-23
ΔList(*list***)**	Returns a list containing the differences between consecutive elements in *list*.	[2nd] [LIST] OPS **7:ΔList(**	11-12
List▸matr(*listname1*,..., *listname n*,*matrixname***)**	Fills *matrixname* column by column with the elements from each specified *listname*.	[2nd] [LIST] OPS **0:List▸matr(**	10-14 11-15
ln(*value***)**	Returns the natural logarithm of a real or complex number, expression, or list.	[LN]	2-4
LnReg [*Xlistname*, *Ylistname*,*freqlist*, *regequ*]	Fits a logarithmic regression model to *Xlistname* and *Ylistname* with frequency *freqlist*, and stores the regression equation to *regequ*.	[STAT] CALC **9:LnReg**	12-26
log(*value***)**	Returns logarithm of a real or complex number, expression, or list.	[LOG]	2-4

Function or Instruction/ Arguments	Result	Key or Keys/ Menu or Screen/Item
Logistic [*Xlistname,* *Ylistname,freqlist,* *regequ*]	Fits a logistic regression model to *Xlistname* and *Ylistname* with frequency *freqlist*, and stores the regression equation to *regequ.*	[STAT] CALC **B:Logistic** 12-27
Matr▶list(*matrix,* *listnameA,...,listname n***)**	Fills each *listname* with elements from each column in *matrix.*	[2nd] [LIST] OPS 10-14 **A:Matr▶list(** 11-16
Matr▶list(*matrix,* *column#,listname***)**	Fills a *listname* with elements from a specified *column#* in *matrix.*	[2nd] [LIST] OPS 10-14 **A:Matr▶list(** 11-16
max(*valueA,valueB***)**	Returns the larger of *valueA* and *valueB.*	[MATH] NUM **7:max(** 2-15
max(*list***)**	Returns largest real or complex element in *list.*	[2nd] [LIST] MATH **2:max(** 11-17
max(*listA,listB***)**	Returns a real or complex list of the larger of each pair of elements in *listA* and *listB.*	[2nd] [LIST] MATH **2:max(** 11-17
max(*value,list***)**	Returns a real or complex list of the larger of *value* or each *list* element.	[2nd] [LIST] MATH **2:max(** 11-17
mean(*list*[*,freqlist*]**)**	Returns the mean of *list* with frequency *freqlist.*	[2nd] [LIST] MATH **3:mean(** 11-17
median(*list*[*,freqlist*]**)**	Returns the median of *list* with frequency *freqlist.*	[2nd] [LIST] MATH **4:median(** 11-17
Med-Med [*Xlistname,* *Ylistname,freqlist,* *regequ*]	Fits a median-median model to *Xlistname* and *Ylistname* with frequency *freqlist*, and stores the regression equation to *regequ.*	[STAT] CALC **3:Med-Med** 12-25
Menu("*title***","***text1***",***label1* [,...,"*text7*",*label7*]**)**	Generates a menu of up to seven items during program execution.	† [PRGM] CTL **C:Menu(** 16-15

Function or Instruction/ Arguments	Result	Key or Keys/ Menu or Screen/Item	
min(*valueA,valueB***)**	Returns smaller of *valueA* and *valueB*.	[MATH] NUM **6:min(**	2-15
min(*list***)**	Returns smallest real or complex element in *list*.	[2nd] [LIST] MATH **1:min(**	11-17
min(*listA,listB***)**	Returns real or complex list of the smaller of each pair of elements in *listA* and *listB*.	[2nd] [LIST] MATH **1:min(**	11-17
min(*value,list***)**	Returns a real or complex list of the smaller of *value* or each *list* element.	[2nd] [LIST] MATH **1:min(**	11-17
valueA **nCr** *valueB*	Returns the number of combinations of *valueA* taken *valueB* at a time.	[MATH] PRB **3:nCr**	2-21
value **nCr** *list*	Returns a list of the combinations of *value* taken each element in *list* at a time.	[MATH] PRB **3:nCr**	2-21
list **nCr** *value*	Returns a list of the combinations of each element in *list* taken *value* at a time.	[MATH] PRB **3:nCr**	2-21
listA **nCr** *listB*	Returns a list of the combinations of each element in *listA* taken each element in *listB* at a time.	[MATH] PRB **3:nCr**	2-21
nDeriv(*expression,variable, value*[,ε]**)**	Returns approximate numerical derivative of *expression* with respect to *variable* at *value*, with specified ε.	[MATH] MATH **8:nDeriv(**	2-7
▸Nom(*effective rate, compounding periods***)**	Computes the nominal interest rate.	[APPS] **1:Finance** CALC **B:▸Nom(**	14-13
Normal	Sets normal display mode.	† [MODE] **Normal**	1-10

Function or Instruction/ Arguments	Result	Key or Keys/ Menu or Screen/Item	
normalcdf(*lowerbound*, *upperbound*[,μ,σ]**)**	Computes the normal distribution probability between *lowerbound* and *upperbound* for the specified μ and σ.	[2nd] [DISTR] DISTR **2:normalcdf(**	13-29
normalpdf(*x*[,μ,σ]**)**	Computes the probability density function for the normal distribution at a specified x value for the specified μ and σ.	[2nd] [DISTR] DISTR **1:normalpdf(**	13-28
not(*value***)**	Returns **0** if *value* is ≠ 0. *value* can be a real number, expression, or list.	[2nd] [TEST] LOGIC **4:not(**	2-26
valueA **nPr** *valueB*	Returns the number of permutations of *valueA* taken *valueB* at a time.	[MATH] PRB **2:nPr**	2-21
value **nPr** *list*	Returns a list of the permutations of *value* taken each element in *list* at a time.	[MATH] PRB **2:nPr**	2-21
list **nPr** *value*	Returns a list of the permutations of each element in *list* taken *value* at a time.	[MATH] PRB **2:nPr**	2-21
listA **nPr** *listB*	Returns a list of the permutations of each element in *listA* taken each element in *listB* at a time.	[MATH] PRB **2:nPr**	2-21
npv(*interest rate*,*CF0*, *CFList*[,*CFFreq*]**)**	Computes the sum of the present values for cash inflows and outflows.	[APPS] **1:Finance** CALC **7:npv(**	14-9
valueA **or** *valueB*	Returns 1 if *valueA* or *valueB* is ≠ 0. *valueA* and *valueB* can be real numbers, expressions, or lists.	[2nd] [TEST] LOGIC **2:or**	2-26

Function or Instruction/ Arguments	Result	Key or Keys/ Menu or Screen/Item	
Output(*row*,*column*,"*text*"**)**	Displays *text* beginning at specified *row* and *column*.	† [PRGM] I/O **6:Output(**	16-20
Output(*row*,*column*,*value***)**	Displays *value* beginning at specified *row* and *column*.	† [PRGM] I/O **6:Output(**	16-20
Param	Sets parametric graphing mode.	† [MODE] **Par**	1-11
Pause	Suspends program execution until you press [ENTER].	† [PRGM] CTL **8:Pause**	16-13
Pause [*value*]	Displays *value*; suspends program execution until you press [ENTER].	† [PRGM] CTL **8:Pause**	16-13
Plot#(*type*,*Xlistname*, *Ylistname*,*mark***)**	Defines **Plot#** (**1**, **2**, or **3**) of *type* **Scatter** or **xyLine** for *Xlistname* and *Ylistname* using *mark*.	† [2nd] [STAT PLOT] PLOTS **1:Plot1-** **2:Plot2-** **3:Plot3-**	12-34
Plot#(*type*,*Xlistname*, *freqlist***)**	Defines **Plot#** (**1**, **2**, or **3**) of *type* **Histogram** or **Boxplot** for *Xlistname* with frequency *freqlist*.	† [2nd] [STAT PLOT] PLOTS **1:Plot1-** **2:Plot2-** **3:Plot3-**	12-34
Plot#(*type*,*Xlistname*, *freqlist*,*mark***)**	Defines **Plot#** (**1**, **2**, or **3**) of *type* **ModBoxplot** for *Xlistname* with frequency *freqlist* using *mark*.	† [2nd] [STAT PLOT] PLOTS **1:Plot1-** **2:Plot2-** **3:Plot3-**	12-34
Plot#(*type*,*datalistname*, *data axis*,*mark***)**	Defines **Plot#** (**1**, **2**, or **3**) of *type* **NormProbPlot** for *datalistname* on *data axis* using *mark*. *data axis* can be **X** or **Y**.	† [2nd] [STAT PLOT] PLOTS **1:Plot1-** **2:Plot2-** **3:Plot3-**	12-34
PlotsOff [1,2,3]	Deselects all stat plots or one or more specified stat plots (**1**, **2**, or **3**).	[2nd] [STAT PLOT] STAT PLOTS **4:PlotsOff**	12-35
PlotsOn [1,2,3]	Selects all stat plots or one or more specified stat plots (**1**, **2**, or **3**).	[2nd] [STAT PLOT] STAT PLOTS **5:PlotsOn**	12-35

Function or Instruction/ Arguments	Result	Key or Keys/ Menu or Screen/Item	
Pmt_Bgn	Specifies an annuity due, where payments occur at the beginning of each payment period.	[APPS] **1:Finance** CALC **F:Pmt_Bgn**	14-14
Pmt_End	Specifies an ordinary annuity, where payments occur at the end of each payment period.	[APPS] **1:Finance** CALC **E:Pmt_End**	14-14
poissoncdf(μ,*x***)**	Computes a cumulative probability at *x* for the discrete Poisson distribution with specified mean μ.	[2nd] [DISTR] DISTR **C:poissoncdf(**	13-33
poissonpdf(μ,*x***)**	Computes a probability at *x* for the discrete Poisson distribution with the specified mean μ.	[2nd] [DISTR] DISTR **B:poissonpdf(**	13-32
Polar	Sets polar graphing mode.	† [MODE] **Pol**	5-3
complex value **▸Polar**	Displays *complex value* in polar format.	[MATH] CPX **7:▸Polar**	2-19
PolarGC	Sets polar graphing coordinates format.	† [2nd] [FORMAT] **PolarGC**	3-13
prgm*name*	Executes the program *name*.	† [PRGM] CTRL **D:prgm**	16-16
ΣPrn(*pmt1*,*pmt2* [,*roundvalue*]**)**	Computes the sum, rounded to *roundvalue*, of the principal amount between *pmt1* and *pmt2* for an amortization schedule.	[APPS] **1:Finance** CALC **0:ΣPrn(**	14-10
prod(*list*[,*start*,*end*]**)**	Returns product of *list* elements between *start* and *end*.	[2nd] [LIST] MATH **6:prod(**	11-18
Prompt *variableA* [,*variableB*,...,*variable n*]	Prompts for value for *variableA*, then *variableB*, and so on.	† [PRGM] I/O **2:Prompt**	16-19

Function or Instruction/ Arguments	Result	Key or Keys/ Menu or Screen/Item
1-PropZInt(*x*,*n* [,*confidence level*]**)**	Computes a one-proportion *z* confidence interval.	† [STAT] TESTS **A:1-PropZInt(** 13-20
2-PropZInt(*x1*,*n1*,*x2*,*n2* [,*confidence level*]**)**	Computes a two-proportion *z* confidence interval.	† [STAT] TESTS **B:2-PropZInt(** 13-20
1-PropZTest(*p0*,*x*,*n* [,*alternative*,*drawflag*]**)**	Computes a one-proportion *z* test. *alternative*=**⁻1** is <; *alternative*=**0** is ≠; *alternative*=**1** is >. *drawflag*=**1** draws results; *drawflag*=**0** calculates results.	† [STAT] TESTS **5:1-PropZTest(** 13-14
2-PropZTest(*x1*,*n1*,*x2*,*n2* [,*alternative*,*drawflag*]**)**	Computes a two-proportion *z* test. *alternative*=**⁻1** is <; *alternative*=**0** is ≠; *alternative*=**1** is >. *drawflag*=**1** draws results; *drawflag*=**0** calculates results.	† [STAT] TESTS **6:2-PropZTest(** 13-15
Pt-Change(*x*,*y***)**	Reverses a point at (*x*,*y*).	[2nd] [DRAW] POINTS **3:Pt-Change(** 8-15
Pt-Off(*x*,*y*[,*mark*]**)**	Erases a point at (*x*,*y*) using *mark*.	[2nd] [DRAW] POINTS **2:Pt-Off(** 8-15
Pt-On(*x*,*y*[,*mark*]**)**	Draws a point at (*x*,*y*) using *mark*.	[2nd] [DRAW] POINTS **1:Pt-On(** 8-14
PwrReg [*Xlistname*, *Ylistname*,*freqlist*, *regequ*]	Fits a power regression model to *Xlistname* and *Ylistname* with frequency *freqlist*, and stores the regression equation to *regequ*.	[STAT] CALC **A:PwrReg** 12-27

Function or Instruction/ Arguments	Result	Key or Keys/ Menu or Screen/Item	
Pxl-Change(*row,column***)**	Reverses pixel at (*row,column*); 0 ≤ *row* ≤ 62 and 0 ≤ *column* ≤ 94.	[2nd] [DRAW] POINTS **6:Pxl-Change(**	8-16
Pxl-Off(*row,column***)**	Erases pixel at (*row,column*); 0 ≤ *row* ≤ 62 and 0 ≤ *column* ≤ 94.	[2nd] [DRAW] POINTS **5:Pxl-Off(**	8-16
Pxl-On(*row,column***)**	Draws pixel at (*row,column*); 0 ≤ *row* ≤ 62 and 0 ≤ *column* ≤ 94.	[2nd] [DRAW] POINTS **4:Pxl-On(**	8-16
pxl-Test(*row,column***)**	Returns 1 if pixel (*row, column*) is on, 0 if it is off; 0 ≤ *row* ≤ 62 and 0 ≤ *column* ≤ 94.	[2nd] [DRAW] POINTS **7:pxl-Test(**	8-16
P▸Rx(*r*,*θ***)**	Returns **X**, given polar coordinates *r* and θ or a list of polar coordinates.	[2nd] [ANGLE] ANGLE **7:P▸Rx(**	2-24
P▸Ry(*r*,*θ***)**	Returns **Y**, given polar coordinates *r* and θ or a list of polar coordinates.	[2nd] [ANGLE] ANGLE **8:P▸Ry(**	2-24
QuadReg [*Xlistname, Ylistname,freqlist, regequ*]	Fits a quadratic regression model to *Xlistname* and *Ylistname* with frequency *freqlist*, and stores the regression equation to *regequ*.	[STAT] CALC **5:QuadReg**	12-25
QuartReg [*Xlistname, Ylistname,freqlist, regequ*]	Fits a quartic regression model to *Xlistname* and *Ylistname* with frequency *freqlist*, and stores the regression equation to *regequ*.	[STAT] CALC **7:QuartReg**	12-26
Radian	Sets radian angle mode.	† [MODE] **Radian**	1-11
rand[(*numtrials*)]	Returns a random number between 0 and 1 for a specified number of trials *numtrials*.	[MATH] PRB **1:rand**	2-20
randBin(*numtrials,prob* [*,numsimulations*]**)**	Generates and displays a random real number from a specified Binomial distribution.	[MATH] PRB **7:randBin(**	2-22

Function or Instruction/ Arguments	Result	Key or Keys/ Menu or Screen/Item	
randInt(*lower*,*upper* [,*numtrials*]**)**	Generates and displays a random integer within a range specified by *lower* and *upper* integer bounds for a specified number of trials *numtrials*.	[MATH] PRB **5:randInt(**	2-22
randM(*rows*,*columns***)**	Returns a random matrix of *rows* (**1–99**) × *columns* (**1–99**).	[2nd] [MATRX] MATH **6:randM(**	10-13
randNorm(μ,σ[,*numtrials*]**)**	Generates and displays a random real number from a specified Normal distribution specified by μ and σ for a specified number of trials *numtrials*.	[MATH] PRB **6:randNorm(**	2-22
re^θ*i*	Sets the mode to polar complex number mode (**re^θ*i***).	† [MODE] **re^θ*i***	1-12
Real	Sets mode to display complex results only when you enter complex numbers.	† [MODE] **Real**	1-12
real(*value***)**	Returns the real part of a complex number or list of complex numbers.	[MATH] CPX **2:real(**	2-18
RecallGDB *n*	Restores all settings stored in the graph database variable **GDB***n*.	[2nd] [DRAW] STO **4:RecallGDB**	8-20
RecallPic *n*	Displays the graph and adds the picture stored in **Pic***n*.	[2nd] [DRAW] STO **2:RecallPic**	8-18
complex value **▸Rect**	Displays *complex value* or list in rectangular format.	[MATH] CPX **6:▸Rect**	2-19
RectGC	Sets rectangular graphing coordinates format.	† [2nd] [FORMAT] **RectGC**	3-13
ref(*matrix***)**	Returns the row-echelon form of a *matrix*.	[2nd] [MATRX] MATH **A:ref(**	10-15

Function or Instruction/ Arguments	Result	Key or Keys/ Menu or Screen/Item	
:Repeat *condition* :*commands* **:End** :*commands*	Executes *commands* until *condition* is true.	† PRGM CTL **6:Repeat**	16-12
Return	Returns to the calling program.	† PRGM CTL **E:Return**	16-16
round(*value*[,#*decimals*]**)**	Returns a number, expression, list, or matrix rounded to #*decimals* (≤ 9).	MATH NUM **2:round(**	2-13
row(value*,*matrix*,*row***)**	Returns a matrix with *row* of *matrix* multiplied by *value* and stored in *row*.	2nd [MATRX] MATH **E:*row(**	10-16
row+(*matrix*,*rowA*,*rowB***)**	Returns a matrix with *rowA* of *matrix* added to *rowB* and stored in *rowB*.	2nd [MATRX] MATH **D:row+(**	10-16
row+(value*,*matrix*, *rowA*,*rowB***)**	Returns a matrix with *rowA* of *matrix* multiplied by *value*, added to *rowB*, and stored in *rowB*.	2nd [MATRX] MATH **F:*row+(**	10-16
rowSwap(*matrix*,*rowA*, *rowB***)**	Returns a matrix with *rowA* of *matrix* swapped with *rowB*.	2nd [MATRX] MATH **C:rowSwap(**	10-16
rref(*matrix***)**	Returns the reduced row-echelon form of a *matrix*.	2nd [MATRX] MATH **B:rref(**	10-15
R▶Pr(x,y**)**	Returns **R**, given rectangular coordinates x and y or a list of rectangular coordinates.	2nd [ANGLE] ANGLE **5:R▶Pr(**	2-24
R▶Pθ(x,y**)**	Returns θ, given rectangular coordinates x and y or a list of rectangular coordinates.	2nd [ANGLE] ANGLE **6:R▶Pθ(**	2-24

Function or Instruction/ Arguments	Result	Key or Keys/ Menu or Screen/Item	
2-SampFTest [*listname1*, *listname2*,*freqlist1*, *freqlist2*,*alternative*, *drawflag*] (Data list input)	Performs a two-sample F test. *alternative*=⁻**1** is **<**; *alternative*=**0** is ≠; *alternative*=**1** is **>**. *drawflag*=**1** draws results; *drawflag*=**0** calculates results.	† [STAT] TESTS **D:2-SampFTest**	13-22
2-SampFTest *Sx1*,*n1*, *Sx2*,*n2*[,*alternative*, *drawflag*] (Summary stats input)	Performs a two-sample F test. *alternative*=⁻**1** is **<**; *alternative*=**0** is ≠; *alternative*=**1** is **>**. *drawflag*=**1** draws results; *drawflag*=**0** calculates results.	† [STAT] TESTS **D:2-SampFTest**	13-22
2-SampTInt [*listname1*, *listname2*, *freqlist1*,*freqlist2*, *confidence level*,*pooled*] (Data list input)	Computes a two-sample *t* confidence interval. *pooled*=**1** pools variances; *pooled*=**0** does not pool variances.	† [STAT] TESTS **0:2-SampTInt**	13-19
2-SampTInt $\bar{x}$*1*,*Sx1*,*n1*, $\bar{x}$*2*,*Sx2*,*n2* [,*confidence level*,*pooled*] (Summary stats input)	Computes a two-sample *t* confidence interval. *pooled*=**1** pools variances; *pooled*=**0** does not pool variances.	† [STAT] TESTS **0:2-SampTInt**	13-19
2-SampTTest [*listname1*, *listname2*,*freqlist1*, *freqlist2*,*alternative*, *pooled*,*drawflag*] (Data list input)	Computes a two-sample *t* test. *alternative*=⁻**1** is **<**; *alternative*=**0** is ≠; *alternative*=**1** is **>**. *pooled*=**1** pools variances; *pooled*=**0** does not pool variances. *drawflag*=**1** draws results; *drawflag*=**0** calculates results.	† [STAT] TESTS **4:2-SampTTest**	13-13

Function or Instruction/ Arguments	Result	Key or Keys/ Menu or Screen/Item
2-SampTTest $\bar{x}1$,*Sx1*,*n1*, $\bar{x}2$,*Sx2*,*n2*[,*alternative*, *pooled*,*drawflag*] (Summary stats input)	Computes a two-sample *t* test. *alternative*=⁻**1** is **<**; *alternative*=**0** is ≠; *alternative*=**1** is **>**. *pooled*=**1** pools variances; *pooled*=**0** does not pool variances. *drawflag*=**1** draws results; *drawflag*=**0** calculates results.	† [STAT] TESTS **4:2-SampTTest** 13-13
2-SampZInt($\sigma 1$,σ_2 [,*listname1*,*listname2*, *freqlist1*,*freqlist2*, *confidence level*]**)** (Data list input)	Computes a two-sample *z* confidence interval.	† [STAT] TESTS **9:2-SampZInt(** 13-18
2-SampZInt($\sigma 1$,σ_2, $\bar{x}1$,*n1*,$\bar{x}2$,*n2* [,*confidence level*]**)** (Summary stats input)	Computes a two-sample *z* confidence interval.	† [STAT] TESTS **9:2-SampZInt(** 13-18
2-SampZTest($\sigma 1$,σ_2 [,*listname1*,*listname2*, *freqlist1*,*freqlist2*, *alternative*,*drawflag*]**)** (Data list input)	Computes a two-sample *z* test. *alternative*=⁻**1** is **<**; *alternative*=**0** is ≠ ; *alternative*=**1** is **>**. *drawflag*=**1** draws results; *drawflag*=**0** calculates results.	† [STAT] TESTS **3:2-SampZTest(** 13-12
2-SampZTest($\sigma 1$,σ_2, $\bar{x}1$,*n1*,$\bar{x}2$,*n2* [,*alternative*,*drawflag*]**)** (Summary stats input)	Computes a two-sample *z* test. *alternative*=⁻**1** is **<**; *alternative*=**0** is ≠; *alternative*=**1** is **>**. *drawflag*=**1** draws results; *drawflag*=**0** calculates results.	† [STAT] TESTS **3:2-SampZTest(** 13-12
Sci	Sets scientific notation display mode.	† [MODE] **Sci** 1-10
Select(*Xlistname*, *Ylistname***)**	Selects one or more specific data points from a scatter plot or xyLine plot (only), and then stores the selected data points to two new lists, *Xlistname* and *Ylistname*.	[2nd] [LIST] OPS **8:Select(** 11-12

Function or Instruction/ Arguments	Result	Key or Keys/ Menu or Screen/Item	
Send(*variable***)**	Sends contents of *variable* to the CBL 2/CBL or CBR System.	† [PRGM] I/O **B:Send(**	16-22
seq(*expression*,*variable*, *begin*,*end*[,*increment*]**)**	Returns list created by evaluating *expression* with regard to *variable*, from *begin* to *end* by *increment*.	[2nd] [LIST] OPS **5:seq(**	11-12
Seq	Sets sequence graphing mode.	† [MODE] **Seq**	1-11
Sequential	Sets mode to graph functions sequentially.	† [MODE] **Sequential**	1-12
SetUpEditor	Removes all list names from the stat list editor, and then restores list names **L1** through **L6** to columns **1** through **6**.	[STAT] EDIT **5:SetUpEditor**	12-21
SetUpEditor *listname1* [,*listname2*,..., *listname20*]	Removes all list names from the stat list editor, then sets it up to display one or more *listnames* in the specified order, starting with column **1**.	[STAT] EDIT **5:SetUpEditor**	12-21
Shade(*lowerfunc*, *upperfunc*[,*Xleft*,*Xright*, *pattern*,*patres*]**)**	Draws *lowerfunc* and *upperfunc* in terms of **X** on the current graph and uses *pattern* and *patres* to shade the area bounded by *lowerfunc*, *upperfunc*, *Xleft*, and *Xright*.	[2nd] [DRAW] DRAW **7:Shade(**	8-10
Shadeχ^2(*lowerbound*, *upperbound*,*df***)**	Draws the density function for the χ^2 distribution specified by degrees of freedom *df* and shades the area between *lowerbound* and *upperbound*.	[2nd] [DISTR] DRAW **3:Shadeχ^2(**	13-35

Function or Instruction/ Arguments	Result	Key or Keys/ Menu or Screen/Item	
ShadeF(*lowerbound, upperbound, numerator df, denominator df***)**	Draws the density function for the F distribution specified by *numerator df* and *denominator df* and shades the area between *lowerbound* and *upperbound.*	[2nd] [DISTR] DRAW **4:ShadeF(**	13-36
ShadeNorm(*lowerbound, upperbound*[,μ,σ]**)**	Draws the normal density function specified by μ and σ and shades the area between *lowerbound* and *upperbound.*	[2nd] [DISTR] DRAW **1:ShadeNorm(**	13-34
Shade_t(*lowerbound, upperbound,df***)**	Draws the density function for the Student-t distribution specified by degrees of freedom df, and shades the area between *lowerbound* and *upperbound.*	[2nd] [DISTR] DRAW **2:Shade_t(**	13-35
Simul	Sets mode to graph functions simultaneously.	† [MODE] **Simul**	1-12
sin(*value***)**	Returns the sine of a real number, expression, or list.	[SIN]	2-3
sin^{-1}(*value***)**	Returns the arcsine of a real number, expression, or list.	[2nd] [SIN^{-1}]	2-3
sinh(*value***)**	Returns the hyperbolic sine of a real number, expression, or list.	[2nd] [CATALOG] **sinh(**	15-10
$sinh^{-1}$(*value***)**	Returns the hyperbolic arcsine of a real number, expression, or list.	[2nd] [CATALOG] **$sinh^{-1}$-**	15-10

Function or Instruction/ Arguments	Result	Key or Keys/ Menu or Screen/Item	
SinReg [*iterations*, *Xlistname*,*Ylistname*, *period*,*regequ*]	Attempts *iterations* times to fit a sinusoidal regression model to *Xlistname* and *Ylistname* using a *period* guess, and stores the regression equation to *regequ*.	[STAT] CALC **C:SinReg**	12-27
solve(*expression*,*variable*, *guess*,{*lower*,*upper*}**)**	Solves *expression* for *variable*, given an initial *guess* and *lower* and *upper* bounds within which the solution is sought.	† [MATH] MATH **0:solve(**	2-12
SortA(*listname***)**	Sorts elements of *listname* in ascending order.	[2nd] [LIST] OPS **1:SortA(**	11-10 12-20
SortA(*keylistname*, *dependlist1*[,*dependlist2*, ...,*dependlist n*]**)**	Sorts elements of *keylistname* in ascending order, then sorts each *dependlist* as a dependent list.	[2nd] [LIST] OPS **1:SortA(**	11-10 12-20
SortD(*listname***)**	Sorts elements of *listname* in descending order.	[2nd] [LIST] OPS **2:SortD(**	11-10 12-20
SortD(*keylistname*, *dependlist1*[,*dependlist2*,..., *dependlist n*]**)**	Sorts elements of *keylistname* in descending order, then sorts each *dependlist* as a dependent list.	[2nd] [LIST] OPS **2:SortD(**	11-10 12-20
stdDev(*list*[,*freqlist*]**)**	Returns the standard deviation of the elements in *list* with frequency *freqlist*.	[2nd] [LIST] MATH **7:stdDev(**	11-18
Stop	Ends program execution; returns to home screen.	† [PRGM] CTL **F:Stop**	16-16
Store: *value*→*variable*	Stores *value* in *variable*.	[STO▸]	1-14
StoreGDB *n*	Stores current graph in database **GDB***n*.	[2nd] [DRAW] STO **3:StoreGDB**	8-19

Function or Instruction/ Arguments	Result	Key or Keys/ Menu or Screen/Item	
StorePic *n*	Stores current picture in picture **Pic***n*.	[2nd] [DRAW] STO **1:StorePic**	8-17
String▸Equu(*string*,Y= *var*)	Converts *string* into an equation and stores it in **Y=** *var*.	[2nd] [CATALOG] **String▸Equ(**	15-8
sub(*string*,*begin*,*length***)**	Returns a string that is a subset of another *string*, from *begin* to *length*.	[2nd] [CATALOG] **sub(**	15-9
sum(*list*[,*start*,*end*]**)**	Returns the sum of elements of *list* from *start* to *end*.	[2nd] [LIST] MATH **5:sum(**	11-18
tan(*value***)**	Returns the tangent of a real number, expression, or list.	[TAN]	2-3
tan^{-1}(*value***)**	Returns the arctangent of a real number, expression, or list.	[2nd] [TAN^{-1}]	2-3
Tangent(*expression*,*value***)**	Draws a line tangent to *expression* at **X**=*value*.	[2nd] [DRAW] DRAW **5:Tangent(**	8-8
tanh(*value***)**	Returns hyperbolic tangent of a real number, expression, or list.	[2nd] [CATALOG] **tanh(**	15-10
tanh^{-1}(*value***)**	Returns the hyperbolic arctangent of a real number, expression, or list.	[2nd] [CATALOG] **tanh^{-1}-**	15-10
tcdf(*lowerbound*, *upperbound*,*df***)**	Computes the Student-*t* distribution probability between *lowerbound* and *upperbound* for the specified degrees of freedom *df*.	[2nd] [DISTR] DISTR **5:tcdf(**	13-30
Text(*row*,*column*,*text1*, *text2*,...,*text n***)**	Writes *text* on graph beginning at pixel (*row*,*column*), where $0 \le row \le 57$ and $0 \le column \le 94$.	[2nd] [DRAW] DRAW **0:Text(**	8-12
Then *See* **If:Then**			

Function or Instruction/ Arguments	Result	Key or Keys/ Menu or Screen/Item	
Time	Sets sequence graphs to plot with respect to time.	† [2nd] [FORMAT] **Time**	6-8
TInterval [*listname*, *freqlist*,*confidence level*] (Data list input)	Computes a *t* confidence interval.	† [STAT] TESTS **8:TInterval**	13-17
TInterval $\bar{x}$,*Sx*,*n* [,*confidence level*] (Summary stats input)	Computes a *t* confidence interval.	† [STAT] TESTS **8:TInterval**	13-17
tpdf(x,*df***)**	Computes the probability density function (pdf) for the Student-*t* distribution at a specified x value with specified degrees of freedom *df*.	[2nd] [DISTR] DISTR **4:tpdf(**	13-29
Trace	Displays the graph and enters TRACE mode.	[TRACE]	3-18
T-Test *μ0*[,*listname*, *freqlist*,*alternative*, *drawflag*] (Data list input)	Performs a *t* test with frequency *freqlist*. *alternative*=⁻**1** is **<**; *alternative*=**0** is ≠; *alternative*=**1** is **>**. *drawflag*=**1** draws results; *drawflag*=**0** calculates results.	† [STAT] TESTS **2:T-Test**	13-11
T-Test *μ0*, $\bar{x}$,*Sx*,*n* [,*alternative*,*drawflag*] (Summary stats input)	Performs a *t* test with frequency *freqlist*. *alternative*=⁻**1** is < ; *alternative*=**0** is ≠ ; *alternative*=**1** is >. *drawflag*=**1** draws results; *drawflag*=**0** calculates results.	† [STAT] TESTS **2:T-Test**	13-11

Function or Instruction/ Arguments	Result	Key or Keys/ Menu or Screen/Item	
tvm_FV[(***N***,*I%*,*PV*,*PMT*, *P/Y*,*C/Y*)]	Computes the future value.	[APPS] **1:Finance** CALC **6:tvm_FV**	14-8
tvm_I%[(***N***,*PV*,*PMT*,*FV*, *P/Y*,*C/Y*)]	Computes the annual interest rate.	[APPS] **1:Finance** CALC **3:tvm_I%**	14-8
tvm_N[(*I%*,*PV*,*PMT*,*FV*, *P/Y*,*C/Y*)]	Computes the number of payment periods.	[APPS] **1:Finance** CALC **5:tvm_N**	14-8
tvm_Pmt[(***N***,*I%*,*PV*,*FV*, *P/Y*,*C/Y*)]	Computes the amount of each payment.	[APPS] **1:Finance** CALC **2:tvm_Pmt**	14-7
tvm_PV[(***N***,*I%*,*PMT*,*FV*, *P/Y*,*C/Y*)]	Computes the present value.	[APPS] **1:Finance** CALC **4:tvm_PV**	14-8
UnArchive	Moves the specified variables from the user data archive memory to RAM. To archive variables, use **Archive**.	[2nd] [MEM] **6:UnArchive**	18-10
uvAxes	Sets sequence graphs to plot **u(*n*)** on the x-axis and **v(*n*)** on the y-axis.	† [2nd] [FORMAT] **uv**	6-8
uwAxes	Sets sequence graphs to plot **u(*n*)** on the x-axis and **w(*n*)** on the y-axis.	† [2nd] [FORMAT] **uw**	6-8
1-Var Stats [*Xlistname*, *freqlist*]	Performs one-variable analysis on the data in *Xlistname* with frequency *freqlist*.	[STAT] CALC **1:1-Var Stats**	12-25
2-Var Stats [*Xlistname*, *Ylistname*,*freqlist*]	Performs two-variable analysis on the data in *Xlistname* and *Ylistname* with frequency *freqlist*.	[STAT] CALC **2:2-Var Stats**	12-25
variance(*list*[,*freqlist*]**)**	Returns the variance of the elements in *list* with frequency *freqlist*.	[2nd] [LIST] MATH **8:variance(**	11-18
Vertical *x*	Draws a vertical line at *x*.	[2nd] [DRAW] DRAW **4:Vertical**	8-6
vwAxes	Sets sequence graphs to plot **v(*n*)** on the x-axis and **w(*n*)** on the y-axis.	† [2nd] [FORMAT] **vw**	6-8
Web	Sets sequence graphs to trace as webs.	† [2nd] [FORMAT] **Web**	6-8

Function or Instruction/ Arguments	Result	Key or Keys/ Menu or Screen/Item	
:While *condition* *:commands* **:End** *:command*	Executes *commands* while *condition* is true.	† [PRGM] CTL **5:While**	16-12
valueA **xor** *valueB*	Returns 1 if only *valueA* or *valueB* = 0. *valueA* and *valueB* can be real numbers, expressions, or lists.	[2nd] [TEST] LOGIC **3:xor**	2-26
ZBox	Displays a graph, lets you draw a box that defines a new viewing window, and updates the window.	† [ZOOM] ZOOM **1:ZBox**	3-20
ZDecimal	Adjusts the viewing window so that **ΔX=0.1** and **ΔY=0.1**, and displays the graph screen with the origin centered on the screen.	† [ZOOM] ZOOM **4:ZDecimal**	3-21
ZInteger	Redefines the viewing window using these dimensions: **ΔX=1** **Xscl=10** **ΔY=1** **Yscl=10**	† [ZOOM] ZOOM **8:ZInteger**	3-22
ZInterval σ[,*listname*, *freqlist*,*confidence level*] (Data list input)	Computes a *z* confidence interval.	† [STAT] TESTS **7:ZInterval**	13-16
ZInterval σ,$\bar{x}$,*n* [,*confidence level*] (Summary stats input)	Computes a *z* confidence interval.	† [STAT] TESTS **7:ZInterval**	13-16
Zoom In	Magnifies the part of the graph that surrounds the cursor location.	† [ZOOM] ZOOM **2:Zoom In**	3-21
Zoom Out	Displays a greater portion of the graph, centered on the cursor location.	† [ZOOM] ZOOM **3:Zoom Out**	3-21

Function or Instruction/ Arguments	Result	Key or Keys/ Menu or Screen/Item	
ZoomFit	Recalculates **Ymin** and **Ymax** to include the minimum and maximum **Y** values, between **Xmin** and **Xmax**, of the selected functions and replots the functions.	† [ZOOM] ZOOM **0:ZoomFit**	3-22
ZoomRcl	Graphs the selected functions in a user-defined viewing window.	† [ZOOM] MEMORY **3:ZoomRcl**	3-23
ZoomStat	Redefines the viewing window so that all statistical data points are displayed.	† [ZOOM] ZOOM **9:ZoomStat**	3-22
ZoomSto	Immediately stores the current viewing window.	† [ZOOM] MEMORY **2:ZoomSto**	3-23
ZPrevious	Replots the graph using the window variables of the graph that was displayed before you executed the last ZOOM instruction.	† [ZOOM] MEMORY **1:ZPrevious**	3-23
ZSquare	Adjusts the **X** or **Y** window settings so that each pixel represents an equal width and height in the coordinate system, and updates the viewing window.	† [ZOOM] ZOOM **5:ZSquare**	3-21
ZStandard	Replots the functions immediately, updating the window variables to the default values.	† [ZOOM] ZOOM **6:ZStandard**	3-22

Function or Instruction/ Arguments	Result	Key or Keys/ Menu or Screen/Item	
Z-Test($\mu 0$,σ[,*listname*, *freqlist*,*alternative*, *drawflag*]**)** (Data list input)	Performs a *z* test with frequency *freqlist*. *alternative*=⁻**1** is **<**; *alternative*=**0** is ≠; *alternative*=**1** is **>**. *drawflag*=**1** draws results; *drawflag*=**0** calculates results.	† [STAT] TESTS **1:Z-Test(**	13-10
Z-Test($\mu 0$,σ,x̄,*n* [,*alternative*,*drawflag*]**)** (Summary stats input)	Performs a *z* test. *alternative*=⁻**1** is **<**; *alternative*=**0** is ≠; *alternative*=**1** is **>**. *drawflag*=**1** draws results; *drawflag*=**0** calculates results.	† [STAT] TESTS **1:Z-Test(**	13-10
ZTrig	Replots the functions immediately, updating the window variables to preset values for plotting trig functions.	† [ZOOM] ZOOM **7:ZTrig**	3-22
Factorial: *value***!**	Returns factorial of *value*.	[MATH] PRB **4:!**	2-21
Factorial: *list***!**	Returns factorial of *list* elements.	[MATH] PRB **4:!**	2-21
Degrees notation: *value*°	Interprets *value* as degrees; designates degrees in DMS format.	[2nd] [ANGLE] ANGLE **1:°**	2-23
Radian: *angle*ʳ	Interprets *angle* as radians.	[2nd] [ANGLE] ANGLE **3:ʳ**	2-24
Transpose: *matrix*ᵀ	Returns a matrix in which each element (row, column) is swapped with the corresponding element (column, row) of *matrix*.	[2nd] [MATRX] MATH **2:ᵀ**	10-12

Function or Instruction/ Arguments	Result	Key or Keys/ Menu or Screen/Item	
x^{th}root$^x\sqrt{}$*value*	Returns *x^{th}root* of *value*.	[MATH] MATH 5:$^x\sqrt{}$	2-6
x^{th}root$^x\sqrt{}$*list*	Returns *x^{th}root* of *list* elements.	[MATH] MATH 5:$^x\sqrt{}$	2-6
list$^x\sqrt{}$*value*	Returns *list* roots of *value*.	[MATH] MATH 5:$^x\sqrt{}$	2-6
listA$^x\sqrt{}$*listB*	Returns *listA* roots of *listB*.	[MATH] MATH 5:$^x\sqrt{}$	2-6
Cube: *value*^{3}	Returns the cube of a real or complex number, expression, list, or square matrix.	[MATH] MATH 3:^{3}	2-6 10-10
Cube root: $^3\sqrt{}$(*value*)	Returns the cube root of a real or complex number, expression, or list.	[MATH] MATH 4:$^3\sqrt{}$	2-6
Equal: *valueA*=*valueB*	Returns 1 if *valueA* = *valueB*. Returns 0 if *valueA* ≠ *valueB*. *valueA* and *valueB* can be real or complex numbers, expressions, lists, or matrices.	[2nd] [TEST] TEST 1:=	2-25 10-11
Not equal: *valueA*≠*valueB*	Returns 1 if *valueA* ≠ *valueB*. Returns 0 if *valueA* = *valueB*. *valueA* and *valueB* can be real or complex numbers, expressions, lists, or matrices.	[2nd] [TEST] TEST 2:≠	2-25 10-11
Less than: *valueA*<*valueB*	Returns 1 if *valueA* < *valueB*. Returns 0 if *valueA* ≥ *valueB*. *valueA* and *valueB* can be real or complex numbers, expressions, or lists.	[2nd] [TEST] TEST 5:<	2-25

Function or Instruction/ Arguments	Result	Key or Keys/ Menu or Screen/Item	
Greater than: *valueA>valueB*	Returns 1 if *valueA* > *valueB*. Returns 0 if *valueA* ≤ *valueB*. *valueA* and *valueB* can be real or complex numbers, expressions, or lists.	[2nd] [TEST] TEST **3:>**	2-25
Less than or equal: *valueA≤valueB*	Returns 1 if *valueA* ≤ *valueB*. Returns 0 if *valueA* > *valueB*. *valueA* and *valueB* can be real or complex numbers, expressions, or lists.	[2nd] [TEST] TEST **6:≤**	2-25
Greater than or equal: *valueA≥valueB*	Returns 1 if *valueA* ≥ *valueB*. Returns 0 if *valueA* < *valueB*. *valueA* and *valueB* can be real or complex numbers, expressions, or lists.	[2nd] [TEST] TEST **4:≥**	2-25
Inverse: *value*$^{-1}$	Returns 1 divided by a real or complex number or expression.	[x^{-1}]	2-3
Inverse: *list*$^{-1}$	Returns 1 divided by *list* elements.	[x^{-1}]	2-3
Inverse: *matrix*$^{-1}$	Returns *matrix* inverted.	[x^{-1}]	10-10
Square: *value*^{2}	Returns *value* multiplied by itself. *value* can be a real or complex number or expression.	[x^2]	2-3
Square: *list*^{2}	Returns *list* elements squared.	[x^2]	2-3
Square: *matrix*^{2}	Returns *matrix* multiplied by itself.	[x^2]	10-10
Powers: *value*^*power*	Returns *value* raised to *power*. *value* can be a real or complex number or expression.	[^]	2-3
Powers: *list*^*power*	Returns *list* elements raised to *power*.	[^]	2-3
Powers: *value*^*list*	Returns *value* raised to *list* elements.	[^]	2-3

Function or Instruction/ Arguments	Result	Key or Keys/ Menu or Screen/Item	
Powers: *matrix*^*power*	Returns *matrix* elements raised to *power*.	[^]	10-10
Negation: ⁻*value*	Returns the negative of a real or complex number, expression, list, or matrix.	[(-)]	2-4 10-9
Power of ten: 10^(*value*)	Returns 10 raised to the *value* power. *value* can be a real or complex number or expression.	[2nd] [10^x]	2-4
Power of ten: 10^(*list*)	Returns a list of 10 raised to the *list* power.	[2nd] [10^x]	2-4
Square root: √(*value*)	Returns square root of a real or complex number, expression, or list.	[2nd] [√]	2-3
Multiplication: *valueA***valueB*	Returns *valueA* times *valueB*.	[×]	2-3
Multiplication: *value***list*	Returns *value* times each *list* element.	[×]	2-3
Multiplication: *list***value*	Returns each *list* element times *value*.	[×]	2-3
Multiplication: *listA***listB*	Returns *listA* elements times *listB* elements.	[×]	2-3
Multiplication: *value***matrix*	Returns value times *matrix* elements.	[×]	10-9
Multiplication: *matrixA***matrixB*	Returns *matrixA* times *matrixB*.	[×]	10-9
Division: *valueA*/*valueB*	Returns *valueA* divided by *valueB*.	[÷]	2-3
Division: *list*/*value*	Returns *list* elements divided by value.	[÷]	2-3
Division: *value*/*list*	Returns value divided by *list* elements.	[÷]	2-3
Division: *listA*/*listB*	Returns *listA* elements divided by *listB* elements.	[÷]	2-3

Function or Instruction/ Arguments	Result	Key or Keys/ Menu or Screen/Item	
Addition: *valueA+valueB*	Returns *valueA* plus *valueB*.	[+]	2-3
Addition: *list+value*	Returns list in which *value* is added to each *list* element.	[+]	2-3
Addition: *listA+listB*	Returns *listA* elements plus *listB* elements.	[+]	2-3
Addition: *matrixA+matrixB*	Returns *matrixA* elements plus *matrixB* elements.	[+]	10-9
Concatenation: *string1+string2*	Concatenates two or more strings.	[+]	15-6
Subtraction: *valueA−valueB*	Subtracts *valueB* from *valueA*.	[−]	2-3
Subtraction: *value−list*	Subtracts *list* elements from *value*.	[−]	2-3
Subtraction: *list−value*	Subtracts *value* from *list* elements.	[−]	2-3
Subtraction: *listA−listB*	Subtracts *listB* elements from *listA* elements.	[−]	2-3
Subtraction: *matrixA−matrixB*	Subtracts *matrixB* elements from *matrixA* elements.	[−]	10-9
Minutes notation: *degrees°minutes' seconds"*	Interprets *minutes* angle measurement as minutes.	[2nd] [ANGLE] ANGLE **2:'**	2-23
Seconds notation: *degrees°minutes' seconds"*	Interprets *seconds* angle measurement as seconds.	[ALPHA] ["]	2-23

The TI-83 Plus Menu Map begins at the top-left corner of the keyboard and follows the keyboard layout from left to right. Default values and settings are shown.

[Y=]

(Func mode)	(Par mode)	(Pol mode)	(Seq mode)
Plot1 Plot2	Plot1 Plot2	Plot1 Plot2	Plot1 Plot2
Plot3	Plot3	Plot3	Plot3
\Y1=	\X1T=	\r1=	*n*Min=1
\Y2=	Y1T=	\r2=	⋱u(*n*)=
\Y3=	\X2T=	\r3=	u(*n*Min)=
\Y4=	Y2T=	\r4=	⋱v(*n*)=
...	...	\r5=	v(*n*Min)=
\Y9=	\X6T=	\r6=	⋱w(*n*)=
\Y0=	Y6T=		w(*n*Min)=

[2nd] [STAT PLOT]

STAT PLOTS
1:Plot1…Off
L1 L2 □
2:Plot2…Off
L1 L2 □
3:Plot3…Off
L1 L2 □
4:PlotsOff
5:PlotsOn

[2nd] [STAT PLOT]

(PRGM editor)	(PRGM editor)	(PRGM editor)
PLOTS	TYPE	MARK
1:Plot1(	1:Scatter	1:□
2:Plot2(	2:xyLine	2:+
3:Plot3(	3:Histogram	3:•
4:PlotsOff	4:ModBoxplot	
5:PlotsOn	5:Boxplot	
	6:NormProbPlot	

[WINDOW]

(Func mode)	(Par mode)	(Pol mode)	(Seq mode)
WINDOW	WINDOW	WINDOW	WINDOW
Xmin=-10	Tmin=0	θmin=0	*n*Min=1
Xmax=10	Tmax=π*2	θmax=π*2	*n*Max=10
Xscl=1	Tstep=π/24	θstep=π/24	PlotStart=1
Ymin=-10	Xmin=-10	Xmin=-10	PlotStep=1
Ymax=10	Xmax=10	Xmax=10	Xmin=-10
Yscl=1	Xscl=1	Xscl=1	Xmax=10
Xres=1	Ymin=-10	Ymin=-10	Xscl=1
	Ymax=10	Ymax=10	Ymin=-10
	Yscl=1	Yscl=1	Ymax=10
			Yscl=1

[2nd] [TBLSET]

TABLE SETUP
TblStart=0
ΔTbl=1
Indpnt:Auto Ask
Depend:Auto Ask

[2nd] [TBLSET]

(PRGM editor)
TABLE SETUP
Indpnt:Auto Ask
Depend:Auto Ask

ZOOM

ZOOM	MEMORY	MEMORY
1:ZBox	1:ZPrevious	(Set Factors...)
2:Zoom In	2:ZoomSto	ZOOM FACTORS
3:Zoom Out	3:ZoomRcl	XFact=4
4:ZDecimal	4:SetFactors...	YFact=4
5:ZSquare		
6:ZStandard		
7:ZTrig		
8:ZInteger		
9:ZoomStat		
0:ZoomFit		

2nd [FORMAT]

(Func/Par/Pol modes)	(Seq mode)
RectGC PolarGC	Time Web uv vw uw
CoordOn CoordOff	RectGC PolarGC
GridOff GridOn	CoordOn CoordOff
AxesOn AxesOff	GridOff GridOn
LabelOff LabelOn	AxesOn AxesOff
ExprOn ExprOff	LabelOff LabelOn
	ExprOn ExprOff

2nd [CALC]

(Func mode)	(Par mode)	(Pol mode)	(Seq mode)
CALCULATE	CALCULATE	CALCULATE	CALCULATE
1:value	1:value	1:value	1:value
2:zero	2:dy/dx	2:dy/dx	
3:minimum	3:dy/dt	3:dr/dθ	
4:maximum	4:dx/dt		
5:intersect			
6:dy/dx			
7:∫f(x)dx			

MODE

Normal Sci Eng
Float 0123456789
Radian Degree
Func Par Pol Seq
Connected Dot
Sequential Simul
Real a+b*i* re^θ*i*
Full Horiz G-T

2nd [LINK]

SEND	RECEIVE
1:All+…	1:Receive
2:All-…	
3:Prgm…	
4:List…	
5:Lists to TI82…	
6:GDB…	
7:Pic…	
8:Matrix…	
9:Real…	
0:Complex…	
A:Y-Vars…	
B:String…	
C:Apps…	
D:AppVars…	
E:Group…	
F:SendId	
G:SendSW	
H:Back Up…	

STAT

EDIT	CALC	TESTS
1:Edit…	1:1-Var Stats	1:Z-Test…
2:SortA(	2:2-Var Stats	2:T-Test…
3:SortD(	3:Med-Med	3:2-SampZTest…
4:ClrList	4:LinReg(ax+b)	4:2-SampTTest…
5:SetUpEditor	5:QuadReg	5:1-PropZTest…
	6:CubicReg	6:2-PropZTest…
	7:QuartReg	7:ZInterval…
	8:LinReg(a+bx)	8:TInterval…
	9:LnReg	9:2-SampZInt…
	0:ExpReg	0:2-SampTInt…
	A:PwrReg	A:1-PropZInt…
	B:Logistic	B:2-PropZInt…
	C:SinReg	C:χ^2-Test…
		D:2-SampFTest…
		E:LinRegTTest…
		F:ANOVA(

[2nd] [LIST]

NAMES	OPS	MATH
1:listname	1:SortA(	1:min(
2:listname	2:SortD(	2:max(
3:listname	3:dim(	3:mean(
...	4:Fill(	4:median(
	5:seq(	5:sum(
	6:cumSum(	6:prod(
	7:ΔList(	7:stdDev(
	8:Select(	8:variance(
	9:augment(	
	0:List▶matr(	
	A:Matr▶list(	
	B:ʟ	

[MATH]

MATH	NUM	CPX	PRB
1:▶Frac	1:abs(	1:conj(	1:rand
2:▶Dec	2:round(	2:real(	2:nPr
3:^{3}	3:iPart(	3:imag(	3:nCr
4:$^3\sqrt{}$(	4:fPart(	4:angle(	4:!
5:$^x\sqrt{}$	5:int(	5:abs(	5:randInt(
6:fMin(	6:min(	6:▶Rect	6:randNorm(
7:fMax(	7:max(	7:▶Polar	7:randBin(
8:nDeriv(	8:lcm(		
9:fnInt(	9:gcd(		
0:Solver...			

[2nd] [TEST]

TEST	LOGIC
1:=	1:and
2:≠	2:or
3:>	3:xor
4:≥	4:not(
5:<	
6:≤	

2nd [MATRX]

NAMES	MATH	EDIT
1:[A]	1:det(	1:[A]
2:[B]	2:^{T}	2:[B]
3:[C]	3:dim(	3:[C]
4:[D]	4:Fill(	4:[D]
5:[E]	5:identity(	5:[E]
6:[F]	6:randM(	6:[F]
7:[G]	7:augment(	7:[G]
8:[H]	8:Matr▸list(	8:[H]
9:[I]	9:List▸matr(	9:[I]
0:[J]	0:cumSum(	0:[J]
	A:ref(	
	B:rref(	
	C:rowSwap(	
	D:row+(	
	E:*row(	
	F:*row+(	

2nd [ANGLE]

ANGLE
1:°
2:'
3:^{r}
4:▸DMS
5:R▸Pr(
6:R▸Pθ(
7:P▸Rx(
8:P▸Ry(

PRGM

EXEC	EDIT	NEW
1:*name*	1:*name*	1:Create New
2:*name*	2:*name*	
...	. .	

PRGM

(PRGM editor) CTL	(PRGM editor) I/O	(PRGM editor) EXEC
1:If	1:Input	1:name
2:Then	2:Prompt	2:name
3:Else	3:Disp	...
4:For(	4:DispGraph	
5:While	5:DispTable	
6:Repeat	6:Output(	
7:End	7:getKey	
8:Pause	8:ClrHome	
9:Lbl	9:ClrTable	
0:Goto	0:GetCalc(	
A:IS>(	A:Get(	
B:DS<(	B:Send(	
C:Menu(		
D:prgm		
E:Return		
F:Stop		
G:DelVar		
H:GraphStyle(		

[2nd] [DRAW]

DRAW	POINTS	STO
1:ClrDraw	1:Pt-On(	1:StorePic
2:Line(	2:Pt-Off(	2:RecallPic
3:Horizontal	3:Pt-Change(	3:StoreGDB
4:Vertical	4:Pxl-On(	4:RecallGDB
5:Tangent(	5:Pxl-Off(	
6:DrawF	6:Pxl-Change(	
7:Shade(	7:pxl-Test(	
8:DrawInv		
9:Circle(		
0:Text(		
A:Pen		

[VARS]

VARS	Y-VARS
1:Window…	1:Function…
2:Zoom…	2:Parametric…
3:GDB…	3:Polar…
4:Picture…	4:On/Off…
5:Statistics…	
6:Table…	
7:String…	

VARS

(Window…) X/Y	(Window…) T/θ	(Window…) U/V/W
1:Xmin	1:Tmin	1:u(*n*Min)
2:Xmax	2:Tmax	2:v(*n*Min)
3:Xscl	3:Tstep	3:w(*n*Min)
4:Ymin	4:θmin	4:*n*Min
5:Ymax	5:θmax	5:*n*Max
6:Yscl	6:θstep	6:PlotStart
7:Xres		7:PlotStep
8:ΔX		
9:ΔY		
0:XFact		
A:YFact		

VARS

(Zoom...)	(Zoom...)	(Zoom...)
ZX/ZY	ZT/Zθ	ZU
1:ZXmin	1:ZTmin	1:Zu(*n*Min)
2:ZXmax	2:ZTmax	2:Zv(*n*Min)
3:ZXscl	3:ZTstep	3:Zw(*n*Min)
4:ZYmin	4:Zθmin	4:Z*n*Min
5:ZYmax	5:Zθmax	5:Z*n*Max
6:ZYscl	6:Zθstep	6:ZPlotStart
7:ZXres		7:ZPlotStep

VARS

(GDB...)	(Picture...)
GRAPH DATABASE	PICTURE
1:GDB1	1:Pic1
2:GDB2	2:Pic2
...	...
9:GDB9	9:Pic9
0:GDB0	0:Pic0

VARS

(Statistics...)	(Statistics...)	(Statistics...)	(Statistics...)	(Statistics...)
XY	Σ	EQ	TEST	PTS
1:n	1:Σx	1:RegEQ	1:p	1:x1
2:$\bar{x}$	2:Σx^2	2:a	2:z	2:y1
3:Sx	3:Σy	3:b	3:t	3:x2
4:σx	4:Σy^2	4:c	4:χ^2	4:y2
5:$\bar{y}$	5:Σxy	5:d	5:F	5:x3
6:Sy		6:e	6:df	6:y3
7:σy		7:r	7:$\hat{p}$	7:Q1
8:minX		8:r^2	8:$\hat{p}1$	8:Med
9:maxX		9:R^2	9:$\hat{p}2$	9:Q3
0:minY			0:s	
A:maxY			A:$\bar{x}1$	
			B:$\bar{x}2$	
			C:Sx1	
			D:Sx2	
			E:Sxp	
			F:n1	
			G:n2	
			H:lower	
			I:upper	

VARS

(Table…)	(String…)
TABLE	STRING
1:TblStart	1:Str1
2:∆Tbl	2:Str2
3:TblInput	3:Str3
	4:Str4
	...
	9:Str9
	0:Str0

Y-VARS

(Function…)	(Parametric…)	(Polar…)	(On/Off…)
FUNCTION	PARAMETRIC	POLAR	ON/OFF
1:Y_1	1:X_{1T}	1:r_1	1:FnOn
2:Y_2	2:Y_{1T}	2:r_2	2:FnOff
3:Y_3	3:X_{2T}	3:r_3	
4:Y_4	4:Y_{2T}	4:r_4	
...	...	5:r_5	
9:Y_9	A:X_{6T}	6:r_6	
0:Y_0	B:Y_{6T}		

2nd [DISTR]

DISTR	DRAW
1:normalpdf(	1:ShadeNorm(
2:normalcdf(	2:Shade_t(
3:invNorm(	3:Shadeχ^2(
4:tpdf(	4:ShadeF(
5:tcdf(	
6:χ^2pdf(	
7:χ^2cdf(	
8:Fpdf(	
9:Fcdf(	
0:binompdf(	
A:binomcdf(	
B:poissonpdf(	
C:poissoncdf(	
D:geometpdf(	
E:geometcdf(	

APPS

1:Finance — 2:CBL/CBR

Finance

CALC	VARS
1:TVM Solver…	1:**N**
2:tvm_Pmt	2:I%
3:tvm_I%	3:PV
4:tvm_PV	4:PMT
5:tvm_**N**	5:FV
6:tvm_FV	6:P/Y
7:npv(	7:C/Y
8:irr(	
9:bal(	
0:ΣPrn(	
A:ΣInt(	
B:▸Nom(	
C:▸Eff(	
D:dbd(	
E:Pmt_End	
F:Pmt_Bgn	

CBL/CBR

1:GAUGE
2:DATA LOGGER
3:CBR
4:QUIT

2nd [MEM]

```
MEMORY
1:About
2:Mem Mgmt/Del…
3:Clear Entries
4:ClrAllLists
5:Archive
6:UnArchive
7:Reset…
8:Group
```

MEMORY

```
(Mem Mgmt/Del…)
RAM FREE  25631
ARC FREE 131069
1:All…
2:Real…
3:Complex…
4:List…
5:Matrix…
6:Y-Vars…
7:Prgm…
8:Pic…
9:GDB…
0:String…
A:Apps…
B:AppVars…
C:Group…
```

MEMORY (Reset…)

```
RAM
1:All RAM…
2:Defaults…

Resetting RAM erases
all data and programs
from RAM.
```

```
ARCHIVE
1:Vars…
2:Apps…
B:Both…

Resetting Both erases
all data, programs
and Apps from
Archive.
```

```
ALL
1:All Memory…

Resetting ALL erases
all data, programs
and Apps from RAM and
Archive.
```

RAM

```
RESET RAM
1:No
2:Reset
Resetting RAM erases
all data and programs
from RAM.
```

```
RESET DEFAULTS
1:No
2:Reset
```

ARCHIVE

```
RESET ARC VARS
1:No
2:Reset
Resetting Vars erases
all data and programs
from Archive.
```

```
RESET ARC APPS
1:No
2:Reset
Resetting Apps erases
all Apps from
Archive.
```

```
RESET ARC BOTH
1:No
2:Reset
Resetting Both erases
all data, programs
and Apps from
Archive.
```

ALL

```
RESET MEMORY
1:No
2:Reset
Resetting ALL will
delete all data,
programs & Apps from
RAM & Archive.
```

MEMORY (GROUP...)

```
GROUP UNGROUP
1:Create New
```

MEMORY (UNGROUP...)

```
1:name
2:name
...
```

[2nd] [CATALOG]

```
CATALOG
cosh(
cosh⁻¹(
...
Equ▸String(
expr(
...
inString(
...
length(
...
sinh(
sinh⁻¹(
...
String▸Equ(
sub(
...
tanh(
tanh⁻¹(
```

User Variables

The TI-83 Plus uses the variables listed below in various ways. Some variables are restricted to specific data types.

The variables **A** through **Z** and θ are defined as real or complex numbers. You may store to them. The TI-83 Plus can update **X**, **Y**, **R**, θ, and **T** during graphing, so you may want to avoid using these variables to store nongraphing data.

The variables (list names) **L1** through **L6** are restricted to lists; you cannot store another type of data to them.

The variables (matrix names) **[A]** through **[J]** are restricted to matrices; you cannot store another type of data to them.

The variables **Pic1** through **Pic9** and **Pic0** are restricted to pictures; you cannot store another type of data to them.

The variables **GDB1** through **GDB9** and **GDB0** are restricted to graph databases; you cannot store another type of data to them.

The variables **Str1** through **Str9** and **Str0** are restricted to strings; you cannot store another type of data to them.

Except for system variables, you can store any string of characters, functions, instructions, or variables to the functions **Y***n*, (**1** through **9**, and **0**), **X***n***T**/**Y***n***T** (**1** through **6**), **r***n* (**1** through **6**), **u(***n***)**, **v(***n***)**, and **w(***n***)** directly or through the **Y=** editor. The validity of the string is determined when the function is evaluated.

Archive Variables

You can store data, programs or any variable from RAM to user data archive memory where they cannot be edited or deleted inadvertantly. Archiving also allows you to free up RAM for variables that may require additional memory. The names of archived variables are preceded by an "*" indicating they are in user data archive.

System Variables

The variables below must be real numbers. You may store to them. Since the TI-83 Plus can update some of them, as the result of a ZOOM, for example, you may want to avoid using these variables to store nongraphing data.

- **Xmin**, **Xmax**, **Xscl**, **ΔX**, **XFact**, **Tstep**, **PlotStart**, ***n*Min**, and other window variables.
- **ZXmin**, **ZXmax**, **ZXscl**, **ZTstep**, **ZPlotStart**, **Zu(*n*Min)**, and other ZOOM variables.

The variables below are reserved for use by the TI-83 Plus. You cannot store to them.

n, **$\bar{x}$**, **Sx**, **σx**, **minX**, **maxX**, **Σy**, **Σy^2**, **Σxy**, **a**, **b**, **c**, **RegEQ**, **x1**, **x2**, **y1**, **z**, **t**, **F**, **χ^2**, **$\hat{p}$**, **$\bar{x}$1**, **Sx1**, **n1**, **lower**, **upper**, **r^2**, **R^2** and other statistical variables.

This section contains statistics formulas for the **Logistic** and **SinReg** regressions, **ANOVA**, **2-SampFTest**, and **2-SampTTest**.

Logistic

The logistic regression algorithm applies nonlinear recursive least-squares techniques to optimize the following cost function:

$$J = \sum_{i=1}^{N} \left(\frac{c}{1 + ae^{-bx_i}} - y_i \right)^2$$

which is the sum of the squares of the residual errors,

where: x = the independent variable list
y = the dependent variable list
N = the dimension of the lists

This technique attempts to estimate the constants *a*, *b*, and *c* recursively to make *J* as small as possible.

SinReg

The sine regression algorithm applies nonlinear recursive least-squares techniques to optimize the following cost function:

$$J = \sum_{i=1}^{N} \left[a \sin(bx_i + c) + d - y_i \right]^2$$

which is the sum of the squares of the residual errors,

where: x = the independent variable list
y = the dependent variable list
N = the dimension of the lists

This technique attempts to recursively estimate the constants *a*, *b*, *c*, and *d* to make *J* as small as possible.

ANOVA(

The **ANOVA** F statistic is:

$$F = \frac{Factor\ MS}{Error\ MS}$$

The mean squares (*MS*) that make up F are:

$$Factor\ MS = \frac{Factor\ SS}{Factor\ df}$$

$$Error\ MS = \frac{Error\ SS}{Error\ df}$$

The sum of squares (*SS*) that make up the mean squares are:

$$Factor\ SS = \sum_{i=1}^{I} n_i(\overline{x}_i - \overline{x})^2$$

$$Error\ SS = \sum_{i=1}^{I} (n_i - 1)Sx_i^2$$

The degrees of freedom *df* that make up the mean squares are:

$$Factor\ df = I - 1 = \text{numerator } df \text{ for F}$$

$$Error\ df = \sum_{i=1}^{I} (n_i - 1) = \text{denominator } df \text{ for F}$$

where:

- I = number of populations
- $\overline{x}_i$ = the mean of each list
- Sxi = the standard deviation of each list
- ni = the length of each list
- $\overline{x}$ = the mean of all lists

2-SampFTest

Below is the definition for the **2-SampFTest**.

$Sx1, Sx2$ = Sample standard deviations having n_1-1 and n_2-1 degrees of freedom df, respectively.

$$\text{F} = \text{F-statistic} = \left(\frac{Sx1}{Sx2}\right)^2$$

$df(x, n_1-1, n_2-1)$ = F*pdf*( ) with degrees of freedom df, n_1-1, and n_2-1

p = reported p value

2-SampFTest for the alternative hypothesis $\sigma_1 > \sigma_2$.

$$p = \int_F^{\infty} f(x, n_1 - 1, n_2 - 1)dx$$

2-SampFTest for the alternative hypothesis $\sigma_1 < \sigma_2$.

$$p = \int_0^{F} f(x, n_1 - 1, n_2 - 1)dx$$

2-SampFTest for the alternative hypothesis $\sigma_1 \neq \sigma_2$. Limits must satisfy the following:

$$\frac{p}{2} = \int_0^{Lbnd} f(x, n_1 - 1, n_2 - 1)dx = \int_{Ubnd}^{\infty} f(x, n_1 - 1, n_2 - 1)dx$$

where:
[*Lbnd*, *Ubnd*] = lower and upper limits

The F-statistic is used as the bound producing the smallest integral. The remaining bound is selected to achieve the preceding integral's equality relationship.

2-SampTTest

The following is the definition for the **2-SampTTest**. The two-sample t statistic with degrees of freedom df is:

$$t = \frac{\bar{x}_1 - \bar{x}_2}{S}$$

where the computation of S and df are dependent on whether the variances are pooled. If the variances are not pooled:

$$S = \sqrt{\frac{Sx_1^2}{n_1} + \frac{Sx_2^2}{n_2}}$$

$$df = \frac{\left(\frac{Sx_1^2}{n_1} + \frac{Sx_2^2}{n_2}\right)^2}{\frac{1}{n_1 - 1}\left(\frac{Sx_1^2}{n_1}\right)^2 + \frac{1}{n_2 - 1}\left(\frac{Sx_2^2}{n_2}\right)^2}$$

---

otherwise:

$$Sx_p = \frac{(n_1 - 1)Sx_1^2 + (n_2 - 1)Sx_2^2}{df}$$

$$S = \sqrt{\frac{1}{n_1} + \frac{1}{n_2}}\, Sx_p$$

$$df = n_1 + n_2 - 2$$

and Sxp is the pooled variance.

This section contains financial formulas for computing time value of money, amortization, cash flow, interest-rate conversions, and days between dates.

Time Value of Money

$$i = \left[e^{(y \times \ln(x+1))}\right] - 1$$

where: $PMT \neq 0$

$y = C/Y \div P/Y$

$x = (.01 \times I\%) \div C/Y$

C/Y = compounding periods per year

P/Y = payment periods per year

$I\%$ = interest rate per year

$$i = (^{-}FV \div PV)^{(1 \div N)} - 1$$

where: $PMT = 0$

The iteration used to compute i:

$$0 = PV + PMT \times G_i \left[\frac{1-(1+i)^{-N}}{i}\right] + FV \times (1+i)^{-N}$$

$$I\% = 100 \times C/Y \times \left[e^{(y \times \ln(x+1))} - 1\right]$$

where: $x = i$

$y = P/Y \div C/Y$

$$G_i = 1 + i \times k$$

where: $k = 0$ for end-of-period payments

$k = 1$ for beginning-of-period payments

$$N = \frac{\ln\left(\dfrac{PMT \times G_i - FV \times i}{PMT \times G_i + PV \times i}\right)}{\ln(1+i)}$$

where: $i \neq 0$

$$N = {}^{-}(PV + FV) \div PMT$$

where: $i = 0$

Time Value of Money (Continued)

$$PMT = \frac{-i}{G_i} \times \left[PV + \frac{PV + FV}{(1+i)^N - 1} \right]$$

where: $i \neq 0$

$$PMT = {}^{-}(PV + FV) \div N$$

where: $i = 0$

$$PV = \left[\frac{PMT \times G_i}{i} - FV \right] \times \frac{1}{(1+i)^N} - \frac{PMT \times G_i}{i}$$

where: $i \neq 0$

$$PV = {}^{-}(FV + PMT \times N)$$

where: $i = 0$

$$FV = \frac{PMT \times G_i}{i} - (1+i)^N \times \left(PV + \frac{PMT \times G_i}{i} \right)$$

where: $i \neq 0$

$$FV = {}^{-}(PV + PMT \times N)$$

where: $i = 0$

Amortization

If computing $bal()$, $pmt2 = npmt$

Let $bal(0) = RND(PV)$

Iterate from $m = 1$ to $pmt2$

$$\begin{cases} I_m = RND[RND12(-i \times bal(m-1))] \\ bal(m) = bal(m-1) - I_m + RND(PMT) \end{cases}$$

then:

$$bal(\,) = bal(pmt2)$$

$$\Sigma Prn(\,) = bal(pmt2) - bal(pmt1)$$

$$\Sigma Int(\,) = (pmt2 - pmt1 + 1) \times RND(PMT) - \Sigma Prn(\,)$$

where:

RND = round the display to the number of decimal places selected

RND12 = round to 12 decimal places

Balance, principal, and interest are dependent on the values of **PMT, PV, I%,** and *pmt*1 and *pmt*2.

Cash Flow

$$npv() = CF_0 + \sum_{j=1}^{N} CF_j(1+i)^{-S_{j-1}} \frac{(1-(1+i)^{-n_j})}{i}$$

$$\text{where: } S_j = \begin{cases} \sum_{i=1}^{j} n_i & j \geq 1 \\ 0 & j = 0 \end{cases}$$

Net present value is dependent on the values of the initial cash flow (CF_0), subsequent cash flows (CFj), frequency of each cash flow (nj), and the specified interest rate (i).

$$irr() = 100 \times i, \text{ where } i \text{ satisfies } npv() = 0$$

Internal rate of return is dependent on the values of the initial cash flow (CF_0) and subsequent cash flows (CFj).

$$i = I\% \div 100$$

Interest Rate Conversions

$$\blacktriangleright\text{Eff}() = 100 \times (e^{CP \times ln(x+1)} - 1)$$

where: $x = .01 \times NOM \div CP$

$$\blacktriangleright\text{Nom}() = 100 \times CP \times \left[e^{1 \div CP \times ln(x+1)} - 1\right]$$

where:
$x = .01 \times EFF$
EFF = *effective rate*
CP = *compounding periods*
NOM = *nominal rate*

Days between Dates

With the **dbd(** function, you can enter or compute a date within the range Jan. 1, 1950, through Dec. 31, 2049.

Actual/actual day-count method (assumes actual number of days per month and actual number of days per year):

dbd((days between dates) = Number of Days II - Number of Days I

$$\text{Number of Days I} = (Y1-YB) \times 365 + (\text{number of days } MB \text{ to } M1) + DT1 + \frac{(Y1-YB)}{4}$$

$$\text{Number of Days II} = (Y2-YB) \times 365 + (\text{number of days } MB \text{ to } M2) + DT2 + \frac{(Y2-YB)}{4}$$

where:

$M1$ = month of first date
$DT1$ = day of first date
$Y1$ = year of first date
$M2$ = month of second date
$DT2$ = day of second date
$Y2$ = year of second date
MB = base month (January)
DB = base day (1)
YB = base year (first year after leap year)

B General Information

Contents

When to Replace the Batteries

The TI-83 Plus uses five batteries: four AAA alkaline batteries and one lithium battery. The lithium battery provides auxiliary power to retain memory while you replace the AAA batteries.

When the battery voltage level drops below a usable level, the TI-83 Plus:

Displays this message when you turn on the unit

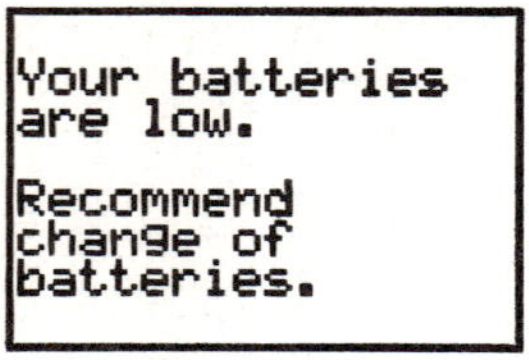
Your batteries
are low.
Recommend
change of
batteries.

Message A

Displays this message when you attempt to download an application.

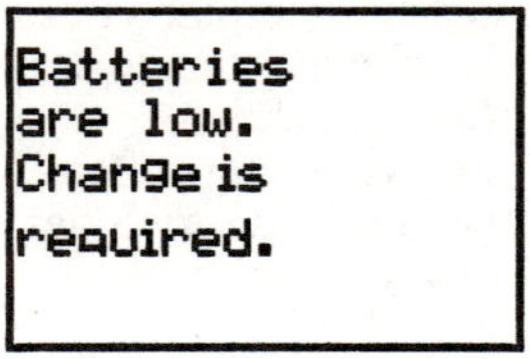
Batteries
are low.
Change is
required.

Message B

After Message A is first displayed, you can expect the batteries to function for about one or two weeks, depending on usage. (This one-week to two-week period is based on tests with alkaline batteries; the performance of other kinds of batteries may vary.)

The low-battery message continues to be displayed each time you turn on the unit until you replace the batteries. If you do not replace the batteries within about two weeks, the calculator may turn off by itself or fail to turn on until you install new batteries.

If Message B is displayed, you must to replace the batteries immediately to successfully download an application.

Replace the lithium battery every three or four years.

Effects of Replacing the Batteries

Do not remove both types of batteries (AAA and lithium auxiliary) at the same time. **Do not** allow the batteries to lose power completely. If you follow these guidelines and the steps for replacing batteries on page B-3, you can replace either type of battery without losing any information in memory.

Battery Precautions

Take these precautions when replacing batteries.

- Do not mix new and used batteries. Do not mix brands (or types within brands) of batteries.
- Do not mix rechargeable and nonrechargeable batteries.
- Install batteries according to polarity (+ and −) diagrams.
- Do not place nonrechargeable batteries in a battery recharger.
- Properly dispose of used batteries immediately. Do not leave them within the reach of children.
- Do not incinerate batteries.

Replacing the Batteries

To replace the batteries, follow these steps.

1. Turn off the calculator. Replace the slide cover over the keyboard to avoid inadvertently turning on the calculator. Turn the back of the calculator toward you.

2. Hold the calculator upright, push downward on the latch on the top of the battery cover with your finger, and then pull the cover toward you.

   Note: To avoid loss of information stored in memory, you must turn off the calculator. Do not remove the AAA batteries and the lithium battery simultaneously.

3. Replace all four AAA alkaline batteries simultaneously. Or, replace the lithium battery.

   - To replace the AAA alkaline batteries, remove all four discharged AAA batteries and install new ones according to the polarity (+ and −) diagram in the battery compartment.
   - To replace the lithium battery, remove the screw from the lithium-battery cover, and then remove the cover. Install the new battery, + side up. Replace the cover and secure it with the screw. Use a CR1616 or CR1620 (or equivalent) lithium battery.

4. Replace the battery compartment cover. Turn the calculator on and adjust the display contrast, if necessary (step 1; page B-4).

Handling a Difficulty

To handle a difficulty, follow these steps.

1. If you cannot see anything on the screen, the contrast may need to be adjusted.

   To darken the screen, press and release [2nd], and then press and hold [▲] until the display is sufficiently dark.

   To lighten the screen, press and release [2nd], and then press and hold [▼] until the display is sufficiently light.

2. If an error menu is displayed, follow the steps in Chapter 1. Refer to pages B-5 through B-10 for details about specific errors, if necessary.

3. If the busy indicator (dotted line) is displayed, a graph or program has been paused; the TI-83 Plus is waiting for input. Press [ENTER] to continue or press [ON] to break.

4. If a checkerboard cursor (▦) is displayed, then either you have entered the maximum number of characters in a prompt, or memory is full. If memory is full:

   - Press [2nd] [MEM] **2** to display the MEMORY MANAGEMENT DELETE menu.
   - Select the type of data you want to delete, or select **1:All** for a list of all variables of all types. A screen is displayed listing each variable of the type you selected and the number of bytes each variable is using.
   - Press [▲] and [▼] to move the selection cursor (▶) next to the item you want to delete, and then press [DEL].(Chapter 18).

5. If the calculator does not seem to work at all, be sure the batteries are fresh and that they are installed properly. Refer to battery information on pages B-2 and B-3.

6. If the calculator still doesn't work even though you are sure the batteries are sufficiently charged, you can try the two solutions in the order they are presented.

   - Download calculator system software as follows:
     1. Remove one battery from the calculator and then press and hold the [DEL] key while re-installing the battery. This will force the calculator to accept a download of system software.
     2. Connect your calculator to a personal computer with the TI-GRAPH LINK™ (optional) accessory to download current or new software code to your calculator.
   - II. If the above solution does not work, reset all memory as follows:
     1. Remove one battery from the calculator and then press and hold down the [CLEAR] key while re-installing the battery. While continuing to hold down the [CLEAR] key, press and hold down the [ON] key. When the home screen is displayed, release the keys.
     2. Press [2nd] [MEM] to display the MEMORY menu.
     3. Select **7:Reset** to display the RAM ARCHIVE ALL menu.
     4. Press [▶] [▶] to display the ALL menu.
     5. Select **1:All Memory** to display the RESET MEMORY menu.
     6. To continue with the reset, select **2:Reset**. The message **MEM cleared** is displayed on the home screen.

When the TI-83 Plus detects an error, it displays **ERR:***message* and an error menu. Chapter 1 describes the general steps for correcting errors. This table contains each error type, possible causes, and suggestions for correction.

Error Type	Possible Causes and Suggested Remedies
ARCHIVED	You have attempted to use, edit, or delete an archived variable. For example, dim(L1) is an error if L1 is archived.
ARCHIVE FULL	You have attempted to archive a variable and there is not enough space in archive to receive it.
ARGUMENT	A function or instruction does not have the correct number of arguments. See Appendix A and the appropriate chapter.
BAD ADDRESS	You have attempted to send or receive an application and an error (e.g. electrical interference) has occurred in the transmission.
BAD GUESS	• In a CALC operation, you specified a **Guess** that is not between **Left Bound** and **Right Bound.** • For the **solve(** function or the equation solver, you specified a *guess* that is not between *lower* and *upper*. • Your guess and several points around it are undefined. Examine a graph of the function. If the equation has a solution, change the bounds and/or the initial guess.
BOUND	• In a CALC operation or with **Select(**, you defined **Left Bound > Right Bound.** • In **fMin(**, **fMax(**, **solve(**, or the equation solver, you entered *lower* ≥ *upper*.
BREAK	You pressed the [ON] key to break execution of a program, to halt a DRAW instruction, or to stop evaluation of an expression.
DATA TYPE	You entered a value or variable that is the wrong data type. • For a function (including implied multiplication) or an instruction, you entered an argument that is an invalid data type, such as a complex number where a real number is required. See Appendix A and the appropriate chapter. • In an editor, you entered a type that is not allowed, such as a matrix entered as an element in the stat list editor. See the appropriate chapter. • You attempted to store to an incorrect data type, such as a matrix, to a list.
DIM MISMATCH	You attempted to perform an operation that references more than one list or matrix, but the dimensions do not match.
DIVIDE BY 0	• You attempted to divide by zero. This error is not returned during graphing. The TI-83 Plus allows for undefined values on a graph. • You attempted a linear regression with a vertical line.

Error Type	Possible Causes and Suggested Remedies
DOMAIN	• You specified an argument to a function or instruction outside the valid range. This error is not returned during graphing. The TI-83 Plus allows for undefined values on a graph. See Appendix A and the appropriate chapter. • You attempted a logarithmic or power regression with a **-X** or an exponential or power regression with a **-Y**. • You attempted to compute **ΣPrn(** or **ΣInt(** with *pmt2* < *pmt1*.
DUPLICATE	You attempted to create a duplicate group name.
Duplicate Name	A variable you attempted to transmit cannot be transmitted because a variable with that name already exists in the receiving unit.
EXPIRED	You have attempted to run an application with a limited trial period which has expired.
Error in Xmit	• The TI-83 Plus was unable to transmit an item. Check to see that the cable is firmly connected to both units and that the receiving unit is in receive mode. • You pressed [ON] to break during transmission. • You attempted to perform a backup from a TI-82 to a TI-83 Plus. • You attempted to transfer data (other than **L1** through **L6**) from a TI-83 Plus to a TI-82. • You attempted to transfer **L1** through **L6** from a TI-83 Plus to a TI-82 without using **5:Lists to TI82** on the LINK SEND menu.
ID NOT FOUND	This error occurs when the SendID command is executed but the proper calculator ID cannot be found.
ILLEGAL NEST	You attempted to use an invalid function in an argument to a function, such as **seq(** within *expression* for **seq(**.
INCREMENT	• The increment in **seq(** is 0 or has the wrong sign. This error is not returned during graphing. The TI-83 Plus allows for undefined values on a graph. • The increment in a **For(** loop is 0.
INVALID	• You attempted to reference a variable or use a function where it is not valid. For example, **Y*n*** cannot reference **Y**, **Xmin**, **ΔX**, or **TblStart**. • You attempted to reference a variable or function that was transferred from the TI-82 and is not valid for the TI-83 Plus. For example, you may have transferred **U*n*−1** to the TI-83 Plus from the TI-82 and then tried to reference it. • In **Seq** mode, you attempted to graph a phase plot without defining both equations of the phase plot.

Error Type	Possible Causes and Suggested Remedies
INVALID (cont.)	• In **Seq** mode, you attempted to graph a recursive sequence without having input the correct number of initial conditions. • In **Seq** mode, you attempted to reference terms other than **(*n*−1)** or **(*n*−2)**. • You attempted to designate a graph style that is invalid within the current graph mode. • You attempted to use **Select(** without having selected (turned on) at least one xyLine or scatter plot.
INVALID DIM	• You specified dimensions for an argument that are not appropriate for the operation. • You specified a list dimension as something other than an integer between 1 and 999. • You specified a matrix dimension as something other than an integer between 1 and 99. • You attempted to invert a matrix that is not square.
ITERATIONS	• The **solve(** function or the equation solver has exceeded the maximum number of permitted iterations. Examine a graph of the function. If the equation has a solution, change the bounds, or the initial guess, or both. • **irr(** has exceeded the maximum number of permitted iterations. • When computing **I%**, the maximum number of iterations was exceeded.
LABEL	The label in the **Goto** instruction is not defined with a **Lbl** instruction in the program.
MEMORY	Memory is insufficient to perform the instruction or function. You must delete items from memory (Chapter 18) before executing the instruction or function. Recursive problems return this error; for example, graphing the equation **Y1=Y1**. Branching out of an **If/Then**, **For(**, **While**, or **Repeat** loop with a **Goto** also can return this error because the **End** statement that terminates the loop is never reached.

Error Type	Possible Causes and Suggested Remedies
MemoryFull	• You are unable to transmit an item because the receiving unit's available memory is insufficient. You may skip the item or exit receive mode. • During a memory backup, the receiving unit's available memory is insufficient to receive all items in the sending unit's memory. A message indicates the number of bytes the sending unit must delete to do the memory backup. Delete items and try again.
MODE	You attempted to store to a window variable in another graphing mode or to perform an instruction while in the wrong mode; for example, **DrawInv** in a graphing mode other than **Func**.
NO SIGN CHNG	• The **solve(** function or the equation solver did not detect a sign change. • You attempted to compute I% when **FV**, (**N*PMT**), and **PV** are all ≥ 0, or when **FV**, (**N*PMT**), and **PV** are all ≤ 0. • You attempted to compute **irr(** when neither *CFList* nor *CFO* is > 0, or when neither *CFList* nor *CFO* is < 0.
NONREAL ANS	In **Real** mode, the result of a calculation yielded a complex result. This error is not returned during graphing. The TI-83 Plus allows for undefined values on a graph.
OVERFLOW	You attempted to enter, or you have calculated, a number that is beyond the range of the calculator. This error is not returned during graphing. The TI-83 Plus allows for undefined values on a graph.
RESERVED	You attempted to use a system variable inappropriately. See Appendix A.
SINGULAR MAT	• A singular matrix (determinant = 0) is not valid as the argument for $^{-1}$. • The **SinReg** instruction or a polynomial regression generated a singular matrix (determinant = 0) because it could not find a solution, or a solution does not exist. This error is not returned during graphing. The TI-83 Plus allows for undefined values on a graph.

Error Type	Possible Causes and Suggested Remedies
SINGULARITY	*expression* in the **solve(** function or the equation solver contains a singularity (a point at which the function is not defined). Examine a graph of the function. If the equation has a solution, change the bounds or the initial guess or both.
STAT	You attempted a stat calculation with lists that are not appropriate. • Statistical analyses must have at least two data points. • **Med-Med** must have at least three points in each partition. • When you use a frequency list, its elements must be ≥ 0. • (**Xmax** − **Xmin**) / **Xscl** must be ≤ 47 for a histogram.
STAT PLOT	You attempted to display a graph when a stat plot that uses an undefined list is turned on.
SYNTAX	The command contains a syntax error. Look for misplaced functions, arguments, parentheses, or commas. See Appendix A and the appropriate chapter.
TOL NOT MET	You requested a tolerance to which the algorithm cannot return an accurate result.
UNDEFINED	You referenced a variable that is not currently defined. For example, you referenced a stat variable when there is no current calculation because a list has been edited, or you referenced a variable when the variable is not valid for the current calculation, such as **a** after **Med-Med**.
VALIDATION	Electrical interference caused a link to fail or this calculator is not authorized to run the application.
VARIABLE	You have tried to archive a variable that cannot be archived or you have have.tried to unarchive an application or group. Examples of variables that cannot be archived include: • Real numbers **LRESID, R, T, X, Y, Theta**, Statistic variables under **Vars**, STATISTICS menu, **Yvars**, and the **AppIdList.**
VERSION	You have attempted to receive an incompatible variable version from another calculator.
WINDOW RANGE	A problem exists with the window variables. • You defined **Xmax** ≤ **Xmin** or **Ymax** ≤ **Ymin.** • You defined **θmax** ≤ **θmin** and **θstep** > **0** (or vice versa). • You attempted to define **Tstep=0**. • You defined **Tmax** ≤ **Tmin** and **Tstep** > **0** (or vice versa). • Window variables are too small or too large to graph correctly. You may have attempted to zoom in or zoom out to a point that exceeds the TI-83 Plus's numerical range.
ZOOM	• A point or a line, instead of a box, is defined in **ZBox.** • A ZOOM operation returned a math error.

Computational Accuracy

To maximize accuracy, the TI-83 Plus carries more digits internally than it displays. Values are stored in memory using up to 14 digits with a two-digit exponent.

- You can store a value in the window variables using up to 10 digits (12 for **Xscl**, **Yscl**, **Tstep**, and **θstep**).
- Displayed values are rounded as specified by the mode setting with a maximum of 10 digits and a two-digit exponent.
- **RegEQ** displays up to 14 digits in **Float** mode. Using a fixed-decimal setting other than **Float** causes **RegEQ** results to be rounded and stored with the specified number of decimal places.

Graphing Accuracy

Xmin is the center of the leftmost pixel, **Xmax** is the center of the next-to-the-rightmost pixel. (The rightmost pixel is reserved for the busy indicator.) **ΔX** is the distance between the centers of two adjacent pixels.

- In **Full** screen mode, **ΔX** is calculated as (**Xmax** − **Xmin**) / 94. In **G-T** split-screen mode, **ΔX** is calculated as (**Xmax** − **Xmin**) / 46.
- If you enter a value for **ΔX** from the home screen or a program in **Full** screen mode, **Xmax** is calculated as **Xmin** + **ΔX** ∗ 94. In **G-T** split-screen mode, **Xmax** is calculated as **Xmin** + **ΔX** ∗ 46.

Ymin is the center of the next-to-the-bottom pixel; **Ymax** is the center of the top pixel. **ΔY** is the distance between the centers of two adjacent pixels.

- In **Full** screen mode, **ΔY** is calculated as (**Ymax** − **Ymin**) / 62. In **Horiz** split-screen mode, **ΔY** is calculated as (**Ymax** − **Ymin**) / 30. In **G-T** split-screen mode, **ΔY** is calculated as (**Ymax** − **Ymin**) / 50.
- If you enter a value for **ΔY** from the home screen or a program in **Full** screen mode, **Ymax** is calculated as **Ymin** + **ΔY** ∗ 62. In **Horiz** split-screen mode, **Ymax** is calculated as **Ymin** + **ΔY** ∗ 30. In **G-T** split-screen mode, **Ymax** is calculated as **Ymin** + **ΔY** ∗ 50.

Graphing Accuracy (continued)

Cursor coordinates are displayed as eight-character numbers (which may include a negative sign, decimal point, and exponent) when **Float** mode is selected. **X** and **Y** are updated with a maximum accuracy of eight digits.

minimum and **maximum** on the CALCULATE menu are calculated with a tolerance of 1E-5; **∫f(x)dx** is calculated at 1E-3. Therefore, the result displayed may not be accurate to all eight displayed digits. For most functions, at least five accurate digits exist. For **fMin(**, **fMax(**, and **fnInt(** on the MATH menu and **solve(** in the CATALOG, the tolerance can be specified.

Function Limits

Function	Range of Input Values
sin x, **cos** x, **tan** x	$0 \le \lvert x \rvert < 10^{12}$ (radian or degree)
sin^{-1} x, **cos**$^{-1}$ x	$^{-}1 \le x \le 1$
ln x, **log** x	$10^{-100} < x < 10^{100}$
e^{x}	$^{-}10^{100} < x \le 230.25850929940$
10^{x}	$^{-}10^{100} < x < 100$
sinh x, **cosh** x	$\lvert x \rvert \le 230.25850929940$
tanh x	$\lvert x \rvert < 10^{100}$
sinh^{-1} x	$\lvert x \rvert < 5 \times 10^{99}$
cosh^{-1} x	$1 \le x < 5 \times 10^{99}$
tanh^{-1} x	$^{-}1 < x < 1$
$\sqrt{x}$ (real mode)	$0 \le x < 10^{100}$
$\sqrt{x}$ (complex mode)	$\lvert x \rvert < 10^{100}$
x**!**	$^{-}.5 \le x \le 69$, where x is a multiple of .5

Function Results

Function	Range of Result	
sin^{-1} x, **tan**$^{-1}$ x	$^{-}90°$ to $90°$	or $^{-}\pi/2$ to $\pi/2$ (radians)
cos^{-1} x	$0°$ to $180°$	or 0 to π (radians)

Product Support

Customers in the U.S., Canada, Puerto Rico, and the Virgin Islands

For general questions, contact Texas Instruments Customer Support:

phone: **1-800-TI-CARES (1-800-842-2737)**
e-mail: **ti-cares@ti.com**

For technical questions, call the Programming Assistance Group of Customer Support:

phone: **1-972-917-8324**

Customers outside the U.S., Canada, Puerto Rico, and the Virgin Islands

Contact TI by e-mail or visit the TI **Calculator** home page on the World Wide Web.

e-mail: **ti-cares@ti.com**
Internet: **education.ti.com**

Product Service

Customers in the U.S. and Canada Only

Always contact Texas Instruments Customer Support before returning a product for service.

Customers outside the U.S. and Canada

Refer to the leaflet enclosed with this product or contact your local Texas Instruments retailer/distributor.

Other TI Products and Services

Visit the TI **Calculator** home page on the World Wide Web.

education.ti.com

Customers in the U.S. and Canada Only

One-Year Limited Warranty for Electronic Product

This Texas Instruments ("TI") electronic product warranty extends only to the original purchaser and user of the product.

Warranty Duration. This TI electronic product is warranted to the original purchaser for a period of one (1) year from the original purchase date.

Warranty Coverage. This TI electronic product is warranted against defective materials and construction. **THIS WARRANTY IS VOID IF THE PRODUCT HAS BEEN DAMAGED BY ACCIDENT OR UNREASONABLE USE, NEGLECT, IMPROPER SERVICE, OR OTHER CAUSES NOT ARISING OUT OF DEFECTS IN MATERIALS OR CONSTRUCTION.**

Warranty Disclaimers. ANY IMPLIED WARRANTIES ARISING OUT OF THIS SALE, INCLUDING BUT NOT LIMITED TO THE IMPLIED WARRANTIES OF MERCHANTABILITY AND FITNESS FOR A PARTICULAR PURPOSE, ARE LIMITED IN DURATION TO THE ABOVE ONE-YEAR PERIOD. TEXAS INSTRUMENTS SHALL NOT BE LIABLE FOR LOSS OF USE OF THE PRODUCT OR OTHER INCIDENTAL OR CONSEQUENTIAL COSTS, EXPENSES, OR DAMAGES INCURRED BY THE CONSUMER OR ANY OTHER USER.

Some states/provinces do not allow the exclusion or limitation of implied warranties or consequential damages, so the above limitations or exclusions may not apply to you.

Legal Remedies. This warranty gives you specific legal rights, and you may also have other rights that vary from state to state or province to province.

Warranty Performance. During the above one (1) year warranty period, your defective product will be either repaired or replaced with a reconditioned model of an equivalent quality (at TI's option) when the product is returned, postage prepaid, to Texas Instruments Service Facility. The warranty of the repaired or replacement unit will continue for the warranty of the original unit or six (6) months, whichever is longer. Other than the postage requirement, no charge will be made for such repair and/or replacement. TI strongly recommends that you insure the product for value prior to mailing.

Software. Software is licensed, not sold. TI and its licensors do not warrant that the software will be free from errors or meet your specific requirements. **All software is provided "AS IS."**

Copyright. The software and any documentation supplied with this product are protected by copyright.

Australia & New Zealand Customers only

One-Year Limited Warranty for Commercial Electronic Product

This Texas Instruments electronic product warranty extends only to the original purchaser and user of the product.

Warranty Duration. This Texas Instruments electronic product is warranted to the original purchaser for a period of one (1) year from the original purchase date.

Warranty Coverage. This Texas Instruments electronic product is warranted against defective materials and construction. This warranty is void if the product has been damaged by accident or unreasonable use, neglect, improper service, or other causes not arising out of defects in materials or construction.

Warranty Disclaimers. Any implied warranties arising out of this sale, including but not limited to the implied warranties of merchantability and fitness for a particular purpose, are limited in duration to the above one-year period. Texas Instruments shall not be liable for loss of use of the product or other incidental or consequential costs, expenses, or damages incurred by the consumer or any other user.

Some jurisdictions do not allow the exclusion or limitation of implied warranties or consequential damages, so the above limitations or exclusions may not apply to you.

Legal Remedies. This warranty gives you specific legal rights, and you may also have other rights that vary from jurisdiction to jurisdiction.

Warranty Performance. During the above one (1) year warranty period, your defective product will be either repaired or replaced with a new or reconditioned model of an equivalent quality (at TI's option) when the product is returned to the original point of purchase. The repaired or replacement unit will continue for the warranty of the original unit or six (6) months, whichever is longer. Other than your cost to return the product, no charge will be made for such repair and/or replacement. TI strongly recommends that you insure the product for value if you mail it.

Software. Software is licensed, not sold. TI and its licensors do not warrant that the software will be free from errors or meet your specific requirements. All software is provided "AS IS."

Copyright. The software and any documentation supplied with this product are protected by copyright.

All Customers Outside the U.S. and Canada

For information about the length and terms of the warranty, refer to your package and/or to the warranty statement enclosed with this product, or contact your local Texas Instruments retailer/distributor.

- A -

- A (continued) -

- B -

- C -

- C (continued) -

- C (continued) -

- D -

- D (continued) -

- D (continued) -

- E -

- F (continued) -

- F (continued) -

-G -

- L (continued) -

- L (continued) -

- M -

- M (continued) -

- N -

- T (continued) -